PROCESS CONTROL FOR THE CHEMICAL AND ALLIED FLUID-PROCESSING INDUSTRIES

HEINEMANN CHEMICAL ENGINEERING SERIES

Process Control

**For the Chemical and
Allied Fluid-Processing Industries**

A. POLLARD
B.Sc., C.ENG., F.I.CHEM.E., F.INST.F.,
M.INST.M.C., F.R.I.C.

*Senior Lecturer in Chemical Engineering
and Process Instrumentation,
University of Leeds*

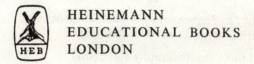

HEINEMANN
EDUCATIONAL BOOKS
LONDON

Heinemann Educational Books Ltd

LONDON EDINBURGH MELBOURNE AUCKLAND TORONTO
HONG KONG SINGAPORE KUALA LUMPUR NEW DELHI
IBADAN NAIROBI LUSAKA JOHANNESBURG
KINGSTON/JAMAICA

ISBN 0 435 72561 0

© A. Pollard 1971

First published 1971

Reprinted 1978

Published by Heinemann Educational Books Ltd
48 Charles Street, London W1X 8AH

Printed by photolithography and bound in
Great Britain by Biddles Ltd, Guildford, Surrey

Contents

Preface

A sound knowledge of the basic principles underlying any particular branch of engineering is fundamental to its practical application in industry and to its further development. A study of the elements of process dynamics and control now forms an essential part of the education of the chemical engineer and is of value also to students in allied fields of fluid-process technology. The present volume covers the basic theory of the dynamics and control of linear process systems and the two main graphical methods (root locus and frequency response) available for process system analysis. The text is based almost entirely on courses given by the author over a number of years to undergraduates in chemical engineering and allied disciplines.

A knowledge of calculus, elementary differential equations, complex numbers, and the unit operations of chemical engineering compatible with the penultimate-year honours course is assumed. It is essential that the reader becomes familiar with the Laplace transform technique as a basic mathematical tool in systems analysis; the basic features of the technique are presented and the simple rules for direct and inverse transformation are easily acquired.

A detailed treatment of specific process systems and the instruments used in their control has not been included; such treatment tends to be descriptive rather than quantitative and often quickly becomes obsolete in this rapidly developing field. The more spectacular recent developments in process control, such as the application of direct digital control and optimalizing and adaptive systems, and the treatment of non-linear and multi-variable systems are given only relatively brief reference in view of the more fundamental material presented. The literature of process control is now so extensive that only a selected bibliography is included.

The author wishes to acknowledge many helpful discussions with his colleagues, particularly Mr F. H. Cass, and with students in the Department of Chemical Engineering in the University of Leeds. He is particularly obliged to Mr P. G. Myatt, of the Leeds Polytechnic, for much helpful advice and criticism during preparation of the manuscript. Finally, the book would not have been produced at all without the encouragement and forebearance of the author's wife, to whom he owes considerable thanks.

1971   A.P.

Nomenclature

A	area, amplitude, input element transfer function, term arising from finite gain in controller action equation
A.R.	normalized amplitude ratio
b	measured (feedback) variable
B	proportional band width
c	controlled variable, concentration, proportionality constant
c_p	specific heat capacity
C	capacitance
C_v	valve flow coefficient
d	differential operator
D	diameter, damping force, derivative action or control
e	base of natural logarithms
e	error, voltage
E	emissivity factor
f	function
f^*	sampled function
f_c	clamped function
F	force, function
g	gravitational constant, bellows spring constant
G	transfer function (forward path), specific gravity
G_c	controller transfer function
G_0	closed-loop M.R. at zero frequency
G_p	process transfer function, closed-loop peak magnitude
G_v	control valve transfer function
h	head or depth of liquid
H	measurement (feedback) transfer function, specific enthalpy
ΔH	heat of reaction
i	current
I	integral action or control, interaction factor
j	$\sqrt{-1}$
k	reaction rate constant, proportionality constant, thermal conductivity
K	proportionality constant, steady-state gain, overall gain, thermal diffusivity
K_c	proportional sensitivity, proportional gain
$K_{c\,max}$	maximum proportional sensitivity (limiting stability)
K_g	controller gain
K_L	load gain
K_m	measurement gain
K_{max}	maximum overall gain (limiting stability)

K_u	ultimate proportional sensitivity ($= K_{c\,max}$)
K_v	valve sensitivity
l	distance, length
L	time delay, length, inductance
$\mathscr{L}$	Laplace transform operator
$\mathscr{L}^{-1}$	inverse Laplace transform operator
m	manipulated variable, mass, valve position
M	change in input variable, closed loop M.R., mass
M_p	closed loop M.R. relative peak height
M.R.	magnitude ratio
n	integer, real number
N	load element transfer function, slope of Ziegler-Nichols reaction curve
p	pressure, controller output signal, pole of function
Δp	pressure difference
P	applied force, period of oscillation
P_u	ultimate period of (continuous) oscillation
q	electrical charge, volumetric flow rate, heat flow rate
Q	heat flow rate
r	reference input signal, root of function or equation
R	resistance, gas law constant, lag ratio
R_d	derivative action resistance
R_i	integral action resistance
s	Laplace transform variable
$s_1, s_2 \ldots$	roots of polynomial or characteristic equation
S	spring force
T	time constant, characteristic time, sampling interval
T_d	derivative (action) time
T_i	integral (action) time
T_m	measurement time constant
u	load variable, disturbance signal
U	overall heat transfer coefficient
v	desired value signal, velocity
V	volume
w	mass flow rate
x	linear displacement, real part of complex number $(x + jy)$
X	thickness, length
y	imaginary part of complex number $(x + jy)$
z	zero of function, z-transform variable
α	damping factor, closed loop phase angle, derivative action phase advance
β	damped frequency
β_0	undamped (natural frequency)
γ	controller phase shift
δ	unit impulse
Δ	finite difference operator

ε	expansion factor, specific internal energy
ζ	damping ratio
θ	temperature
θ_d	desired value temperature
θ_i	input signal, input temperature
θ_o	output signal, output temperature
$\bar{\theta}$	design temperature (normal steady-state)
λ	integral action phase lag
μ	viscosity
π	pi (3.141 . . .)
Π	product
ρ	density
σ	Stefan-Boltzmann radiation constant
Σ	summation
ϕ	angle, phase angie
ψ	phase lag, angular displacement
ω	angular frequency (rads/unit time)
ω_c	critical frequency at 180° phase lag
ω_{co}	cross-over frequency
ω_n	natural frequency
ω_r	resonant frequency
∂	partial differential operator

Abbreviations for Units

SI		British-American	
°C	deg. Celsius	Btu	British thermal unit
J	joule	°F	deg. Farenheit
K	deg. Kelvin	ft	foot
kg	kilogramme	gal	gallon (UK)
m	metre	h	hour
N	newton	in	inch
s	second	lb	pound (mass)
W	watt	min	minute
		lbf/in^2	lb(force)/in^2(gauge)
		°R	deg. Rankine

Prefixes

k	kilo ($\times 10^3$)
m	milli- ($\times 10^{-3}$)

(See Mullin, *The Chemical Engineer*, September 1967)

Chapter 1: Introduction to Process Control

Automatic control is now widely used in many different fields ranging between such diverse applications as the automatic piloting of aircraft to the control of temperature in the domestic oven. In the analysis of problems within this range, an understanding of the basic principles is of primary importance since these same principles can be generally applied to any control problem regardless of variations in physical or mechanical detail. Differences between individual problems arise principally from the number, scale, and complexity of the relationships of the variables involved in the particular situation.

In this introduction the basic principles will be reviewed and ultimately considered with respect to the control of industrial processes typical of the chemical and allied fluid-processing industries, i.e. what is usually referred to as *process control*, as distinct from *position control* or *servo-operation*, which is the other major branch of automatic control engineering.

A process may be defined as simply an advance to some particular end or objective, which is achieved in all but the simplest cases by some orderly pattern or sequence of events or operations. In the chemical industry, where process control finds its widest applications, the process consists of the chemical engineering operations necessary to convert one or more raw materials into finished products, often with the simultaneous formation of by-products which may have to be separated from the desired end-product. Processing action, however, may be given a much wider interpretation on the simple definition given above. The movement of material from storage to a point of use and the transfer of components between stages of manu-facture are processing actions. A process may be carried out on human beings; the training of individuals for some particular task or the education of students are processes in the sense that an orderly pattern of operations must be followed to attain the required objectives. Processing is also carried out on paper-work, as in accounting, design, etc. and in some of these fields the computer is already taking over the data-processing opera-tions from individuals. However, such applications which involve the reaction of the human element in the achievement of the process objective form a natural branch from the purely mechanical field, and are not normally included as branches of process control but are regarded as the special field to which the name *cybernetics* has been applied.

In general, if a process is to be worthwhile it must be carried out efficiently. Efficiency may be defined in a number of ways depending upon the nature of the particular process, but in the final analysis the definition will usually be based on economic factors. To maintain efficiency, control of the process operations is necessary at every stage, and in all but the very simplest of processes this control cannot be exercised by purely manual means—the required efficiency cannot be maintained by human endeavour alone. The manual operator, at the very least, must usually be provided with some instrumental assistance to measure the variables in the process. Process instrumentation, or the measurement of process variables, is a very important division of the field of process control.

The operation of a process calls for certain events to take place; these events were originally manual operations and it is convenient still to refer to them as 'manipulations'. The manipulations necessary for a particular process may be quite simple or very complex. One person or several may be called upon to apply purely mechanical skills to the moving of levers, the adjustment of control valves, etc., or all or part of these operations may be carried out automatically by suitable apparatus.

The important feature common to all processes is that a process is never in a state of static equilibrium for more than a very short period of time. A process is a dynamic entity subject to continual upsets or disturbances which tend to drive it away from the desired state of equilibrium; the process must then be manipulated upon or corrected to drive it back towards the desired state. Manipulation is in itself a disturbance of the process, a correction is simply a manipulation to oppose the effect of some previous disturbance. Both disturbances and manipulation are events arising externally to the process. Some disturbances bring about only transient effects in the process behaviour; these pass away and may never occur again. Others may apply cyclic or periodic forces which make the process respond in a cyclic or periodic fashion. Most disturbances are completely random with respect to time and show no repetitive pattern; thus their occurrence may be expected but cannot be predicted at any particular time.

Process disturbances occur in many and varied ways. Raw materials for manufacturing processes are supplied in batches which may not be truly homogeneous so that the properties of the material may change as it enters the process. Machinery and plant do not run smoothly for indefinite periods but show a gradual deterioration in performance; plant may be affected by climatic conditions such as changes in ambient temperature and pressure, changes in cooling water temperature, and so on.

As has already been pointed out, an operating process is never static. Material or energy is always flowing and whilst it is possible to visualize a steady-state equilibrium, as is usually assumed in the design of continuously operating process plant, in practice this ideal state is rarely attained in operation except for very short periods. The reaction of the process under operating conditions is dynamic, and consequently it is necessary to study the dynamic reactions of processes which are to be controlled.

The Control System

The starting point in any control system is the particular variable which it is necessary to control to achieve some desired measure of the efficiency in the particular process. A process, of course, is never designed with the object of controlling a particular variable; rather more logically a process is designed for a particular objective or output and it is then found, sometimes by trial and error but more usually by inference from previous experience, that control of a particular variable associated with some stage of the process is necessary to achieve the desired efficiency.

The output of the controlled system is the value of the particular variable being controlled, and this will not necessarily be the same as the output of the process. The process output is obviously a dependent variable, since its value depends on the operation of the process and the manipulation of the inputs to the process. Whilst the process output may be almost any physical quantity, the output of the control system associated with the process need not be a product from the process in the usual sense. The output of the control system is more usually a variable associated with the process operation which determines in some way the efficiency of the process in reaching the desired end. For example, the temperature or pressure in a chemical reactor may determine the efficiency of operation of the reactor in producing products of the required specifications; hence one or both of these variables would be used to control the operation and so would be the outputs of one or two control systems. In the latter case the two systems might be completely independent of each other or they may interact with each other in certain ways. Generally in process control, the controlled variable is one of the more usual process variables, i.e. temperature, pressure, fluid flow rate, or liquid level, but could quite conceivably be almost any other physical property of the materials or the environment such as the composition of a material stream, density, pH, viscosity, speed, etc.

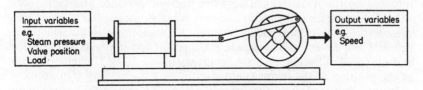

FIGURE 1-1. Steam engine input and output variables

It is more illuminating at this stage to look at some practical processes where the properties and effects are fairly obvious, rather than to discuss generalities. The first process to be considered is a steam engine, illustrated in Figure 1-1. Here is a process—the engine—and associated with it a number of variables which can be grouped initially into two categories, 'input' and 'output' variables. The former are independent variables; their values or magnitudes are determined by factors independent of the process. Some of the input variables can be manipulated, i.e. the values can be changed at will by a process operator (it is immaterial at this stage whether the operator is manual or automatic). Other input variables are not capable of being manipulated but their values are determined by outside agencies or natural phenomena. A change in the value of any input variable will affect the operation of the process in some way or other and will ultimately lead to changes in the values of some or all of the output variables.

These latter are not independent variables but are determined by the values of the input variables and the conservation balance of the material and/or energy flows into and out of the process. Generally one of the dependent output variables will represent, either directly or indirectly, the primary objective of the process operation, and this would be selected as the controlled variable, i.e. the process variable which must be held at a certain value or within certain limits if the process is to be operated efficiently and the required objective attained. In a rather similar way, in the majority of cases one of the input variables will have a more direct and immediate effect on the operation of the process and so also on the values of the output variables than will any of the others, and this particular variable would be chosen as the one to be manipulated to control the process through the selected output variable.

The steam engine presents a relatively straightforward picture. The object of the process is to convert the pressure energy of the steam into mechanical energy for a particular purpose which may be to raise a load or to propel a vehicle. The output variable immediately indicative of the engine's performance as a power unit is the speed of the output shaft, and the operation of the process of power conversion is then controlled on the basis of the engine speed. The primary input variable which has the most direct influence on the engine speed is the rate of flow of steam to the cylinders, but this in turn is determined by the restriction to the flow imposed by the usual regulating valve and the pressure drop across the valve, which again is a function of two other variables, the steam supply pressure and the engine back pressure. Of these, the variable which it is simplest and easiest to manipulate is the restriction imposed by the valve, and the speed of the engine is almost invariably controlled by manipulation of the position of the steam inlet valve.

It must be noted that other variables can also change the engine speed and these, by definition, must also be input variables. The steam supply pressure which partly determines the pressure drop across the valve and hence the steam flow through the valve has already been mentioned. Similarly, if the engine is required to do more work by an increase in the load on the output shaft, it will tend to slow down. An increase in friction in the piston or bearings due to lack of lubrication will have a similar effect. These are all input variations even though the load on the engine is concerned with the output of the process. For convenience a distinction can be drawn between *supply* and *demand* disturbances, i.e. those on the input and output sides of the process. A change affecting the steam pressure from the boiler is a supply disturbance, changes in the load on the output shaft are demand disturbances.

It will be noted that changes in the value of any input variable, as defined, will lead to a subsequent change in the output variable, but in each case the selected manipulated variable, i.e. the position of the steam valve, can be used to correct for the disturbance by admitting more or less steam as required. It will also be noted that selection of the steam valve position as the manipulated variable is not unique, and that other input variables could be manipulated to provide the necessary correction to a disturbance. For example, the steam valve could be dispensed with and the steam flow regulated by manipulation of the steam supply pressure through the boiler. This would be a less effective method as it would involve operating the boiler as part of the engine process and would introduce more input variables affecting the performance of the boiler. In a similar way the speed could be controlled by manipulating part of the load, by effectively applying a brake to the output shaft to use up any excess power following a reduction in the working load or an excess flow of steam. Again this can be seen to be less efficient and obviously less economic than manipulation of the steam flow by means of the valve.

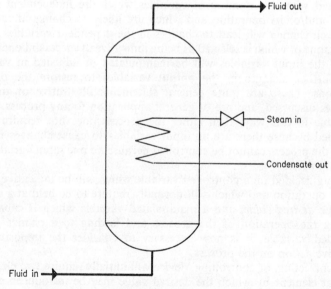

FIGURE 1-2. Fluid heater

A second process example is the fluid heater shown in Figure 1-2, in which the objective is to maintain a supply of heated fluid. Here again is a fairly obvious choice of both controlled and manipulated variables. That which is most indicative of the aim of the process and the efficiency of achievement is the temperature of the fluid leaving the heater, and the process would then be controlled by control of the output temperature. The input variable with the fastest and most direct reaction on the outlet temperature is the heat flow rate into the heater but, as in the previous example, two or more input variables may again be involved. If steam or some other heat transfer medium is used, the manipulated variable may again be the position of a flow regulating valve with the supply and back pressures as additional input variables. If the heating is by electricity then the position of a rheostat or variable transformer would be manipulated with the supply voltage as an additional variable. Rather less practical would be to control the output temperature by manipulating the flow of the fluid through the heater, with the heat supply as a completely independent variable. Supply disturbances already noted are the heating medium pressure or voltage; additionally the fluid inlet temperature is an additional input variable, and also in this case ambient conditions on the outside of the heater (temperature, air movement, etc.) are other possible sources of disturbance since these determine the magnitude of the heat losses from the apparatus. On the demand side the major and most probable disturbance is the flow rate of the heated fluid through the heater.

To sum up at this point, each process or process operation will have associated with it a number of variables which are independent of the process and/or its operation and which are likely to change at random. Each such change will lead to changes in the dependent variables of the process, one of which is selected as being indicative of successful operation. One of the input variables will be manipulated or adjusted in value to cause further changes in the output variables to restore the original conditions. These are quite general statements illustrative of the two examples discussed, and are of general application to any process. If not applicable in this way, then either the process does not require to be controlled because there are no input variables to cause changes in output, or the process cannot be controlled because no one input variable can be manipulated.

Having decided on a controlled variable which will be indicative of the process operation and which will normally require to be held at a certain level, the *desired value*, and a manipulated variable which is capable of affecting the operation of the process and leading to a change in the controlled variable, it is now necessary to consider the imposition of corrective action on the process.

The correcting, or controlling, device will initially require two elements, an input element by which the desired value may be introduced and an output element which will manipulate the appropriate variable. The former is basically nothing more than a pointer which can be set on a scale of desired values of the controlled variable; the output element is usually some form of valve used to regulate the rate of flow of material or energy which is the manipulated variable, as in the two examples considered. The two elements thus permit the pointer to be set on the scale to define a certain value of the controlled variable, and this in turn will set the valve at a certain position, permitting a certain flow into the process. Thus, in the case of the steam engine, a certain flow of steam is passed by the valve which permits the engine to run at a certain speed as dictated by the flow of steam and the other input variables. The only further requirement is the calibration of the scale of the input element to correlate with the actual values of the controlled variable.

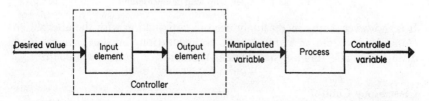

FIGURE 1–3. Open-loop control

This forms a type of control system as shown in Figure 1–3, which, for reasons which will be apparent later, is known as an *open-loop* system. As a control system it cannot, however, be very effective except in certain special cases. The system will fail in most applications because no account is obviously taken of the other input variables apart from the manipulated variable. All other input variables are perfectly free and can change in a completely random manner; any such change will upset the correlation between the controlled variable and the input element. A little consideration will show that any such correlation can only exist under one condition of load, i.e. if all the input variables are constant. If any input variable changes there will be a consequential change in the controlled variable and in this system the manipulated variable will not be adjusted to apply any correction. Another line of development must then be pursued.

There are two possible ways of improving the situation. From the previous discussion it will be appreciated that a change in any input variable can be corrected by a change in the manipulated variable. If the load on the engine is increased, the steam valve can be opened to admit more steam and thus maintain the engine speed. It is therefore possible in principle to use another input element to effectively measure the load on the engine and to use this to make an appropriate correction to the output element (the valve), so making the required change in the manipulated variable when the load changes to maintain the desired value of the controlled variable. This principle of *load-change compensation* is, however, applicable to changes in only one source of disturbance, and the method would strictly have to be applied separately to every individual input variable which could affect the process. This would require a very complex system and in practice this technique is rarely used in the form described.

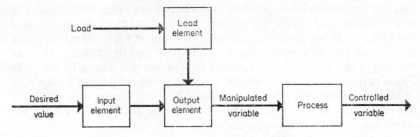

FIGURE 1–4. Open-loop load compensation

It is, however, being increasingly used in conjunction with the alternative *closed-loop* system of control as a compensation for the primary load disturbance affecting a system, under the name of *feedforward control*.

Closed-loop Control

The alternative method of dealing with the problem of external disturbances is relatively simple, so much so that it is the technique which is almost universally employed in practice to control any operation. The principle is to use the effects of the disturbances on the controlled variable to adjust the manipulated variable and so correct for the disturbance. In other words, since the disturbances are bound to happen, let them affect the process and so cause changes in the controlled variable but now use these changes, i.e. the departures from the desired value of the variable, to adjust the manipulated variable and so correct for the effect of the disturbance. The source or cause of the disturbance is ignored; only the effects on the controlled variable are used to apply the correction.

This now introduces the vitally important principle of *feedback*. The desired value of the controlled variable is, as before, set on the scale of an input element to provide an input signal, which may be symbolized by θ_i. This has now to be compared to an output signal, θ_o, which is effectively the actual value of the controlled variable at the time so that any difference between the two signals, which represents the departure of the controlled variable from the desired value, can be used to adjust the output element to change the manipulated variable and so apply the correction to restore the controlled variable to the desired value. Initially then, before the signals can be compared, the value of the output signal—the controlled variable— must be 'fed back' from the output side of the process to the controlling device, and this brings into being a *closed-loop* system, as shown in Figure 1–5. The purpose of the controller is now to make the output signal, θ_o, equal to the input, θ_i, and the logical step is to use the *difference* between the two at any time, which may be termed the *error* or *deviation* in the system, to generate the corrective action. A further element is now required in the controller in the form of an error discriminator (ε), which effectively subtracts θ_o from θ_i. The input to a system is conventionally regarded as positive, and hence the feedback of the output must be negative with respect to the input; this step thus adds 'negative feedback' to the system.

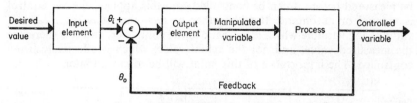

FIGURE 1-5. Closed-loop control

Quite often, and especially in process control, a further additional element is required in the control loop which is not strictly a part of the controller as such, but is often included in the same instrument case, purely for convenience. This additional element is required when the input signal to the controller is of a different physical nature from the controlled variable. The input signal is most often the mechanical position of a pointer on a scale of desired values, i.e. a mechanical displacement, but it may equally be an air loading-pressure on a diaphragm in a pneumatically-actuated instrument, or an electrical signal such as current or voltage in an electrical or electronic instrument. The controlled variable will be a conventional process variable such as rate of flow, pressure, temperature, etc. which cannot be fed back and compared directly with the input signal. Instead, the controlled variable has to be converted or *transduced* in either range or physical nature, or both, so that both input and output signals are of the same physical form and in the same range of magnitude and one can be subtracted from the other. This transducing function is, of course, the basic function of the conventional measuring instrument which converts the particular process variable into, usually, the movement of a pointer over a scale, but equally can be arranged to produce a proportional pneumatic or electrical signal in the required range. The process control loop will thus normally include a measurement transducer in the feedback link. It should be noted that the transducer is strictly that alone, it is only required to convert the form of the signal and need not perform any other function of a conventional measuring instrument such as the display or recording of the actual value of the controlled variable. In practice it costs very little extra to do this and the information so obtained is obviously useful as a monitor of the control system performance. It is therefore more usual in process control for the measurement transducer to be provided with some form of visual display such as the usual pointer and scale or recording pen and chart. These may again be included in the same case as the controller, but there is an increasing tendency to the use of centralized control areas when it is often more convenient to transmit the measured value of the variable from the transducer to a separate indicator or recorder.

The above discussion serves to emphasize the importance of measurement as an adjunct to the art of control. It is axiomatic that a variable must be measured before it can be controlled, and this applies whether control is manual or automatic. It will further be seen that it is the measured value of the variable which is fed back to the controller, and it is therefore the measured value, and not the actual value, which is effectively being controlled. The importance of this point will be discussed later.

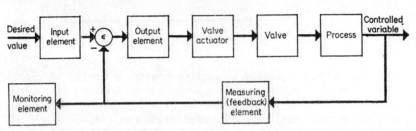

FIGURE 1–6. Process control loop

The process control system in Figure 1–6 shows a further difference from the generalized closed-loop system of Figure 1–5. A regulating valve is a standard piece of ironmongery or 'hardware' and is not then usually included in or regarded as part of the controller. Instead it is usually considered as an item of plant external to the controller. It must, of course, be provided with an operating device or actuator which in turn responds to a signal from the output element of the controller.

The control system now differs fundamentally from that previously discussed. The use of the feedback principle brings into being a closed loop, a cyclic system of elements around which information, in the form of signals, is passed. The action of the controller on the process now, in principle, determines its own input. The original sequence, as in Figure 1–3, was a chain of elements and is thus referred to as an *open-loop* system. The closed loop of Figure 1–5 must always require the use of negative feedback to literally 'close the loop'.

A fundamental difference between the open- and closed-loop systems which is worthy of note at this stage is that the power available in the open loop depends only on the input to the system. In a closed loop, power can be fed back into the loop from the process via the output signal feedback and the system can thus gain energy and become unstable in the classical sense.

Regulator and Servo Controls

There are two general problems in automatic control systems, the first of which is to reduce the effect of changes in load (i.e. input variables). The desired value of the controlled variable in such cases is usually fixed and significantly invariable with time; the load variables may change sporadically or continuously at random, both in time and in the magnitude of the changes. The purpose of the control system is to minimize the effect of the load disturbances on the value of the controlled variable. A load disturbance control is illustrated by the example of the fluid heater, in which the purpose of the control is to maintain a constant outlet temperature in spite of fluctuations in the load variables, e.g. the inlet temperature, rate of flow, heat supply, etc. as shown in Figure 1–7.

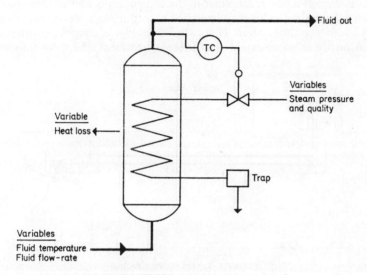

FIGURE 1–7. Fluid heater temperature control

The alternative control problem is the control of a system to follow a changing desired value or input signal. In most cases of this type, the load variables are constant or changes are of little significance. The desired value now changes unpredictably and the control system is designed to make the value of the controlled variable follow these changes as closely as possible. An example is provided by the steering mechanism of a large ship where it is impossible to position the rudder by hand and a power-amplifying or servo-system is used. A small wheel at the steering station is turned by hand, and this provides the required desired value signal to the steering engine; this latter is usually hydraulic, and the actual position of the rudder is fed back to the controller so that any error between the input and the feedback operates the engine to re-position the rudder and reduce the error (Figure 1–8).

These two systems are basically very similar in principle, both being dependent on a closed loop, but they also represent the two main branches into which automatic control has been artificially and somewhat arbitrarily divided. Load-disturbance or regulator control is typical of process control where the desired values are typically constant and the loads variable. Follow-up and position controls or servo-systems have variable desired values with more or less constant loads. This division of automatic control is largely due to historical development and study of each of these categories separately and in isolation from the other, and this is the explanation of the different approaches in the literature and the confusing and sometimes illogical nomenclature. To draw a distinct line between the two fields is often difficult; for example, process controls often include a servo-system as a power amplifier in the control loop, and in some process systems the desired value may be varied according to a pre-set programme or by another control system. There are, however, some practical differences which justify to some extent the different treatment, the most important

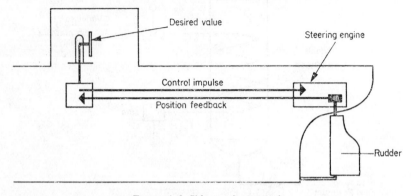

FIGURE 1–8. Ship steering control

being the difference in time scales. Process control systems normally operate on a longer time scale than position-control or servo-systems; the responses of process systems are slower and are measured in minutes or even hours rather than in seconds or fractions of a second, as is usually the case with servo-systems. This slower response is mainly due to the much greater inertia in process systems which limits the rate of change of the variables. Further differences are found in design technique and philosophy apart from the wide variation in apparatus employed. Servo-systems are generally designed and manufactured for a specific application and are normally required in sufficient numbers for this to be justified. Process plants, on the other hand, are more in the nature of 'one-off' jobs and it would not be economic to design a different process controller for each specific plant. Instead, a basic design of controller with variable functions and parameters is employed which can be adjusted to suit the particular characteristics of the plant to which it is applied.

Returning to the closed-loop system, the fundamental feature of a cyclic system of a number of elements or stages has already been noted.

It can be seen that the controller operates the process through the regulating unit; but it is equally true to say that the process now operates the controller through the measuring device and the feedback link. Since this is a cyclic system, the overall performance is necessarily dependent on every part of the closed loop, and every part of the loop must be included in considering the overall operation. Although it is often convenient to study various parts of the loop in isolation, eventually these parts must be assembled into the complete loop before the performance of the whole can be assessed.

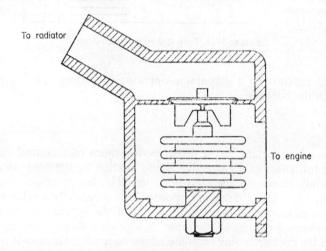

FIGURE 1–9. I.C. engine thermostat

Practical examples corresponding to the process loop generalized in Figure 1–6 can vary considerably in detail. Any of the units external to the plant can have more than one stage, or they may be combined with one another in part or even entirely. A simple example in which all three functions of measurement, control, and regulation are combined into one piece of equipment is the I.C. engine thermostat shown in Figure 1–9, in which the temperature sensitive bellows fulfills the measurement function by its response to changes in the temperature of the water circulating in the cylinder jackets. The expansion and contraction of the bellows operates directly on the regulating valve which diverts the hot water to the radiator-cooler. The controller in this instance is nothing more than the simple mechanical link between the thermostat bellows and the valve stem. A more conventional example of process control is the oil-fired tube still shown in Figure 1–10, in which the temperature of the oil leaving the furnace is measured by a conventional thermometer, of which the detecting element is inserted into a pocket in the oil line and the measuring and indicating elements are included in the same case as the controlling instrument. The latter is pneumatic in actuation and supplies a corrective signal

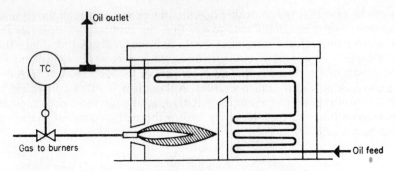

FIGURE 1–10. Tube still temperature control

as an air pressure to a diaphragm-operated regulating valve in the fuel supply to the burners.

Lag

The most important factor in the performance of a control system is that the full effect of a corrective action applied by the controller is not immediately shown by the controlled variable, but is almost inevitably subject to some lag or delay in response. This is basically a delay between cause and effect due to the natural inertia of the system, and that which exists between a change in the input to a process through the regulating unit and the corresponding response of the controlled variable is the *plant* or *process lag*.

To establish the nature and magnitude of such lags, it is useful to consider first the behaviour of a process without lag. In the fluid heater of Figure 1–2, under conditions of constant load each position of the regulating valve will yield a certain flow of heat into the process which, by virtue of the heat balance over the system, will correspond to a certain outlet temperature of the fluid leaving the heater. If the process is initially at a steady state with a certain outlet temperature and the valve is then opened by a given amount, the outlet temperature will rise to a new value determined by the new heat flow into the system. Except in a perfect system without lag, this new temperature will not be attained instantaneously as the valve is moved to its new position. In any practical system opening the valve will initially lead to an almost instantaneous increase in the flow of the heating medium, but this increased flow has then to build up a higher pressure inside the heater coils and the latter heated to a higher temperature before there is any significant increase in the heat transfer to the fluid passing through the heater. Only then will the temperature of the fluid near the coils begin to rise. The increase has next to be dissipated through the mass of fluid in the heater to increase the bulk temperature and finally the hotter fluid has to be transferred to the thermometer before an increase in outlet temperature is registered by the measuring instrument. The response of the outlet temperature will not then be an instantaneous change matching that of the valve position, but will be of the form shown in Figure 1–11. Following the opening of the valve there is very little initial response, and some time may elapse before any detectable change occurs; there is then a slow increase, building up to a maximum rate of change, which is followed by a decreasing rate as the final temperature is approached.

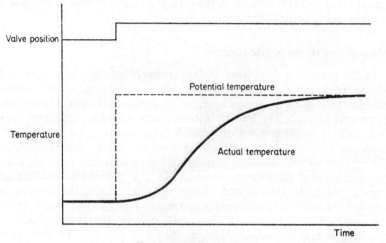

FIGURE 1–11. Process lag

In this instance the process possesses the property of *self-regulation*, i.e. a capacity to regulate its own potential level; the process is to some extent self-controlling since any change in load will ultimately lead to a new steady-state value of the temperature. This final steady-state value may be termed a *potential value* which corresponds to the new load condition, i.e. to the position of the valve, since all other load variables in the system are assumed to be constant. This potential value may be considered to exist from the instant that the valve position is changed or a change in load is applied. The instantaneous difference between the actual value of the variable and the potential value at any time may be regarded as a measure of the driving force causing the variable to change; if the change in load is due to the corrective action of a controller, this difference may be termed the *potential correction* applied by the controller at that time. The potential value is at all times a valid concept, even though the actual value may not be physically attainable.

Process lag is usually the most important factor in the control loop and is entirely a property of the plant or process. Lags may also exist in other parts of the control loop and it is, in fact, basically impossible to transmit a signal instantaneously through any part of the system. In practice most of these delays are extremely short compared to the lag in the plant and can often be ignored for this reason. However, the lag between a change in the actual value of the controlled variable and its indication by the measuring instrument may be significant. This *measuring lag* is fundamentally of the same nature as the process lag, and is particularly undesirable since not only does it delay the imposition of control action by delaying the feedback signal, but the measuring instrument gives an incorrect account of the system performance. The indicated value always lags behind the actual value and it is, of course, the former which is displayed and which is actually being controlled.

Manual *versus* Automatic Control

In the preceding discussion the control problem has been approached as one of purely automatic control, but it is, of course, still common practice to include a human operator in the control system, particularly in process control. The human operator may perform a relatively simple task such as pressing a button when a light flashes or turning a knob to keep one pointer over another. In process control he generally observes the reading of an indicating measuring instrument and operates a hand regulating-valve accordingly to maintain the instrument reading at a desired value. In essence the human operator closes the control loop between the output signal from the process and the element which regulates the manipulated variable, i.e. he carries out the functions of error discrimination and correction. Note, as shown in Figure 1–12, that there is still a closed-loop system in a manual control.

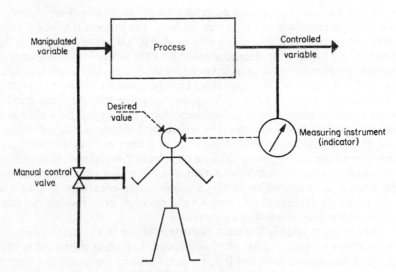

FIGURE 1–12. Manual control

A system which includes a human operator is naturally subject to the limitations of the human element; these are principally the human reaction time to a given stimulus, which varies between individuals and in the one individual, depending upon conditions of health and environment and the onset of fatigue. There can be no dispute that a skilled manual operator can control some processes with considerable success. Concentration is, however, difficult to maintain over long periods, and fatigue is bound to occur sooner or later when efficiency will suffer. The more complex the process and the faster the process reactions, the more difficult it becomes for a human operator to maintain satisfactory control for even limited periods.

Replacement of the human operator by a mechanical device to carry out the same basic control function makes the control automatic and immediately replaces the limitations of the human element by those of the mechanical design. The automatic controller has the advantages of acting logically with the same reaction to a given stimulus for very long periods; the instrument does not suffer from fatigue in the way of the human operator but does, of course, like the human operator, require some maintenance periodically to prolong its working life. A well-designed automatic system will give potentially better and more consistent results over longer periods than can a manual control, if only because the instrument must react consistently. This means that process variables can be held to closer limits, usually leading to improvements in product quality and more economic operation. The plant can generally be operated closer to its economic or optimum throughput, so increasing product quantity. The elimination of peak values of variables may also permit some economies in the capital cost of the plant as reserve or buffer capacity can be reduced. It is in this greater productivity that automatic control yields its major economic dividends.

The subject of manual *versus* automatic control would not be complete without some discussion of the position of the human operator possibly displaced by the use of an automatic control system. It is unfortunate that the view still exists in some circles that the purpose of automatic control is to eliminate the costs of manual labour. This is just not the case, since in most industries the use of complete automatic control of the processes will never completely replace the human element. This is particularly the case in the process industries, where very few plants can be operated entirely automatically and some human supervision is always required to deal with situations which instruments cannot be designed to handle. Effectively the human element is still required to provide the experience of past events, the memory, and the important capacity for independent action which the instrument lacks. Process workers will also still be required to deal with such factors as plant breakdown, leakages, the provision of raw materials, the removal of finished products, and so on. The most that can be said of the effect of automatic control on process labour is that a re-deployment of the labour force may be effected. Some workers will be released from the monotony of minute-to-minute manual control of the plant and will be diverted to other duties, some will have the opportunity of becoming supervisors or craftsmen, since additional skills are often called for in automatically controlled installations.

The above argument, of course, pre-supposes that a previously manually-controlled plant is changed over to fully automatic operation. In the more progressive countries this change-over was accomplished many years ago as plants became obsolescent and were replaced. The problem today is that of erecting new plants of larger capacity or for new products to replace the less efficient plants of yesterday. These new plants will be designed for automatic control from the outset, and some could not in fact be controlled manually, but the general trend is not to cut down the labour force but to employ it more profitably by increasing the *per capita* production with the aid of automatic control.

Control Diagrams

It is convenient in the study of control systems to make use of conventional diagrams to illustrate the inter-relationship of the systems components. Most readers will be familiar with the flow diagrams used in chemical engineering and similar studies, e.g. Figures 1–2 and 1–7, and there is considerable similarity between this sort of diagram and that used in control studies, the closed loop being readily distinguishable. However, this similarity can be as much misleading as helpful, particularly if it is remembered that the bases of the two diagrams are fundamentally different. In a flow diagram for chemical engineering use, the main feature is the flow of material and/or energy; in the control diagram the prime concern is the transmission of information in the system as a function of the independent time-variable. These latter considerations are best generalized by the concept of a *signal*. When a thermometer is used to measure temperature, the information produced is transmitted as a signal to the next stage, which may be an indicator or the error-generating unit of a controller. For the purpose of the generalization it is not necessary to specify the physical nature of the signal, whether it is a movement of a pointer, an air pressure or voltage, etc., and it is not essential to know how the signal is transmitted. What is necessary is to know how the signal is affected in quality and time, i.e. how long does it take to travel through the particular stage and how much, if any, is lost on the way. Any measurable variable in a process is information which can be regarded as a signal, and any stage which affects the signal in time and magnitude, is a transducer which operates on and transmits the signal.

The process control example of the fluid heater discussed previously is shown in Figure 1–7 as a conventional flow diagram. In Figure 1–13 the same process is shown as a signal flow diagram which shows most of the variables affecting the process. Initially there is the desired value input to the controller, the output signal from which is assumed to be an air pressure fed to the control-valve actuator. This in turn produces a mechanical movement of the valve and so regulates the flow of steam to the heater. The heat content of the steam passes, via the heater tubes, to the process fluid, the temperature change of which is measured by a thermometer. Information from the thermometer is passed as a mechanical displacement into the controller for comparison with the desired value. The flow of steam condensate from the heater does not form part of the control loop, and will have no effect on the outlet temperature of the fluid so long as the

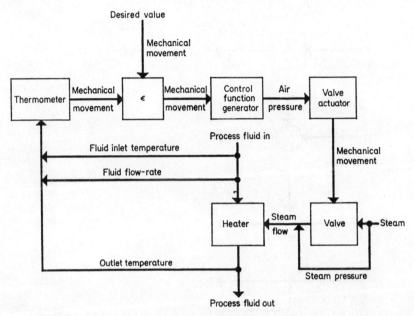

FIGURE 1–13. Fluid heater signal flow diagram

steam trap operates efficiently. The fluid inlet stream does affect the outlet temperature by changes in both rate of flow and inlet temperature, and this is indicated on the diagram. Note that these two variables have opposite effects; an increase in inlet temperature will produce an increase in the outlet temperature, but an increase in the rate of flow will lead to a decreased outlet temperature. The other major variable which it would be useful to show on the diagram is a possible change in steam pressure before the control valve. This will affect the rate of flow of steam through the valve and so change the heat flow rate into the system and ultimately to a change in the fluid outlet temperature. It is difficult, however, to illustrate this effect in the diagram since the effect must be transmitted through the process to influence the outlet temperature.

In a conventional block diagram it is possible to show both the point of incidence of any load variable and its effect on the system. Each transducer of information is represented by a simple rectangular block, with the input signal entering on one side and the output signal leaving on the other. The 'block' is made sufficiently large so that an expression can be written inside it to show the effect on the signal in passing through the particular transducer. This expression is the *system* or *transfer function* of the particular element represented by the block and is effectively the ratio of the output to the input. Thus if the function in the block is G, with input x and output y, then

$$y = G \cdot x$$

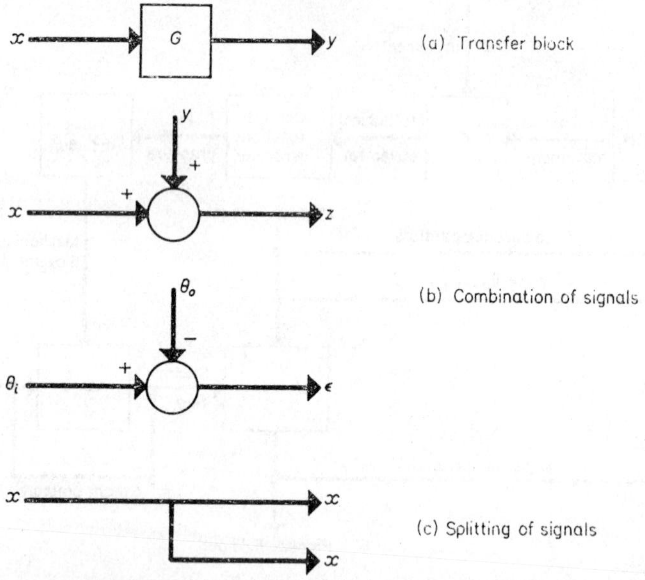

(a) Transfer block

(b) Combination of signals

(c) Splitting of signals

FIGURE 1-14. Conventional symbols for control diagrams

The function may be a dimensionless number if the signal is merely changed in magnitude (amplified or attenuated); it may be dimensional if the signal is changed in physical form, e.g. $lbf/in^2/in$, and it may also be a function of time if the signal is subjected to a lag or delay. As will be seen in the next chapter, most transfer functions involve derivatives of the time variable and it is then generally more convenient to use as the transfer function the ratio of the Laplace transforms of the output and input signals. It should be noted that there can only be one signal entering and one signal leaving each transfer block; this reflects the partial nature of the system analysis, since to determine the transfer function only one variable may be considered at a time.

Two signals can, however, be combined to form a third by addition or subtraction, and this is shown on the diagram by a circle with two lines entering and one leaving. The circle represents the operation of combining the signals, and the use of positive and negative signs indicates the particular arithmetic combination. Thus, in Figure 1–14,

$$\varepsilon = \theta_i - \theta_o$$

$$z = x + y$$

It is also possible to split a signal into two or more simultaneous paths. This implies that the transducers following the split are non-interacting and that each reacts to the signal without affecting its magnitude, i.e. in electrical terms there must be a high input impedance.

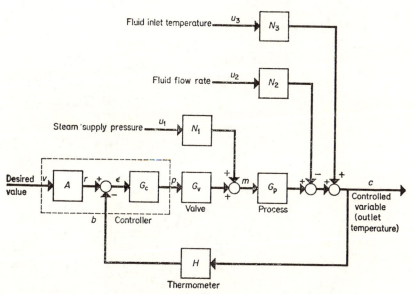

FIGURE 1–15. Conventional block diagram for fluid heater control

It is usual, although not absolutely essential, to indicate the direction of transmission by the use of arrowheads at appropriate points. These symbols are illustrated in Figure 1–14, and Figure 1–15 illustrates their use for the fluid heater control system. It will be seen that the latter figure shows the three load variables discussed above and also their points of incidence on the system. It is clear that the supply disturbance (steam pressure) must pass through the process elements to affect the controlled variable, whereas the demand disturbances (fluid flow-rate and inlet temperature) do not. It is also clear from the positive and negative signs that the latter two disturbances have opposing effects on the outlet temperature. Figure 1–15 also illustrates symbols which may be used for the particular variables and transfer functions in the control system, these are identified in Table 1–1.

TABLE 1–1

Variables	Transfer functions
v = desired value r = input signal e = error signal p = controller output m = manipulated variable c = controlled variable b = feedback variable u = load variable	A = input element G_c = controller elements G_v = control valve G_p = process elements H = feedback elements N = load elements

Chapter 2: Process Dynamics

Traditionally the design of machines and plant has been based on two factors, the *equilibrium* of thermal and mechanical energy and of chemical reactions, and the *kinetics* of moving bodies, flowing fluids, and heat and mass transfer. For manual control, design of the process to 'steady-state' conditions in this way is quite feasible. Automatic control, however, is based on dynamic action. Disturbances occur and variables change with respect to time, the function of the controller is to restore the controlled variable back to its design (or desired) value, and so again the variables are changed with respect to time. A controlled process can never continue in a steady-state condition indefinitely, since if it should, then control is not necessary.

Unsteady-state problems have long been known and studied in many diverse fields: the mechanics of solid bodies, fluid flow, hydraulics, heat transfer, etc. The relationships for the changes with time of the variables in physical systems can be derived from the same physical laws that are used to predict steady-state relationships in kinetic and equilibrium studies. The prediction of the dynamic unsteady-state response of a system is made by calculations based on the familiar principles which can be summarized by three types of equations: conservation, transport-rate and conversion equations. The basic method of determining the dynamic response of a system is to write first the equation for conservation of the material (mass or volume for an incompressible fluid), momentum, force or energy in the system at any given instant of time. The use of the transport-rate equation for the particular system then introduces the time variable and, by use of conversion equations, the response equation can be converted into the particular variables which it is desired to relate.

The objective of the dynamic analysis of the process is to see how conditions (variables) change with time. Many of the properties of the system and hence the process variables will be functions of time, e.g. (dx/dt), and the equation for the dynamic response of a particular variable with respect to changes in the input variables will be a differential equation. As these equations have to be manipulated and solved to determine the temporal response of the output variable, it is desirable that the final equation should be linear and with constant coefficients. In practice many process elements are not strictly linear, and it is thus often necessary to obtain an approximate solution by a process of 'linearization'.

Introduction to Dynamic Response

As an illustration of the development of a dynamic response equation, consider the simple mass-spring system of Figure 2–1, which consists of a weight hung from a spring. The system is initially at rest with the weight at a position x_0 measured from an arbitrary datum; the weight is then displaced and so set in motion. The problem is to determine the dynamic relationship between the position of the weight and the independent time variable. If the weight is at position x at time t, what is the relationship between the displacement $(x - x_0)$ and t?

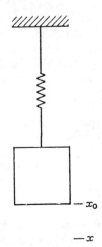

FIGURE 2–1. Mass-spring system

The solution is found by writing first the conservation equation for the system, which in this case is a force balance. At any instant the gravitational force on the weight is constant and the opposing force developed by the spring is determined by the displacement; the difference between the two is the inertial force due to the acceleration of the weight. Thus

$$F_i = F_g - F_s$$

where F_i is the inertial force,
F_g is the gravitational force,
F_s is the spring force.

The time variable is introduced by the transport-rate equation which in this case is Newton's second law of motion, the inertial force is equal to the mass of the weight times the acceleration, i.e.

$$F_i = m(d^2x/dt^2)$$

Having now introduced the linear variable x and the constant mass m, conversion equations are now used to introduce these terms into the other forces. Thus, Hooke's Law relates the spring force to the displacement, i.e.

$$F_s = K(x - x_0)$$

and the gravitational force is given by the mass times the gravitational acceleration, g,

$$F_g = mg$$

If these are substituted in the force balance equation, this now becomes

$$m(d^2x/dt^2) = mg - K(x - x_0)$$

This may be simplified by measuring the displacement of the weight from the zero position, i.e. by putting $x_0 = 0$, and the equation can be re-arranged to

or
$$\left. \begin{aligned} m(d^2x/dt^2) + Kx = mg \\ \ddot{x} + (K/m)x = g \end{aligned} \right\} \tag{2-1}$$

This is a second-order equation with constant coefficients (if certain assumptions are tenable) and is thus capable of solution by normal methods. It is, of course, the equation of simple harmonic motion and this immediately indicates that the analysis is incomplete. The equation implies that the weight would oscillate for ever, whereas in practice the oscillation must gradually subside and the weight come to rest. The derivation does not, in fact, allow for the frictional drag of air resistance; this frictional drag may be assumed to be proportional to the velocity of movement of the weight and the consequent introduction of a (dx/dt) term into the equation would give a damped harmonic solution, which represents the true state of affairs.

This is, however, a good illustration of how to obtain a system equation in a simple case, and it also illustrates that to get the neat equation above some simplifying assumptions have tacitly been made; these are

(a) that friction is negligible;
(b) that the spring is linear;
(c) that the mass of the spring can be neglected;
(d) that elastic deformation of the weight can be ignored.

In (b) a linear relationship has been assumed, otherwise the equation would be non-linear. In (c) and (d) the parameters have been 'lumped'; it is assumed that all the mass in the system is concentrated in the weight and all the elastic deformation in the spring. The lumping of parameters is a widely used technique and is generally permissible; the assumption of a linear relationship between two variables is also widely done, even though in many cases it is not strictly accurate.

Electrical $R–C$ Circuit

A second example of dynamic analysis is taken from the electrical field; consider the electrical circuit consisting of a resistance R and a condenser of capacitance C (Figure 2–2). A voltage e_1 is applied to charge the condenser, but at any intermediate time the voltage across the condenser will

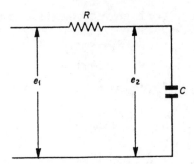

FIGURE 2–2. Electrical circuit with resistance and capacitance

be e_2. By a voltage balance across the resistance and by use of the appropriate transport-rate equation (Ohm's Law),

$$e_1 - e_2 = i \cdot R$$

The current is a variable with time and is the rate of flow of charge
($i = dq/dt$), and the charge on the condenser is given by $q = Ce_2$. The
voltage balance equation can thus be re-written

$$e_1 - e_2 = R(dq/dt) = RC(de_2/dt)$$

or

$$RC(de_2/dt) + e_2 = e_1 \qquad (2\text{–}2)$$

This is a first-order differential equation with constant coefficients relating
the input and output voltages with time. Once again the parameters have
been 'lumped', since all the resistance is assumed to be concentrated at one
point (the resistor) and the all the capacitance at another (the condenser).

Liquid Level

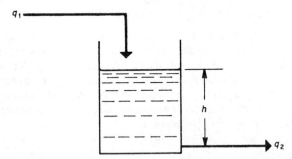

FIGURE 2–3. Liquid level system

An example from the process field is presented by the tank shown in
Figure 2–3, in which liquid enters at the top and leaves from the bottom
under a gravity discharge due to the pressure head of the liquid in the tank.

If the inflow and outflow rates are respectively q_1 and q_2 in units of volume per unit time, then over an interval dt, during which the inflow and outflow rates are not equal, there will be an accumulation or depletion of liquid in the tank given by $(q_1 - q_2)\,dt$. This accumulation or depletion of the contents will cause the level to change by an amount dh_2, and the volume accumulated or depleted will be given by the change in level multiplied by the cross-sectional area of the tank, A. Thus a volume balance can be written

$$(q_1 - q_2)\,dt = A\,dh_2$$

or
$$A(dh_2/dt) = q_1 - q_2$$

To relate the outflow, q_2, to the level in the tank, h_2, and to preserve a linear relationship, it must be assumed that the outflow is laminar, i.e. that the liquid is viscous and that the outflow connection is a smooth pipe of length, L, and diameter D. The outflow can then be given by the Poiseuille-Hagen equation:

$$q_2 = [\pi D^4 g/(128\mu L)]\Delta p$$

where μ is the viscosity of the liquid and Δp is the pressure difference across the outflow pipe. If the latter discharges freely to atmosphere, the pressure difference is the pressure head due to the level of liquid in the tank, i.e. $h_2\rho$, where ρ is the density of the liquid. Hence

$$q_2 = [\pi D^4 g\rho/(128\mu L)]h_2$$

in which the term in brackets is effectively a constant which can be replaced by $1/K$, so that $q_2 = h_2/K$. Substituting now in the volume balance equation,

$$q_1 - h_2/K = A(dh_2/dt)$$

or
$$AK(dh_2/dt) + h_2 = Kq_1 \qquad (2\text{--}3)$$

This equation gives the relationship between the two quantities which vary with time, the inflow, q_1, and the level in the tank, h_2. It will be seen that the relationship is a first-order differential equation with constant coefficients similar in structure to Equation (2–2).

Thermal System

Consider any type of contact temperature-sensing element such as the thermometer bulb shown in Figure 2–4. The temperature indicated by the thermometer is effectively that of the bulb, θ_2. If the temperature of the fluid surrounding the bulb is θ_1, there will be a net flow of heat between the two in a time dt, given by $hA(\theta_1 - \theta_2)\,dt$, where h is the heat transfer coefficient and A is the surface area of the bulb. The amount of heat transferred will be gained or lost by the bulb depending upon the direction of the temperature gradient, and will lead to a change in the bulb temperature of $d\theta_2$. The heat transferred will then also be given by the change in heat content of the bulb, $mc_p \, d\theta_2$, where m is the mass of the bulb and c_p the specific heat capacity (heat capacity per unit mass).

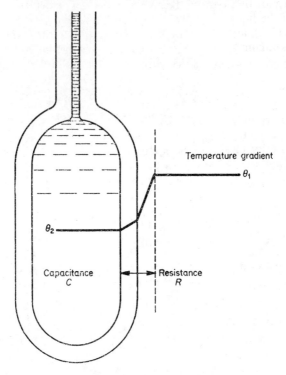

FIGURE 2–4. Thermometer bulb in fluid

A heat balance can then be written

$$hA(\theta_1 - \theta_2) \, dt = mc_p \, d\theta_2$$

whence

$$[mc_p/(hA)] \, d\theta_2/dt + \theta_2 = \theta_1 \qquad (2\text{--}4)$$

which is again a first-order differential equation with constant coefficients. It will be noted that parameters have once again been 'lumped'. All resistance to heat flow is assigned to the thin film of stagnant liquid in contact with the bulb, and all the heat capacity is assigned to the bulb while that of the liquid film is ignored; this is permissible since the film is thin and its mass is negligible.

Analogous Quantities

In deriving Equation (2–3) for the liquid level system, several properties of the system were grouped into one constant K, leaving only the two parameters, A and K, in the final equation. Equation (2–4) for the thermal system contains four constants, two of which can be seen to represent the heat capacity of the thermometer bulb (mc_p) and the other two the conductance of the liquid film surrounding the bulb (hA). Comparing Equations (2–2), (2–3), and (2–4), in the first R is the electrical resistance, in the second K is a combination of seven constants introduced to express the effect of viscous drag (or viscous resistance) on the flow, and in the third hA is a measure of thermal conductance. Since conductance is the reciprocal of resistance, all three terms are a measure of the resistance to flow in the particular system, i.e. K can be regarded as the 'flow resistance' of the pipework in Equation (2–3), and $1/hA$ as the 'thermal resistance' to heat flow in Equation (2–4).

This similarity or analogy between different physical systems can be used to advantage. The three equations refer to three entirely different physical systems but the equations are of basically the same form; the systems are therefore analogous in that their dynamic responses are represented by virtually the same equation. Since this equation is first-order, these are *first-order* systems.

The usefulness of the analogy begins when it is realized that the complex group represented by K in Equation (2–3) is merely the 'resistance' of the system. The second constant, A, in this equation should then be analogous to the electrical capacitance C in Equation (2–2) and to the thermal capacitance, mc_p, in Equation (2–4). The three quantities are dimensionally completely different, but capacitance is defined as the increase in content per unit increase in filling force, i.e. charge/potential, q/e in electrical terms. In terms of liquid level this becomes volume divided by pressure head, i.e. $(A\,dh)/dh = A$, and in the thermal system, heat content divided by temperature, i.e. $(mc_p\theta)/\theta = mc_p$.

There is an obvious similarity between the output voltage e_2, the output pressure head h_2, and the 'output' temperature θ_2, in the three cases, each effectively being a measure of force or potential. The analogy between the input terms e_1, Kq_1, and θ_1 will become apparent later.

Analogies of this sort are, of course, the basis of analogue computation. The behaviour of the liquid level system or the thermometer can be simulated by that of an electrical analogue consisting of a resistor and a condenser. By applying an input voltage equivalent to the liquid inflow or temperature, the output voltage across the condenser becomes a measure of the level in the tank or the temperature of the thermometer bulb, and by proper 'scaling' of the quantities, including time, a quantitative simulation can be obtained. It will also be apparent that a hydraulic analogy to the thermal system could be developed.

Analogue computation lies outside the scope of the present study, although its use in simulation and solution of control problems is a very valuable tool. The resistance and capacitance analogies are introduced as an aid to further process analysis and to simplify further calculations. Similar analogies can be shown to exist in other process systems.

The Time Constant

The three process systems considered can be represented by a general first-order equation:

$$RC(d\theta_2/dt) + \theta_2 = \theta_1$$

The two parameters in this equation, R and C, are analogous to the resistance and capacitance in the electrical circuit and, in general, these two parameters will be constant. Hence their product, RC, will also be constant. The product, which appears in the equation, can thus be replaced by a single parameter which is a fundamental characteristic of the particular system to which the equation is applied. It is a feature of resistance and capacitance, in the electrical definition and also in any process analogies thereto, that the product has the single dimension of time. Thus, in a thermal system, conductance hA has units such as W/K (or J/s K); thermal capacitance mc_p, J/K. The product RC, which is equal to $mc_p/(hA)$, is thus J/K × (J/s K)$^{-1}$, which reduces to the single unit of time (s).

The product of resistance and capacity is thus a fundamental time property of the system which is referred to as the *time constant* and is denoted by T. The first-order equation may then be written in the still more general form

$$T(d\theta_2/dt) + \theta_2 = \theta_1 \qquad (2\text{--}5)$$

It may be noted that some systems may exhibit first-order behaviour and a time constant, but the latter is not readily identified as the product of a resistance and a capacitance. In such cases the time constant may not be a true constant but may change with changes in an input variable, as in the case discussed on pages 36–37.

Transfer Function and Gain

The relationship between the input and output of any process system or individual element of a system may be expressed as a ratio of the output to the input, a quantity which is defined as the *system* or *transfer function*. Thus

$$\text{Output} = \text{Input} \times \text{Transfer function}$$

In cases where the relationship between output and input is a direct proportionality, i.e. when the element is an amplifier or a simple transducer, the transfer function is the numerical ratio of output to input, and will be dimensionless if the input and output are of the same physical form and are expressed in the same dimensional units. The function will be dimensional if the element is a transducer which changes the physical nature of the signal so that the output and input are measured in different units.

When the input and output are related by a differential equation such as that of the first-order system (Equation 2–5), it is convenient to use an alternative mathematical notation, to replace the differential operator (d/dt) by a single symbol which allows the equation to be re-arranged into a more useful form. Whilst the symbols D and p are conventionally used for this purpose, there is some advantage at this stage in using the symbol s, thus $d\theta/dt = s\theta$. Equation (2–5) can then be written

$$Ts\theta_2 + \theta_2 = \theta_1$$

whence
$$(Ts + 1)\theta_2 = \theta_1$$

and
$$\frac{\theta_2}{\theta_1} = \frac{1}{Ts + 1} \tag{2–6}$$

Equation (2–6) now defines the transfer function between the output, θ_2, and the input, θ_1, as $1/(Ts + 1)$, which is characteristic of first-order systems.

As will be seen later (pages 52–53), when the initial value of θ_2 is zero (i.e. $\theta_2 = 0$ at $t = 0$), $s\theta_2$ is the Laplace transform of $d\theta_2/dt$. It then becomes possible to define the transfer function more strictly as the ratio of the Laplace transforms of the output and input signals.

Under steady-state conditions, i.e. when the variables are constant in magnitude either under initial conditions at zero time or subsequently after a transient period following a finite change in magnitude of the input, then $d\theta_2/dt = 0$, or $s = 0$. The term in s thus disappears from the transfer function which, in the cases of Equations (2–2), (2–4), (2–5), and (2–6), reduces to unity and the output is equal to the input ($\theta_2 = \theta_1$). With the liquid level system previously discussed the numerator of the transfer function is not unity since from Equation (2–3),

$$h_2/q_1 = K/(Ts + 1)$$

and at the steady-state when $s = 0$, $h_2 = Kq_1$. The ratio between the output and input signals at the steady state is thus a direct proportionality between the two signals, since no time function is involved. This ratio is defined as the *steady-state gain*, or simply the *gain* of the particular element. In Equations (2–2) and (2–4) the gain is unity and dimensionless, whereas in Equation (2–3) it is the factor K which has the dimensions of s/m^2 (K is actually the flow resistance of the pipework in the level system). This difference arises naturally from the nature of the signals in the particular cases; in Equation (2–2) the signals are both voltages, in Equation (2–4) both signals are temperatures, but in Equation (2–3) the input q_1 is a rate of flow and the output h_2 is a liquid level or head. In practice the input and output signals of an element of a control system are not usually of the same form, and the steady-state gain will therefore usually be a dimensional factor.

Further consideration of the quantity, Kq_1, which appears in Equation (2–3), shows that this will have the dimension of length (or head), i.e.

$$(s/m^2)(m^3/s) = m$$

It is, in fact, the head which would be present in the tank when the out-flow is q_1, i.e. the head when the outflow, q_2, has become equal to the inflow, q_1, which will be the case at the steady state following a finite change in the inflow. This head (Kq_1) may then be regarded as an input signal to the system, or a potential head, h_1, which corresponds to the instantaneous value of the inflow q_1 at any time; it is the head which will ultimately be reached in the tank if the particular value of the inflow is maintained for a sufficiently long period. Substitution of this potential head h_1 for Kq_1 in Equation (2–3) puts this equation into the same form as Equations (2–2) and (2–4).

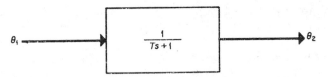

$\theta_1 \qquad\qquad \dfrac{1}{Ts+1} \qquad\qquad \theta_2$

FIGURE 2–5. Block diagram for first-order system with unit gain

Now that the transfer function has been defined and may be evaluated from the system equation, it is possible to represent the first-order system by the conventional block diagram representation of Figure 2–5.

Mixing Processes

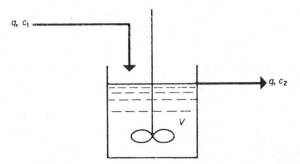

$q, c_1 \qquad\qquad V \qquad\qquad q, c_2$

FIGURE 2–6. Stirred mixing vessel

Consider an agitated tank, as in Figure 2–6, such as may be used to damp fluctuations in some property such as concentration of a solute, pH, density or temperature of the liquid feed to a succeeding item of plant. Assuming perfect mixing, i.e. that the input stream is instantaneously mixed with the liquid in the tank and that the particular property, e.g. concentration, is uniform throughout the tank and is therefore equal to that of the outflow, a conservation equation can be written:

$$qc_1 - qc_2 = V(dc_2/dt)$$

where c_1 and c_2 are respectively the inflow and outflow concentrations per unit volume, q is the volumetric liquid flow rate and V is the hold-up volume in consistent units. This equation can be re-arranged into the usual first-order form

$$(V/q)(dc_2/dt) + c_2 = c_1 \qquad (2\text{--}7)$$

in which V/q has obviously the dimension of time, being the hold-up or residence time of the liquid in the tank. V/q may thus be written as a time constant T, and the usual first-order transfer function can then be obtained:

$$c_2/c_1 = 1/(Ts + 1)$$

It will be apparent that in this case the time constant T is subject to change whenever the flow rate through the system varies, although it will often be the case in practice that changes in the rate of flow from the normal average value are not sufficiently large to make any significant change in the time constant.

The assumption of perfect mixing may not always be tenable in practice, as several seconds may be required to thoroughly mix the contents of a tank and in some applications the mixing lag is often appreciable, particularly if the hold-up time is small. In the present example the hold-up time is considered to be appreciably larger than the mixing lag, so that the latter can be neglected.

The stirred tank may also be used as a chemical reactor, in which case the quantity of material reacting per unit time is an additional term in the material balance. If the reaction is zero-order the time constant is again the hold-up time, since changes in concentration do not affect the rate of reaction. For higher-order reactions the rate is determined by the concentration of one of the reactants, and the system equation and time constant are modified. Thus for a first-order reaction, the rate of reaction is directly proportional to concentration of a reactant, i.e.

$$r = kc_2V$$

where r is the mass reacting per unit time (kg/s),

k is the reaction rate constant (s^{-1}),

c_2 is the mass concentration of reactant (kg/m³),

and V is the hold-up volume (m³),

and the material balance for the reactant becomes:

$$qc_1 - qc_2 - kc_2V = V(dc_2/dt)$$

whence

$$(V/q)(dc_2/dt) + (kV/q + 1)c_2 = c_1$$

(V/q) is again the hold-up or residence time but is not now the time constant since to obtain the transfer function, c_2/c_1, the equation must be further re-arranged to the standard first-order form, i.e.

$$[T'/(kT' + 1)](dc_2/dt) + c_2 = c_1/(kT' + 1) \qquad (2\text{--}8)$$

where $T' = V/q$. The time constant is thus $T'/(kT' + 1)$ and the effect of the reaction is to make the time constant smaller than the hold-up time. The reactor also differs from the mixing vessel in that the output concentration c_2 differs from the input concentration c_1 under steady-state conditions, since the steady state gain is not now unity but $1/(kT' + 1)$. This is, of course, logical since some of the reactant is removed from solution by the reaction and c_2 must be smaller than c_1 if reaction takes place.

For reactions which are second- or fractional-order, an exact response equation cannot be obtained by these methods since the differential equations are not now linear. It is possible, however, to obtain an approximate response to small changes by using a linear approximation of the rate expression, as is considered later under 'linearization'.

Temperature Response of Heated Vessels

In the previous examples first-order equations have been developed for elements with a single input variable; in the following example it will be shown that similar first-order equations can be derived when the system has more than one input variable.

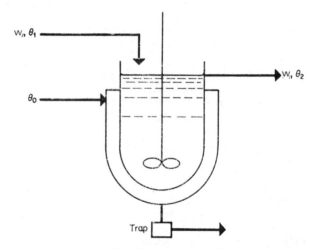

FIGURE 2–7. Steam-heated stirred tank

Consider the steam-heated vessel of Figure 2–7, and assume that the thermal capacity of the vessel wall is negligible and that the temperatures in the steam space and in the liquid in the vessel are both uniform. Assume further there is no heat loss from the system; a heat balance can be written:

$$wc_p\theta_1 - wc_p\theta_2 + UA(\theta_0 - \theta_2) = Mc_p(d\theta_2/dt)$$

where w is the mass flow rate of the liquid,

M is the mass of liquid in the tank,

c_p is the specific heat capacity of the liquid,

U is the heat transfer coefficient,

A is the area of heating surface,

and θ_0, θ_1, θ_2 are the temperatures of steam and liquid input and output respectively.

The equation can re-arranged into general first-order form, thus:

$$[Mc_p/(wc_p + UA)](d\theta_2/dt) + \theta_2$$
$$= [wc_p/(wc_p + UA)]\theta_1 + [UA/(wc_p + UA)]\theta_0$$

or

$$T(d\theta_2/dt) + \theta_2 = K_1\theta_1 + K_2\theta_0 \tag{2-9}$$

In this derivation the temperatures, θ_0, θ_1, and θ_2, are all measured from a common datum such as 0°C. It can readily be shown that the heat balance equation above is correct even though three different base temperatures are used and the temperature variables are regarded as deviations from specified initial values. If the temperatures in the heat balance equation are written as an initial temperature plus the deviation, e.g. $\bar{\theta}_1 + \theta_1$, etc., the equation becomes

$$wc_p(\bar{\theta}_1 + \theta_1) - wc_p(\bar{\theta}_2 + \theta_2) + UA(\bar{\theta}_0 + \theta_0 - \bar{\theta}_2 - \theta_2)$$
$$= Mc_p(d\theta_2/dt)$$

But at the steady-state initial value,

$$d\theta_2/dt = 0$$

and also

$$\theta_0 = \theta_1 = \theta_2 = 0$$

so that

$$wc_p\bar{\theta}_1 - wc_p\bar{\theta}_2 + UA(\bar{\theta}_0 - \bar{\theta}_2) = 0$$

Subtracting this last equation, which is the steady-state heat balance, from the preceding equation, gives the original heat balance equation. The variables in most control system equations are almost invariably written, as in this last example, in the form of deviations from initial or normal values so that at the initial steady-state condition obtaining at zero time, all the variables are zero.

Equation (2–9) now shows that the output temperature is determined by two input variables, the steam temperature, θ_0, and the input temperature, θ_1. In controlling this system the former would almost invariably be the manipulated variable (through the steam supply) and the latter would be a load variable. The system will now have two transfer functions, one defining the effect of steam temperature on the output temperature and the other that of the inlet temperature. By writing s for d/dt in Equation (2–9), the system equation becomes

$$\theta_2 = [K_1/(Ts + 1)]\theta_1 + [K_2/(Ts + 1)]\theta_0 \tag{2-10}$$

The two transfer functions thus have the same time constant,

$$T = Mc_p/(wc_p + UA)$$

but different steady-state gains,

$$K_1 = wc_p/(wc_p + UA)$$

and

$$K_2 = UA/(wc_p + UA)$$

The equation also shows that both input functions are additive, increases in either input or steam temperature lead to an increase in the outlet temperature. A block diagram for this system thus has two blocks, one for each input variable signal, with the addition circle to combine the outputs. However, since the time constant is the same for each variable, Equation (2–10) can also be written:

$$\theta_2 = [K_1/(Ts + 1)][\theta_1 + (K_2/K_1)\theta_0]$$

and the alternative diagram of Figure 2–8 is equally correct.

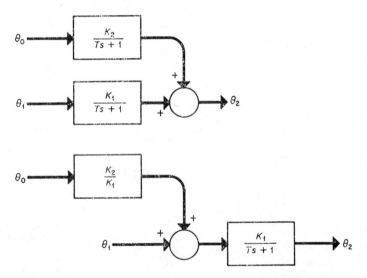

FIGURE 2–8. Alternative block diagrams for steam-heated tank

In the above discussion it has been tacitly assumed that the flow rate of liquid through the tank is constant. This is not necessarily so and the liquid flow rate can be regarded as a further input load variable. Significant changes in this quantity will, however, also change the hold-up time and the time constant of the system, and it is therefore necessary to restrict the change in flow rate to small values so that the effect on the time constant can be neglected. A further complication is now introduced in that the first two terms of the heat balance equation will contain the product of two variables, e.g. the first term becomes $wc_p\theta_1$, where both w and θ_1 are now input variables. The product of two variables is a non-linear expression but for reasonably small changes in the variables the expressions can be linearized by the method discussed in the next section. This effectively requires the average rate of flow, $\bar{w}$, to be defined for the first two terms of the heat balance which allow for changes in inlet and outlet temperature, and a deviation from the average flow rate, w, along with the average temperatures, $\bar{\theta}_1$ and $\bar{\theta}_2$, for the additional term in the heat balance due to the change in the flow rate. The heat balance now becomes

$$\bar{w}c_p\theta_1 - \bar{w}c_p\theta_2 + UA(\theta_0 - \theta_2) + wc_p(\bar{\theta}_1 - \bar{\theta}_2) = Mc_p(d\theta_2/dt)$$

or, by re-arrangement,

$$T(d\theta_2/dt) + \theta_2 = K_1\theta_1 + K_2\theta_0 - K_3w$$

where T, K_1, and K_2 have the same values as in Equation (2–9), but with the average flow rate, $\bar{w}$, in the denominator, and

$$K_3 = c_p(\bar{\theta}_2 - \bar{\theta}_1)/(\bar{w}c_p + UA)$$

The time constant for the response of the outflow temperature is thus the same for changes in both feed and steam temperatures and also for the flow rate, but the steady-state gains are different. This is generally true for most single-capacity first-order systems, the time constant is the same for all input variables (so long as changes in the variables are reasonably small), but the gains are generally different. Recognizing that the time constant is the same for different inputs often saves time and effort in deriving system equations and transfer functions, since once the time constant has been found the gains for the different inputs can be obtained from the conventional steady-state relationships rather than from the differential equations defining the unsteady-state response.

Linearization

All the process examples considered up to this stage have been linear, or a linear relationship between variables has been assumed, in order that the system equation should be a linear differential equation. Linear systems are those to which super-position can be applied, i.e. the response of the system to different disturbances at the same or different points, and simultaneously or at different times, is given by the sum of the individual responses considered independently. This principle does not apply to non-linear systems.

The distinction between linear and non-linear systems is not simply one of convenience; a number of techniques are available for solution of linear problems, but there are no general mathematical techniques for treating non-linear problems and no general analysis is therefore possible. Block diagram algebra, the use of characteristic roots to study stability, techniques based on Laplace transform and s-plane analysis are all limited in application to linear systems with constant coefficients. Only limited techniques are available for certain special cases in non-linear systems, although analogue and digital computers may be used but not always economically.

Many relationships in process analysis are non-linear, e.g. the square-root relationship between turbulent flow and pressure difference, mass and heat transfer correlations based on dimensionless groups with fractional powers, chemical reactions other than first order, the fourth-power radiation law, etc. On the 'hardware' side, control valves are often made with non-linear characteristics such as parabolic and semi-logarithmic relationships; often these non-linearities are desirable in certain applications and the valve is specifically designed to exhibit the desired relationship. Using these relationships in dynamic analysis, a non-linear equation is obtained which is not analogous to the resistance-capacitance equations previously derived.

It is possible, however, to obtain some useful results from such non-linear relationships by 'linearizing' the function. It is important though to understand the degree of error which may be introduced into the analysis by these techniques. Measurement and control are closely related arts; a process cannot be controlled unless the controlled variable can be measured, but the measurement can be made with almost any degree of precision required. Control, on the other hand, is concerned with the adjustment of variables, and extreme precision is rarely required. Process control systems normally operate with the several variables holding fairly steady at normal, constant operating levels. The variables are subject to some departure from these levels due to disturbances, but if the system is well designed these perturbations from the normal values will be relatively small, and the dynamic behaviour of the system can be described reliably by linear differential equations even though the basic process equations are non-linear.

There are several methods of linearizing a function, all rest on the simple fact that any non-linear function is approximately linear if considered over a sufficiently narrow range. Obviously, a non-linear curve cannot be approximated over its entire range by a straight line without considerable error, but at any point on such a curve a linear relationship, which is the slope of the curve, exists at that particular point. For small distances on each side of the point, departures from linearity will be relatively small but will increase as the range is widened. On the assumption that only small perturbations in the variables about the normal operating levels will occur, the range of variation will be narrow and departures from linearity over these narrow ranges will be sufficiently small to be usually negligible.

A standard method linearizing a function is to expand the non-linear relationship by means of a Taylor series but to retain only the linear terms, on the above assumption that only small perturbations about the normal operating values of the variables will occur. This is effectively the same as writing the partial differentials of the function with respect to each variable in turn, i.e. for a relationship such as

$$a = f(b, c, \ldots)$$
$$da = (\partial a/\partial b)\, db + (\partial a/\partial c)\, dc + \ldots$$

and to substitute for the partial differentials those of the original non-linear function at suitable 'mean values', i.e. corresponding values of the variables at their normal operating levels, which effectively are the mid-points of the range of perturbation being considered.

The case of the tank with liquid discharging under laminar flow conditions has already been considered and was shown to be a linear first-order system defined by Equation (2–3). If the discharge from the tank is not laminar but turbulent, as is most likely to be the case in practice, the discharge rate q_2 will not now be proportional to the pressure head h_2 offered by the liquid level in the tank, but will be proportional to the square root of the head, i.e. $q_2 = Kh^{\frac{1}{2}}$.

To obtain a linear differential equation for the system using this function it is necessary to assume that q_2 will be proportional to h_2 over a suitable narrow range, but it is then necessary to define this new relationship, i.e. to define a 'turbulent flow resistance'. Since h_2 is the only independent variable in the relationship, the partial differential equation is

$$dq_2 = (\partial q_2/\partial h_2)\, dh_2$$

The value of the partial differential can be found by differentiating the original function and then evaluating at the mean values of flow and level corresponding to the mid-point of the range over which the linear relationship is assumed to apply. These mean values will be the normal operating values, say $\bar{q}_2$ and $\bar{h}_2$. The partial differential thus becomes

$$(\partial q_2/\partial h_2) = K(\tfrac{1}{2}h^{-\frac{1}{2}})$$

which can be further simplified by substituting for constant K the value given by the mean values from the original function, i.e. $\bar{q}_2 = K\bar{h}_2^{\frac{1}{2}}$, whence $K = \bar{q}_2/\bar{h}_2^{\frac{1}{2}}$, and

$$(\partial q_2/\partial h_2) = (\bar{q}_2/\bar{h}_2^{\frac{1}{2}})(\tfrac{1}{2}\bar{h}_2^{-\frac{1}{2}}) = \bar{q}_2/2\bar{h}_2$$

Substituting in the partial differential equation,

$$dq_2 = (\bar{q}_2/2\bar{h}_2)\, dh_2$$

whence
$$q_2 = (\bar{q}_2/2\bar{h}_2)h_2$$

which gives a linear relationship between the flow and head over a narrow range, the mid-point of which is defined by the normal operating values, $\bar{q}_2$ and $\bar{h}_2$.

This is, of course, equivalent to saying that the graph of q_2 *versus* h_2 can be regarded as a straight line passing through the mean values of $\bar{q}_2$ and

$\bar{h}_2$, with a slope of $\bar{q}_2/2\bar{h}_2$, which is the slope of the actual curve (dq_2/dh_2) at this point, as shown in Figure 2–9.

By analogy to the flow resistance previously discussed, the proportionality between q_2 and h_2 can be written $q_2 = h_2/R$, where R is now the turbulent flow resistance and is equivalent to the laminar flow resistance defined in the derivation of Equation (2–3). The laminar flow resistance was seen to be constant over the whole range of flow, but the turbulent flow resistance has a different value for each mean rate of flow and can thus be defined only for a specified mean rate of flow such as $\bar{q}_2$.

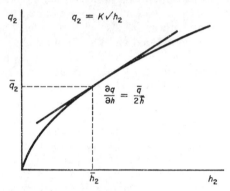

FIGURE 2–9. Linearization of liquid level system with turbulent outflow

By using the linearized flow equation in the material balance for the system, the response equation obtained is identical with that for the linear system (Equation (2–3)).

Example 2–1A: Linearization of Liquid Level System

A tank of 1 m² sectional area has a normal level of 1.44 m when the inflow and outflow are 2.4×10^{-3} m²/s. The inflow is increased suddenly to 3.0×10^{-3} m³/s. Compare the actual response of the level to the linearized response.

Actual response. The outflow is turbulent and $q_2 = kh^{\frac{1}{2}}$. Under normal operating conditions, $h = 1.44$ m when $q_2 = 2.4 \times 10^{-3}$ m³/s. Hence

$$k = 2.4 \times 10^{-3}/(1.44)^{\frac{1}{2}} = 2.0 \times 10^{-3} \text{ m}^3/\text{s m}^{\frac{1}{2}}$$

The final level, when the outflow is equal to the new inflow, is

$$h = (q_2/k)^2 = (3.0 \times 10^{-3}/2.0 \times 10^{-3})^2 = 2.25 \text{ m}$$

Volume balance for the new inflow,

$$A(dh/dt) = q_1 - q_2$$
$$= q_1 - kh^{\frac{1}{2}}$$
$$1.0(dh/dt) = 3.0 \times 10^{-3} - 2.0 \times 10^{-3}h^{\frac{1}{2}}$$

Let $h = X^2$. Then

$$dh = 2X\,dX$$

$$2X(dX/dt) = (3 - 2X) \times 10^{-3}$$

which can be re-arranged to

$$10^{-3}\,dt = X\,dX/(1.5 - X)$$

Integrating, $\qquad 10^{-3}t + C = -X - 1.5\ln(1.5 - X)$

Substituting the initial conditions, $X = (1.44)^{\frac{1}{2}}$ at $t = 0$,

$$C = 0.606$$

The actual change in level is thus given by

$$-[h^{\frac{1}{2}} + 1.5\ln(1.5 - h^{\frac{1}{2}})] = 10^{-3}t + 0.606$$

A plot of this equation is shown in Figure 2–10 with $h = 2.25$ m as the ultimate value.

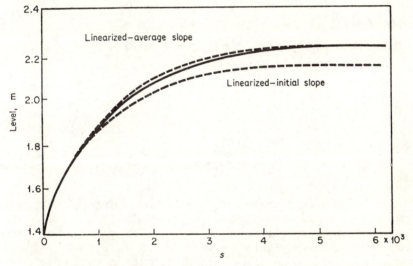

FIGURE 2–10. Actual and linearized response of level following step change in inflow (Example 2–1)

Linearized response. The linearized response may be based on the initial or final slope of the curve relating q_2 to h_2 (i.e. $\partial q_2/\partial h$), or on an average value, a chord slope ($\Delta q_2/\Delta h$).

using the initial slope,

$$\partial q_2/\partial h = \bar{q}_2/2\bar{h}$$

$$= 2.4 \times 10^{-3}/2 \times 1.44$$

$$= 10^{-3}/1.2 \text{ m}^3/\text{s m}$$

The turbulent flow resistance is thus

$$R = \partial h/\partial q_2 = 1.2 \times 10^3 \text{ s/m}^2$$

Since the capacitance (tank area) is 1 m^2, the time constant, $T = RC$, will be 1.2×10^3 s.

Using the initial slope, the change in value of h is given by $\Delta q_2(\partial h/\partial q_2)$, i.e.

$$(3.0 - 2.4) \times 10^{-3} \times 1.2 \times 10^3 = 0.72 \text{ m}$$

The final value of the level is thus

$$1.44 + 0.72 = 2.16 \text{ m}$$

As will be seen later (pages 53–54), the response of a first-order system to a step change in input of magnitude A is given by $A[1 - \exp(-t/T)]$. In the present case A (the magnitude of the final change in level) is 0.72 and the time constant T is 1.2×10^3. The response of the level is thus given by

$$h = 1.44 + 0.72[1 - \exp(-t/T)]$$

where $T = 1.2 \times 10^3$ s.

Using an average slope, the turbulent resistance R is given by

$$R = \Delta h/\Delta q_2$$
$$= (2.25 - 1.44)/[(3.0 - 2.4) \times 10^{-3}]$$
$$= 1.35 \times 10^3 \text{ s/m}^2$$

and the time constant (RC) is thus 1.35×10^3 s. The final change in level is now $2.25 - 1.44 = 0.81$ m, and the response is given by

$$h = 1.44 + 0.81[1 - \exp(-t/T)]$$

where $T = 1.35 \times 10^3$ s.

The two linearized responses are also plotted in Figure 2–10 for comparison with the actual response. It will be seen that the maximum error between the actual and linearized responses is of the order of 3 per cent, even though the change in level is over 50 per cent. The use of the time constant for the initial slope used with the correct final value of the level would further improve the linearization.

Example 2–1B

Derive the response equations for the change in level if the area of the tank is 10 ft^2, the normal level 9 ft when the inflow is 6 ft^3/min, and the inflow is increased to 7.5 ft^3/min.

Proceeding as in Example 2–1A, for the actual response, $q_2 = 2h^{\frac{1}{2}}$, and the final level when the outflow reaches the new inflow of 7.5 ft$^\circ$/min i $h = 14.06$ ft.

The volume balance is then

$$10(dh/dt) = 7.5 - 2h^{\frac{1}{2}}$$

and the actual response is found to be:

$$-[h^{\frac{1}{2}} + 3.75 \ln(3.75 - h^{\frac{1}{2}})] = 10^{-1}] - 1.921$$

Linearized response, using the initial slope, is

$$\partial q_2/\partial h = \tfrac{1}{3} \text{ ft}°/\text{min ft}$$

The turbulent resistance $R = 3$ min/ft^2 and the time constant (RC) is 30 min.

Increase in level = 4.5 ft, final value = 13.5 ft. The response is then given by

$$h = 9.0 + 4.5[1 - \exp(-t/30)]$$

Using the average slope,

$$R = \Delta h/\Delta q_2 = 3.37 \text{ min/ft}^2$$

and the time constant is 33.7 min.

The response is then given by

$$h = 9.0 + 5.06[1 - \exp(-t/33.7)]$$

Example 2–2: Linearization of Radiant Heat Transfer

Consider the radiant heat transfer between the walls of a furnace at a relatively high temperature and the refractory sheath of a thermocouple. When the temperatures of the furnace and sheath differ, a heat balance can be written:

$$q_{in} - q_{out} = C(d\theta_2/dt)$$

where C is the thermal capacitance of the sheath and θ_2 is its temperature.

It is difficult to estimate the heat loss from the sheath by conduction, but this will be almost negligible for small changes of temperature and may be assumed to be zero. Similarly, because of the high temperatures and the relatively small temperature difference between the furnace walls and the sheath, convective heat transfer will also be negligible and the heat interchange will be due to radiative transfer alone. Hence

$$q_{in} = \sigma EA(\theta_1^4 - \theta_2^4)$$

where θ_1 and θ_2 are the furnace and sheath temperatures respectively in absolute units (K), and σ, E, and A have the usual significance.

Linearizing this expression for average temperatures, $\bar{\theta}_1$ and $\bar{\theta}_2$,

$$q_{in} = [\partial(\sigma EA\theta_1^4)/\partial\theta_1]\theta_1 - [\partial(\sigma EA\theta_2^4)/\partial\theta_2]\theta_2$$
$$= (4\sigma EA\bar{\theta}_1^3)\theta_1 - (4\sigma EA\bar{\theta}_2^3)\theta_2$$

Radiant heat transfer resistances can now be defined by

$$1/R_1 = 4\sigma EA\bar{\theta}_1^3 \quad \text{and} \quad 1/R_2 = 4\sigma EA\bar{\theta}_2^3$$

from which

$$R_1/R_2 = \bar{\theta}_2^3/\bar{\theta}_1^3$$

However, both $\bar{\theta}_1$ and $\bar{\theta}_2$ are relatively large numbers, being the absolute temperatures of the furnace and the sheath, which differ by only a small amount; R_1/R_2 will then be very close to unity. R_1 and R_2 can then be replaced by an average value, R, with very little error and the heat interchange can then be written

$$q_{in} = (\theta_1 - \theta_2)/R$$

where $R = 1/(4\sigma EA\bar{\theta}_a^3)$ and θ_a is the average of the temperatures $\bar{\theta}_1$ and $\bar{\theta}_2$. The heat balance then reduces to

$$C(d\theta_2/dt) = (\theta_1 - \theta_2)/R$$

or

$$\theta_2/\theta_1 = 1/(Ts + 1)$$

where $T = RC$.

Example 2–2A

Determine the time constant of a refractory thermocouple sheath, 25 mm outer diameter, 5 mm wall thickness, 0.3 m length over a hemispherical end, specific heat capacity 0.8 kJ/(kg K), density 2.304×10^3 kg/m³, at a furnace temperature of 1400°C. The Stefan-Boltzmann constant,

$$\sigma = 5.6697 \times 10^{-8} \text{ W/(m}^2 \text{ K}^4)$$

The surface area of the sheath is that of the cylindrical section, of length $(300 - 12.5)$ mm, plus that of the hemispherical end, of 25 mm diameter.

Surface area

$$A = \pi \times 25 \times 10^{-3} \times 287.5 \times 10^{-3} + \pi/2 \times (25 \times 10^{-3})^2 \text{ m}^2$$
$$= 23.56 \times 10^{-3} \text{ m}^2$$

At a furnace temperature of 1400°C (1673K), the emissivity of the sheath E will be virtually unity and the difference in temperature of the furnace and sheath will be sufficiently small for the average temperature, θ_a, to be taken as 1673K.

$$1/R = 4\sigma EA\bar{\theta}_a^3$$
$$= 4 \times 5.6697 \times 10^{-8} \times 23.56 \times 10^{-3} \times (1673)^3$$
$$= 25.02 \text{ W/K}$$
$$= 25.02 \times 10^{-3} \text{ kW/K}$$

whence $R = 39.97$ K/kW

The volume of material in the sheath is that of the cylindrical portion (length 287.5 mm, 25 mm outer diameter, and 15 mm inner diameter) plus that of the hemispherical end (25 mm O.D. and 15 mm I.D.).

$$\text{Volume} = \pi/4 \times (25^2 - 15^2) \times 10^{-6} \times 287.5 \times 10^{-3}$$
$$+ \pi/12 \times (25^3 - 15^3) \times 10^{-9} \text{ m}^3$$
$$= 93.72 \times 10^{-6} \text{ m}^3$$

$$\text{Mass} = \text{Volume} \times \text{Density}$$
$$= 93.72 \times 10^{-6} \times 2.304 \times 10^3 \text{ kg}$$
$$= 0.215 \text{ kg}$$

$$\text{Thermal capacitance} = \text{Mass} \times \text{Specific heat capacity}$$
$$= 0.215 \times 0.8$$
$$= 0.172 \text{ kJ/K}$$

Time constant,
$$T = RC$$
$$= 39.97 \times 0.172 \text{ (K/kW)(kJ/K)}$$
$$= 6.88 \text{ s}$$

Example 2–2B

Determine the time constant of a refractory thermocouple sheath, 1 in outer diameter, 0.2 in wall thickness, length 12 in over a hemispherical end, specific heat capacity 0.2 Btu/(lb °F), density 144 lb/ft^3, at a furnace temperature of 1400°C. Stefan-Boltzmann constant, 0.173×10^{-8} Btu/(ft^2 hr °R^4).

$$\text{Length of cylindrical section} = 11.5 \text{ in}$$

$$\text{Surface area} = \pi \times 1/12 \times 11.5/12 + \pi/2 \times (1/12)^2 \text{ ft}^2$$
$$= 0.2617 \text{ ft}^2$$

$$\text{Temperature} = 1400°C$$
$$= 3012°R$$

$$1/R = 4 \times 0.173 \times 10^{-8} \times 1 \times 0.2617 \times (3012)^3$$
$$= 49.49 \text{ Btu/(h °R)}$$

$$R = 3600/49.49$$
$$= 72.74°R \text{ s/Btu}$$

$$\text{Volume} = \pi/4 \times (1 - 0.6^2)/144 \times 11.5/12$$
$$+ \pi/12 \times (1 - 0.6^3)/1728 \text{ ft}^3$$
$$= 0.003\,462 \text{ ft}^3$$

$$\text{Mass} = 0.003\,462 \times 144$$
$$= 0.4896\,\text{lb}$$

$$\text{Thermal capacitance} = 0.4896 \times 0.2$$
$$= 0.099\,72\,\text{Btu/}^{\circ}\text{F}$$

Time constant
$$T\ RC$$
$$= 72.74 \times 0.099\,72$$
$$= 7.25\,\text{s}$$

Pressure Systems

Processes involving the flow of gases through interconnecting pipework and pressure vessels are common in industrial practice. Such pressure systems are similar to level systems in that the flow equations must usually be linearized to obtain a system transfer equation since the flow is usually turbulent. That both inflow and outflow are usually dependent on vessel pressure and that there may be multiple inlets and outlets make these systems generally rather more complex than level systems. Only rarely is the time constant a simple multiple of the hold-up time, and the possibilities of sonic flow and temperature changes are additional complications.

For systems in which the pressure differences are less than about 5 per cent of the upstream pressures, compressibility effects can usually be ignored and the turbulent flow resistance, dp/dq, for a given restriction is then twice the pressure difference divided by the rate of flow at the normal operating levels, i.e. $2\Delta\bar{p}/\bar{q}$ (cf. turbulent flow resistance of the level system). However, in many applications the pressure differences are often substantial, and it is then necessary to introduce an expansion factor, ε, into the flow equation, which then becomes

$$q = KA\varepsilon(2g\Delta p/\rho)^{\frac{1}{2}}$$

The resistance is not now easily determined since the expansion factor is considerably dependent on pressure. It is usually easier to obtain the resistance from the slope of a plot of the flow against pressure or pressure difference.

A special case occurs when the pressure difference is larger than a critical value defined by the critical ratio with the upstream pressure (0.53 for air), in which case the flow velocity in the restriction attains the velocity of sound. Downstream pressure waves cannot then propagate upstream, and the flow is directly proportional to the upstream pressure. The flow resistance is then the ratio of the upstream pressure to the rate of flow.

The capacitance of a pressure vessel is the amount of gas stored per unit change in pressure, and is usually expressed in units of volume per unit of pressure.

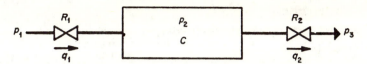

Considering now the vessel of Figure 2-11, in which the pressure p_2 is the dependent variable, a volume balance gives the continuity equation:

$$C(dp_2/dt) = q_1 - q_2$$

where C is the capacitance and q_1 and q_2 are the volumetric rates of flow in and out of the vessel.

Both inflow and outflow are determined by pressure difference; hence

$$C(dp_2/dt) = (p_1 - p_2)/R_1 - (p_2 - p_3)/R_2$$

where R_1 and R_2 are the flow resistances $(d\Delta p/dq)$, linearized if the flow is turbulent. Collecting like terms:

or

$$\left. \begin{aligned} C(dp_2/dt) + p_2(1/R_1 + 1/R_2) = p_1/R_1 + p_3/R_2 \\ p_2(Ts + 1) = K_1 p_1 + K_2 p_3 \end{aligned} \right\} \qquad (2\text{-}11)$$

where $\quad T = C/(1/R_1 + 1/R_2),$
$\quad K_1 = (1/R_1)/(1/R_1 + 1/R_2),$
$\quad K_2 = (1/R_2)/(1/R_1 + 1/R_2).$

Equation (2-11) shows how the output variable (the vessel pressure) depends on the input variables (upstream and downstream pressure). Either of the latter could be a manipulated variable or a load variable, or both could be load variables if the system were controlled by changing R_1 or R_2 as the manipulated variable.

It is relatively easy to show that the system remains first-order even when there are several inlets and/or outlets. Equation (2-11) is simply extended to include the additional upstream and downstream pressures. The time constant is the capacitance divided by the sum of the reciprocals of the resistances, i.e. $C/\Sigma(1/R)$, and the gains for the pressure terms are the reciprocal of the appropriate resistance divided by the sum of the reciprocals, i.e.

$$K_1 = (1/R_1)/\Sigma(1/R)$$

If sonic flow occurs through an inlet resistance, then flow is independent of vessel pressure and the particular resistance does not affect the time constant. If sonic flow occurs through an outlet resistance, again the resistance does not affect the time constant but the gain for the downstream pressure term is zero.

Valve Characteristics

If a valve is used in either the inlet or discharge line of the vessel of Figure 2–11 the position of the valve stem becomes an additional input variable which is usually the manipulated variable, both upstream and downstream pressures then becoming load variables. The flow through the valve is now a function of two variables, the pressure difference and the valve stem position m, i.e. $q = f(\Delta p, m)$. The total differential for the function of the two variables is then

$$dq = (\partial q/\partial \Delta p)\, d\Delta p + (\partial q/\partial m)\, dm$$

The first partial differential is the reciprocal of the flow resistance at the average valve position, the second is the rate of change of flow with valve

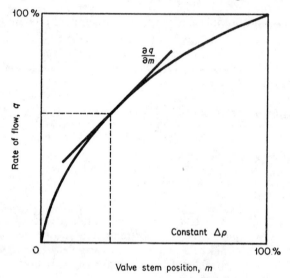

FIGURE 2–12. Linearization of valve characteristic

position at the average pressure difference. The latter can most readily be obtained from a plot of the valve characteristic, i.e. the relationship between the flow rate and valve position at constant pressure difference (Figure 2–12), and this is the *valve sensitivity*, K_v. Considering this to be constant in the operating region, the above equation can be integrated to

$$q = \Delta p/R + K_v m$$

Substituting in the material balance equation for the appropriate flow rate,

$$C(dp_2/dt) = (p_1 - p_2)/R_1 - (p_2 - p_3)/R_2 + K_v m$$

whence

$$p_2(Ts + 1) = K_1 p_1 + K_2 p_3 + K_3 m \tag{2-12}$$

The time constant for small changes in valve position is the same for changes in p_1 or p_3, the gain term K_3 is given by $K_v/\Sigma(1/R)$.

Linearization in Simultaneous Equations

In the section on Mixing Processes (p. 32) the transfer equation for a mixing vessel with a chemical reaction was derived. For simplicity a first-order reaction with a rate proportional to concentration was considered. For reactions of higher order a non-linearity is introduced which must then be linearized, but a more realistic treatment must recognize that the reaction rate will also be a function of temperature which will also probably be non-linear. To derive the transfer functions it is now necessary to consider the material balance with regard to the reacting material and also the heat balance for the thermal effects.

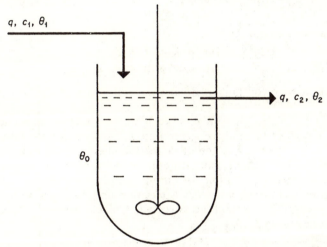

FIGURE 2–13. Stirred tank reactor with reaction rate dependent on temperature

Consider the reaction vessel of Figure 2–13 in which the inflow is a constant rate of flow q, at a temperature θ_1, with a reactant concentration of c_1. The outflow is the same flow q at a temperature θ_2 and reactant concentration c_2. Assuming perfect mixing, the material balance for the reactant will be

$$V(dc_2/dt) = qc_1 - qc_2 - Vr$$

where V is the hold-up volume and r is the reaction rate per unit volume and is a function of c_2 and θ_2. The heat balance equation is similar to those previously derived but a term ΔHr must be added for the heat liberated in the reaction, i.e.

$$\rho V c_p (d\theta_2/dt) = \rho q c_p (\theta_1 - \theta_2) - UA(\theta_2 - \theta_0) + \Delta HVr$$

where θ_0 is the ambient temperature around the vessel.

The non-linear terms in these equations are those including the reaction rate term, r. This may be linearized for the two variables, c_2 and θ_2, thus:

$$r = (\partial r/\partial c_2)c_2 + (\partial r/\partial \theta_2)\theta_2$$

where the partial derivatives are assumed constant over the operating range, and r may then be written as

$$r = k_c c_2 + k_\theta \theta_2$$

Substituting in the heat and material balances produces two simultaneous linear equations:

$$V(dc_2/dt) = q(c_1 - c_2) - Vk_c c_2 - Vk_\theta \theta_2$$

$$\rho V c_p (d\theta_2/dt) = \rho q c_p (\theta_1 - \theta_2) - UA(\theta_2 - \theta_0)$$
$$+ \Delta HVk_c c_2 + \Delta HVk_\theta \theta_2$$

which can be re-arranged to

$$(T_1 s + 1)c_2 = K_1 c_1 + K_2 \theta_2$$

$$(T_2 s + 1)\theta_2 = K_3 \theta_1 + K_4 \theta_0 + K_5 c_2$$

in which the time constants and gains are collections of the various parameters.

It will be seen that the system now exhibits two different time constants, one for the effect of input concentration and output temperature on the output concentration, and the other for the effect of input and ambient temperatures and output concentration on the output temperature. The effect of input concentration on output temperature and of input and ambient temperature on output concentration are both second-order effects as both time constants are involved. A block diagram of this system is shown in Figure 2–14.

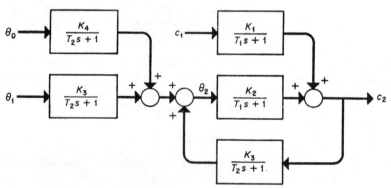

FIGURE 2–14. Block diagram for system of Figure 2–13

Response of First-order Systems

The main reason for deriving system response equations is to investigate the effects which a change in an input variable with respect to time will have on a particular output variable. It should be noted that the analysis of the system response can be applied to a change in only one variable at a time and that if the system has more than one input variable (as Equation 2–9), all other input variables apart from the one being considered must be regarded as constant. Since each variable is measured from a datum corresponding to its normal steady-state value, this is equivalent to making the unchanged variables equal to zero; the appropriate terms then disappear from the equation, leaving for a first-order system the general first-order equation (Equation 2–5) which has only one input and one output variable. This can then be written in terms of the transfer function between the two variables.

If the input variable is now given a time function, the equation is reduced to that of the output variable with respect to time; solution of the equation thus yields the time function of the output variable, i.e. the system response with respect to time for the particular input function assumed.

In practice changes in input variables are usually random, i.e. time functions which are not strictly defined. Much information can, however, be derived by considering the effects of certain standard changes in the input variables for which simple mathematical functions can be written. The two major functions of this type are the *step*, defined by making the input a constant finite value from zero time, and the *harmonic* or *sinusoidal* function, defined by making the input a sine wave of fixed amplitude and frequency from zero time. Although other functions can be and are often used, the step and sinusoidal functions are of most value in systems analysis. The step function is used in both practical and theoretical studies of the transient response of systems, and the sinusoidal function is the basis of frequency response analysis which will be discussed later in more detail.

Before proceeding to general solutions for these functions, it is necessary to define the initial or boundary conditions which exist before the changes in input are applied. These are the usual requirements that both input and output variables are zero up to zero time when the input function is applied, i.e. $\theta_1 = \theta_2 = 0$ at $t = 0$. Since the variables are deviations from the normal operating values, this is equivalent to saying that the system is at a normal steady-state equilibrium condition before the change in input is imposed. Systems analysis is concerned with changes in the value of the variables not with the absolute values.

The general form of response equation for a first-order system is that of Equation (2–5), i.e.

$$T(d\theta_2/dt) + \theta_2 = \theta_1$$

The input variable, θ_1, will now be set equal to the appropriate time function of the change in input, i.e.

$$T(d\theta_2/dt) + \theta_2 = \text{Input time function}$$

and this equation must now be solved for the response of the output, θ_2, as a function of time. This type of differential equation can be solved in a number of ways but the Laplace transform technique is now universally employed.

Use of the Laplace Transform in Process Analysis

For present purposes the Laplace transformation may be regarded as a relatively simple method of converting ('transforming') a linear differential equation into an algebraic one so that the convenience of algebra can be used in the solution. The method is rigorous and generally quicker and easier than the traditional methods.

The essential step is that of 'transforming' the variables which are functions of time in the differential equation. If $y = f(t)$ is a function of the variable t (which in systems analysis usually represents time) defined for $t > 0$, then the integral $\int_0^\infty f(t)e^{-st}\,dt$ is the Laplace transform of $f(t)$. The solution of the integral is a function of s (which may be any number large enough to make the integral finite and convergent). Extensive lists of these transforms have been compiled (of which a representative selection is given in Appendix I) and it is rarely necessary to evaluate the integral. By the use of such tables of transforms each term in the differential equation may be transformed from a function of t into a function of s. At least one of the terms in the equation will be an unknown function of time, the identification of which is the reason for solving the equation, and this will now become an unknown function of s. The transformed equation, however, is no longer a differential equation but is an algebraic equation which can be manipulated by the usual algebraic techniques to yield a solution in s for the unknown function. The final step is to convert or 'inversely transform' this function back into the time domain by using the table of transforms inversely.

It is often customary to designate the transformed, or *s-domain*, variables by upper-case letters and/or the addition of a bracketed (s), to distinguish them from the *time-domain* variables which are designated by lower-case letters and/or the addition of a bracketed (t), e.g. the Laplace transform of a variable $x(t)$ (x being a function of t) is $X(s)$ (X being a function of s). This designation is not absolutely necessary, since s and t cannot occur in the same equation and the presence of s in the equation will indicate that it has been transformed.

Taking the general first-order system equation in terms of a variable y,

$$T(dy_2/dt) + y_2 = y_1$$

and transforming each term,

$$T(sY_2 - Y_{2_{t=0}}) + Y_2 = Y_1$$

where Y_1 and Y_2 are now functions of s. $Y_{2_{t=0}}$ is the initial value of y_2 at

$t = 0$, and since the usual conditions are that $y_1 = y_2 = 0$ at $t = 0$, the equation reduces to:

$$Ts Y_2 + Y_2 = Y_1$$

or

$$Y_2/Y_1 = 1/(Ts + 1)$$

This will be recognized as the transfer function relationship for the first-order system, as shown by Equation (2–6), and the transfer function is thus the ratio of the Laplace transforms of the output and input functions.

It is now possible to illustrate the further utility of the transfer function, which is the ratio of the output to input transforms; substitution of the appropriate transformed function in place of the input immediately yields an identity in s which is the transform of the output function. This has only to be inversely transformed back into a function of time to give the desired result, the time function of the output.

Step Function Response

A 'step' function is developed by a change in the input from a value of zero at zero time to a constant finite value. The Laplace transform of a constant A is A/s; hence if the input θ_1 to a first-order system changes from 0 to A at $t = 0$, the transformed value of θ_1 is $\theta_1(s) = A/s$. Substituting this in the transfer function (Equation 2–6):

$$\theta_2(s) = (A/s)[1/(Ts + 1)]$$

which is the transform of the output function, θ_2

This may be inversely transformed by separation by partial fractions or by re-shaping into a standard transform whose inversion may be taken from the table of transform pairs. In the present case dividing the denominator by T gives

$$\theta_2(s) = \frac{A/T}{s(s + 1/T)}$$

where the denominator is of the form $(s + a)(s + b)$, where $a = 0$ and $b = 1/T$. From the table of Laplace transform pairs (see Appendix I) the function $1/[(s + a)(s + b)]$ is the Laplace transform of the time function $[\exp(-at) - \exp(-bt)]/(b - a)$. The inverse transformation thus yields

$$\theta_2(t) = (A/T)\{[\exp(0) - \exp(-t/T)]/(1/T - 0)\}$$

or

$$\theta_2 = A[1 - \exp(-t/T)] \tag{2–13}$$

The same result is obtained by separating the denominator of the identity for $\theta_2(s)$ into partial fractions and using the inverse transforms of $1/s$ and $1/(s + a)$.

A plot of the response with respect to time is shown in Figure 2–15. The shape of the curve is typical of all first-order systems response to a step function input. The slope of the curve is a maximum at zero time, thus the response is immediate, but then slows down to approach the final value asymptotically. As t increases, $\exp(-t/T)$ approaches zero, hence the final value of the steady-state is A, as would be expected.

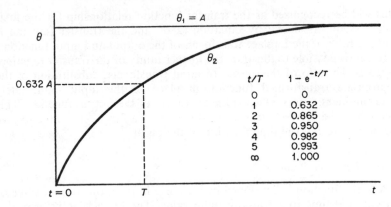

t/T	$1 - e^{-t/T}$
0	0
1	0.632
2	0.865
3	0.950
4	0.982
5	0.993
∞	1.000

FIGURE 2–15. Step function response of first-order system

The shape of the curve can be generalized by considering the value of $[1 - \exp(-t/T)]$, which is the fraction of the total step response accomplished by the response at any particular time. As can readily been seen from the table of values included in Figure 2–15, the step is 63.2 per cent completed when $t/T = 1$, i.e. at the expiry of one time constant. It will also be noted that when $t/T = 4$ the change is more than 98 per cent complete, so that after the expiry of four time constants the output variable is within 2 per cent of the final value. After five time constants the change is virtually completed within the accuracy of most commercial measuring instruments.

Example 2–3A

Flows of 0.1 m³/s of a gas A and 0.025 m³/s of a gas B are passed through a vessel of 5 m³ volume. If the flow of B is suddenly increased to 0.03 m³/s, how soon will the concentration of B in the mixture reach 95 per cent of the final value?

Total flow into the vessel under normal conditions = 0.125 m³/s

Time constant = Hold-up time = $V/q = 5/0.125 = 40$ s

The step function response, $1 - \exp(-t/T)$, reaches a value of 0.95 for unit step input when $t/T = 3$, i.e. after the expiry of 3 time constants.
Hence, the time required to reach 95 per cent of the final value is

$$3 \times 40 = 120 \text{ s}$$

Example 2–3B

A flow of 1000 gal/h of solution containing 1 lb/gal of a reagent passes through a reaction vessel of 500 gal capacity. The first-order reaction rate constant is $5\,h^{-1}$. Determine the concentration of the residual reagent 15 min after the inflow concentration is increased by 10 per cent, and the final steady-state value.

$$\text{Hold-up time} = V/q = 500/1000 = 0.5\,h$$

For normal conditions, the material balance is

$$q(\bar{c}_1 - \bar{c}_2) = k\bar{c}_2 V$$

Hence
$$\bar{c}_2/\bar{c}_1 = 1/(1 + kV/q)$$
$$= 1/(1 + 5 \times 0.5)$$
$$= 0.286$$

Thus when $\bar{c}_1 = 1$ lb/gal, $\bar{c}_2 = 0.286$ lb/gal.

A 10 per cent increase in c_1 will produce 0.0286 lb/gal increase in c_2 at the steady state. Hence the final value of c_2 is 0.3146 lb/gal.

The transfer function (see equation 2–7) is

$$c_2/c_1 = K/(Ts + 1)$$

where $K = 1/(1 + kV/q) = 0.286$

and $\quad T = (V/q)/(1 + kV/q) = 0.5/3.5 = 0.143\,h.$

The response of c_2 to a step change of 10 per cent in c_1 is

$$c_2 = 0.286 + 0.0286\,[1 - \exp{(-t/0.143)}]$$

Thus, when $t = 15$ min,

$$t/T = 0.25/0.143 = 1.75$$

and

$$c_2 = 0.3096\,\text{lb/gal}$$

Sinusoidal Forcing

The second standard forcing function to be considered is the sine wave function, $A \sin \omega t$, from $t = 0$. One reason for considering this particular function is that actual operating processes can develop a semi-stable condition in which the variables cycle about mean values, i.e. the variables oscillate with a more or less pure sine wave pattern. Under these conditions the components of the system are subjected to sinusoidal forcing, and thus the response of a system to such a forcing is of interest. In addition, sine wave inputs can be generated in a reasonably pure form by apparatus such as cranks, oscillators, etc. The injection of a sine wave input of variable frequency into a process element or system is the basis of frequency response analysis which is a powerful tool in systems analysis and design.

The Laplace transform of the sine function $A \sin \omega t$ is $A\omega/(s^2 + \omega^2)$. Substituting this for the input transform θ_1 in the transfer function for the first-order system (Equation 2–6) gives for the value of the output transform, θ_2

$$\theta_2 = A\omega/[(Ts + 1)(s^2 + \omega^2)]$$

This function of s is converted into a form suitable for inversion by resolution into partial fractions, thus

$$\frac{A\omega}{(Ts + 1)(s^2 + \omega^2)} = A\left[\frac{N_1}{(Ts + 1)} + \frac{N_2}{(s^2 + \omega^2)}\right]$$

where

$$N_1 = \left.\frac{\omega}{(s^2 + \omega^2)}\right|_{Ts + 1 = 0} = \frac{\omega}{(-1/T)^2 + \omega^2} = \frac{\omega^2 T^2}{1 + \omega^2 T^2}$$

and

$$N_2 = \left.\frac{\omega}{(Ts + 1)}\right|_{s^2 + \omega^2 = 0} = \left.\frac{\omega(Ts - 1)}{(T^2 s^2 - 1)}\right|_{s^2 = -\omega^2}$$

$$= \frac{-\omega(Ts - 1)}{\omega^2 T^2 + 1} = \frac{\omega - \omega Ts}{1 + \omega^2 T^2}$$

$$\theta_2(s) = \frac{[A\omega T^2/(1 + \omega^2 T^2)]}{(Ts + 1)} + \frac{[A(\omega - \omega Ts)/(1 + \omega^2 T^2)]}{(s^2 + \omega^2)}$$

$$= \left[\frac{A}{(1 + \omega^2 T^2)}\right]\left[\frac{\omega T}{(s + 1/T)} + \frac{\omega}{(s^2 + \omega^2)} - \frac{\omega Ts}{(s^2 + \omega^2)}\right]$$

The transform inversions of the last three terms are respectively $\omega T \exp(-t/T)$, $\sin \omega t$, and $\omega T \cos \omega t$. Hence

$$\theta_2(t) = [A/(1 + \omega^2 T^2)]$$
$$\times [\omega T \exp(-t/T) + \sin \omega t - \omega T \cos \omega t] \qquad (2\text{–}14)$$

Equation (2–14) thus defines the variation with time of the output θ_2 when the input is $A \sin \omega t$ from zero time. The first term in the expansion of Equation (2–14), i.e. $A\omega T \exp(-t/T)/(1 + \omega^2 T^2)$, is a transient since it contains a negative exponential term. This will decrease with time and becomes effectively zero after the expiry of five time constants. The remaining terms define the 'steady-state' solution of the forced response, which is obviously an oscillation of the same frequency, ω, since this is the only frequency term in the sine and cosine terms, but changed in amplitude and phase. A plot of the input and output functions is given in Figure 2–16, in which the transient and forced responses are shown by the dotted lines and the overall response by the solid line. As the transient disappears, the response decays into a continuous oscillation given by

$$\theta_2 = [A/(1 + \omega^2 T^2)] (\sin \omega t - \omega T \cos \omega t) \qquad (2\text{–}15)$$

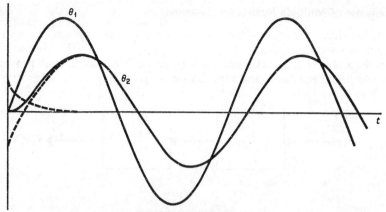

FIGURE 2-16. Response of first-order system to sinusoidal forcing

The most obvious property of the steady-state response is that if the time constant, T, is zero, the response reduces to $A \sin \omega t$, which is the input function, and the response is thus identical with the input. This is so logical as to be little more than a check on the validity of the equation. When T is not zero, it can be seen that the wave is reduced in amplitude and lags in phase behind the input. This can be put into an alternative and more convenient form by combining the sine and cosine terms into the form, $R \sin (\omega t - \psi)$, by the identity

$$R \sin (\omega t - \psi) = R \sin \omega t \cos \psi - R \cos \omega t \sin \psi$$

By comparison with the original terms of Equation (2–16), $R \sin \psi = \omega T$ and $R \cos \psi = 1$, from which it can be deduced that $R = (1 + \omega^2 T^2)^{\frac{1}{2}}$ and $\psi = \tan^{-1} \omega T$. Equation (2–15) can then be written

$$\theta_2 = [A/(1 + \omega^2 T^2)^{\frac{1}{2}}] \sin (\omega t - \psi) \qquad (2\text{-}16)$$

Comparing this result with the input function, $\theta_1 = A \sin \omega t$, it is readily seen that the amplitude of the output is reduced from A to $A/(1 + \omega^2 T^2)^{\frac{1}{2}}$, since the denominator of the response amplitude must be greater than one. In radio parlance, the wave is *attenuated*. Secondly the input sine wave, $\sin \omega t$, becomes an output wave, $\sin (\omega t - \psi)$, which has the same frequency, ω, but a phase lag of ψ is introduced. Both the phase lag and the attentuation are dependent on both the frequency ω and the time constant T; since these are both positive numbers the product ωT can vary between zero and infinity, i.e. the phase angle $(\tan^{-1} \omega T)$ can have values from 0 to 90°, and the output amplitude can be reduced from A to zero. For a given system the value of T is constant, and both attenuation and phase lag are effectively determined by the frequency of the oscillation. As the frequency increases, so also does the phase shift up to a maximum of 90° for a first-order system at $\omega = \infty$, and the response amplitude is reduced from the input amplitude at zero frequency to zero at infinite frequency.

Response of Multiple First-order Systems

It is evident that the output, θ_2, from a first-order system can be the input to a second similar system which will produce a further output, θ_3; e.g. the addition of a pocket or sheath around a thermometer bulb provides an additional resistance and capacitance, so providing a two-stage system

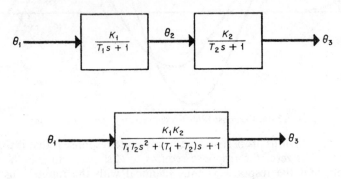

FIGURE 2–17. Alternative block diagram for two non-interacting first-order stages in series

of two first-order elements, although this particular example is a rather special case. In the more general case of two RC stages in series, the second stage will have a response given by the same first-order Equation (2–5), the input now being the output from the first stage, i.e.

$$T_2(\mathrm{d}\theta_3/\mathrm{d}t) + \theta_3 = \theta_2$$

in which T_2 is the time constant of the second stage ($=R_2C_2$). This equation can be combined with that of the first stage (with time constant now T_1) by eliminating θ_2 and $\mathrm{d}\theta_2/\mathrm{d}t$ between the two equations, yielding a second-order equation:

$$T_1T_2(\mathrm{d}^2\theta_3/\mathrm{d}t^2) + (T_1 + T_2)(\mathrm{d}\theta_3/\mathrm{d}t) + \theta_3 = \theta_1 \qquad (2\text{–}17)$$

Transforming this equation term by term with the usual initial conditions of zero values at zero time,

$$T_1T_2s^2\theta_3 + (T_1 + T_2)s\theta_3 + \theta_3 = \theta_1$$

or

$$\theta_3/\theta_1 = 1/[T_1T_2s^2 + (T_1 + T_2)s + 1]$$

The same result is obtained more simply by combining the transfer functions of the two stages, i.e.

$$\theta_2/\theta_1 = 1/(T_1s + 1)$$

$$\theta_3/\theta_2 = 1/(T_2s + 1)$$

and which gives

$$\theta_3/\theta_1 = (\theta_2/\theta_1)(\theta_3/\theta_2)$$
$$= 1/[(T_1s + 1)(T_2s + 1)]$$
$$= 1/[T_1T_2s^2 + (T_1 + T_2)s + 1] \qquad (2\text{--}18)$$

The overall response of the two stages to a given input, θ_1, is now given by the solution of the second-order equation using the Laplace transform method. For the step function input, $\theta_1(t) = A$, $\theta_1(s) = A/s$, and substituting for θ_1 in Equation (2–18):

$$\theta_3(s) = A/[s(T_1s + 1)(T_2s + 1)]$$
$$= \frac{(A/T_1T_2)}{[s(s + 1/T_1)(s + 1/T_2)]}$$

This is now of the form $1/[s(s + a)(s + b)]$, which is the transform of $(1/ab)\{1 + [b \exp(-at) - a \exp(-bt)]/(a - b)\}$, in which $a = 1/T_1$ and $b = 1/T_2$, multiplied by the constant A/T_1T_2. Thus the inverse transform of the above equation is

$$\theta_3(t) = T_1T_2\left[\frac{A}{T_1T_2}\right]\left[1 + \frac{(1/T_2)\exp(-t/T_1) - (1/T_1)\exp(-t/T_2)}{1/T_1 - 1/T_2}\right]$$
$$= A\left[1 - \frac{T_1\exp(-t/T_1) - T_2\exp(-t/T_2)}{T_1 - T_2}\right] \qquad (2\text{--}19)$$

It can be seen that the steady-state response when the two exponential terms have become zero is A, as might be expected, but the transient term is completely different from that of a single stage (first-order) system. The

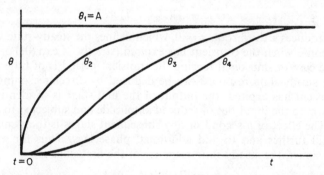

FIGURE 2–18. Step function response of non-interacting first-order stages in series

response θ_3 is plotted against time in Figure 2–18 and is completely different from that of θ_2. The initial slope of the curve at zero time is zero, since the slope depends on the difference $(\theta_2 - \theta_3)$ and both these quantities are zero at zero time. The response therefore starts very slowly, increases to a maximum rate of change and then decays asymptotically to the final value. It will be seen by inspection of the equation that the response is the same irrespective of whether the larger of the two time constants is the first or second in the arrangement.

The addition of further RC stages in the series does not fundamentally alter the shape of the final output response curve, as indicated in Figure 2–18 by the response θ_4, but the response becomes still slower and more sluggish in character. It is interesting to note that as the number of stages becomes larger the final response has some of the characteristics of a *time delay*, since there is finite period before the response can be detected.

The response equation (Equation 2–19) is of the general form:

$$\theta_3 = A[1 + B_1 \exp (s_1 t) + B_2 \exp (s_2 t)] \qquad (2\text{–}20)$$

where $s_1 = -1/T_1$ and $s_2 = -1/T_2$. These quantities can be seen to be the *zeros* of the denominator of the overall transfer function (Equation 2–18), or the roots of a characteristic equation formed by setting the denominator of the overall transfer function equal to zero, i.e.

$$(T_1 s + 1)(T_2 s + 1) = 0$$

This characteristic equation defines the natural, as opposed to the forced, response of the system which is obtained by making the forcing function zero, i.e. $\theta_1 = 0$. In the present case the roots of the characteristic equation, s_1 and s_2, can only be real and negative since the two time constants can only be real and positive. The natural response is thus the transient part of the solution, since both exponential terms will ultimately become zero. In the case of an input step function, as considered above, a third term s is introduced into the denominator of the response transform $\theta_3(s)$, and the characteristic equation becomes:

$$s(T_1 s + 1)(T_2 s + 1) = 0$$

This introduces a third root, $s = 0$, which defines the steady-state value of the response when the transient has expired (strictly, $A \exp (0t)$ or A).

In the case of sinusoidal forcing the transient period is of little interest and the steady-state response can be deduced by direct reasoning. Once the transient has expired, the output of the first stage is of identical sine wave form to the input but of reduced amplitude and subjected to a lag in phase. The effect of a second or any subsequent stages is to attenuate the wave still further and to add additional phase lags. Thus for an input $\theta_1 = A \sin \omega t$, the steady-state response of the first stage is

$$\theta_2 = B \sin (\omega t - \psi_1)$$

where $B = A/(1 + \omega^2 T^2_1)^{\frac{1}{2}}$ and $\psi_1 = \tan^{-1} \omega T_1$. θ_2 now becomes the input to the second stage whose output will be

$$\theta_3 = C \sin (\omega t - \psi_1 - \psi_2)$$

where $C = B/(1 + \omega^2 T_2^2)^{\frac{1}{2}}$ and $\psi_2 = \tan^{-1} \omega T_2$. Since the maximum phase lag is 90° at infinite frequency, for each individual stage, the maximum lag for a two stage system will be 180°, and so on for further first-order stages in series.

Interacting Stages

As mentioned above, the addition of a pocket to a thermometer bulb provides two *RC* stages in series but is also a special case in that the two stages are *interacting*. In the general case of two stages in series discussed in the previous section, it is assumed that the response of the first stage is determined only by the input function and that it is not affected in any way by the value of the output from the second stage. This would be the case with the cascaded level system shown in Figure 2–19(a) where the outflow

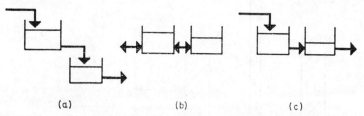

(a)   (b)   (c)

FIGURE 2–19. Interacting and non-interacting stages: (a) non-interacting (cascaded), (b) interacting (dead end), (c) interacting (flow)

from the first tank is independent of the level in the second tank. In Figure 2–19(b) and 2–19(c) the outflow from the first tank is determined by the difference in levels between the two tanks, and these systems are thus interacting. This applies to most thermal systems since heat flow rate is dependent on temperature difference and not on absolute temperature. The value of an intermediate temperature θ_2 thus depends on heat transfer determined by the two temperature differences $(\theta_1 - \theta_2)$ and $(\theta_2 - \theta_3)$. The thermometer bulb and pocket typifies the simplest type of interaction since the system is 'dead-end', the bulb having no outflow heat term, as Figure 2–19(b).

The actual response of an interacting system is not fundamentally altered by the interaction as far as the type of response is concerned, but the time constants are modified as shown by the following example.

Consider the thermometer bulb with pocket as shown in Figure 2–20. The fluid temperature is θ_1, the pocket temperature (assumed uniform through the thickness) is θ_2 and that of the bulb is θ_3. The resistance R_1 is that of the film surrounding the pocket, R_2 is that of the fluid between the pocket and the bulb. C_1 and C_2 are the capacitances of the pocket and bulb respectively.

Initially the heat balance for the bulb is the same as that made previously (Equation 2–4), allowing for the change in symbols:

$$C_2(d\theta_3/dt) = (\theta_2 - \theta_3)/R_2$$

The heat balance for the pocket, however, has both inflow and outflow terms since the pocket gains heat from the fluid and loses heat to the bulb, i.e.

$$C_1(d\theta_2/dt) = (\theta_1 - \theta_2)/R_1 - (\theta_2 - \theta_3)/R_2$$

The overall relationship between the fluid temperature θ_1 and that of the bulb θ_3 is obtained by eliminating the intermediate temperature θ_2 from the two equations, giving

$$R_1C_1R_2C_2(d^2\theta_3/dt^2) + (R_1C_1 + R_2C_2 + R_1C_2)(d\theta_3/dt) + \theta_3 = \theta_1$$

It will be seen that this equation contains the two time constants of the individual stages, T_1 and T_2 (given by R_1C_1 and R_2C_2, respectively), and a

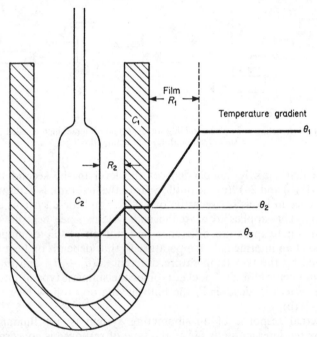

FIGURE 2–20. Thermometer bulb with pocket

third constant, R_1C_2, which, being a product of resistance and capacitance, will also have the dimension of time. Although the third constant has no physical significance, it can be regarded as an additional time constant, an *interdependence* constant, T_{12}, and the equation can be re-written:

$$T_1T_2(d^2\theta_3/dt^2) + (T_1 + T_2 + T_{12})(d\theta_3/dt) + \theta_3 = \theta_1 \quad (2\text{--}21)$$

Comparing this with the equivalent equation for two stages in series without interaction, the two equations are identical apart from the additional time constant in the second coefficient and this is obviously due to the interaction. Equation (2–21) can be restored to the same form as Equation (2–17) by replacing the three constants, T_1, T_2, and T_{12}, by two constants, T_a and T_b, such that

$$T_1 T_2 = T_a T_b$$

and
$$T_1 + T_2 + T_{12} = T_a + T_b$$

This is equivalent to taking the factors of the characteristic equation of Equation (2–21), i.e.

$$(-1/T_a), (-1/T_b) = [-b \pm (b^2 - 4ac)^{\frac{1}{2}}]/2a$$

where $a = T_1 T_2$; $b = T_1 + T_2 + T_{12}$; and $c = 1$.

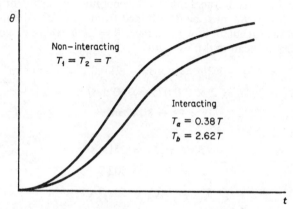

FIGURE 2–21. Effect of interaction on step function response of two first-order stages in series

The two interacting first-order stages with individual time constants of T_1 and T_2 thus have the same performance as shown by the response equation of two non-interacting stages with time constants of T_a and T_b. Effectively one time constant is increased and the other decreased by the interaction, compared to the values of T_1 and T_2, e.g. if the actual time constants are both equal to T, the effective constants are $T_a = 0.38T$ and $T_b = 2.62T$. The responses of the two cases are compared in Figure 2–21, and the practical result is to make the response of the system slower and more sluggish than that of the independent stages. Thus the second-order Equation (2–17) or the transfer function (Equation 2–18) are appropriately descriptive of coupled first-order stages whether or not the stages are interacting, but in the latter case the real time constants are effectively suppressed and replaced by the modified constants.

Example 2–4A

Determine the effective time constants for a system of two tanks, each 2 m diameter, when the outlet from the bottom of the first tank is connected

to the bottom of the second and the latter discharges to atmosphere. The
normal levels are 3 m and 2 m when the flow is 0.025 m³/s.

The two tanks form an interacting second-order system, since the head
on the outflow from the first tank is the difference in level between the
tanks. The characteristic equation for the system will thus be

$$T_1 T_2 s^2 + (T_1 + T_2 + T_{12})s + 1 = 0$$

in which $T_1 = R_1 C_1$, $T_2 = R_2 C_2$, and $T_{12} = R_2 C_1$. (Note that the interdependence constant differs in this case from that of the thermometer
bulb and pocket due to the different order of the resistances and capacitances, this being a 'flow' system as Figure 2–19(c).)

Since the flow from the tanks will generally be turbulent, the flow
resistances R_1 and R_2 must be determined from a linearized equation. As
already seen the turbulent flow resistance, $R = 2\bar{h}/\bar{q}$, where $\bar{h}$ and $\bar{q}$ are the
normal values of head and flow. Thus

$$R_1 = 2(\bar{h}_1 - \bar{h}_2)/\bar{q} = 2 \times (3 - 2)/0.025 = 80 \text{ m s/m}^3$$
$$R_2 = 2\bar{h}_2/\bar{q} = 2 \times 2/0.025 = 160 \text{ m s/m}^3$$
$$C_1 = C_2 = (\pi/4) \times 2^2 = 3.142 \text{ m}^2$$

Hence

$$T_1 = R_1 C_1 = 251.4 \text{ s}$$
$$T_2 = R_2 C_2 = 502.7 \text{ s}$$
$$T_{12} = R_2 C_1 = 502.7 \text{ s}$$

Since the time constants are real, the characteristic equation above has
real roots which define the effective time constants, T_a and T_b, where

$$T_a T_b = T_1 T_2 \quad \text{and} \quad T_a + T_b = T_1 + T_2 + T_{12}$$

Solving these relationships or by application of the quadratic formula
to the equation, the effective time constants are:

$$T_a = 110.3 \text{ s} \quad \text{and} \quad T_b = 1144 \text{ s}$$

Example 2–4B

Determine the effective time constants for a two-tank inter-dependent
system, as described above, when the tank diameters are 6 ft and the levels
are 10 ft and 6 ft for an outflow of 3000 ft³/h.

Proceeding as above,

$$R_1 = 2 \times (10 - 6)/3000 = 0.002\,67 \text{ ft h/ft}^3$$
$$R_2 = 2 \times 6/3000 = 0.0040 \text{ ft h/ft}^3$$
$$C_1 = C_2 = (\pi/4) \times 6^2 = 28.278 \text{ ft}^2$$

Hence

$$T_1 = R_1C_1 = 0.0753 \text{ h} = 4.52 \text{ min}$$

$$T_2 = R_2C_2 = 0.1132 \text{ h} = 6.79 \text{ min}$$

$$T_{12} = R_2C_1 = 6.79 \text{ min}$$

Whence $T_a = 1.89$ min and $T_b = 16.20$ min.

Systems with Resistance Only

Consideration has so far been devoted to process systems containing resistance and capacitance in combination. It is also possible for certain types of systems to exhibit these properties singly and, to complete the picture of the dynamic elements from which process systems are built up, it is necessary to consider those systems which contain either resistance or capacitance alone.

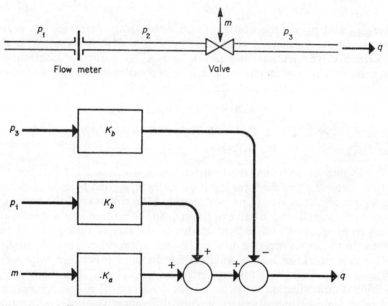

FIGURE 2–22. Flow rate process (resistance element only)

A process system containing only resistance without capacitance is exemplified by the flow of a fluid through a pipe where the resistance is provided by a restriction such as a valve or metering device. It may be assumed that the capacitance of the pipe is zero in the case of an incompressible fluid, or negligible if the fluid is gaseous and the pressure drop within certain limits. Whilst the regulation of fluid flow is basically the manipulation of a valve to attain a desired rate of flow at a given pressure drop, it is necessary also to assume that there are no effects due to the inertia of the flowing fluid. This assumption is close to reality since changes of the variables occur relatively slowly.

Consider first the flow of a liquid which passes through a metering restriction such as an orifice plate, etc. and then through a regulating valve. The flow through the former may be related to the pressure drop by the usual relation for turbulent flow, which may be linearized to

$$q = (p_1 - p_2)/R_1 + K_1$$

where p_1 and p_2 are the upstream and downstream pressures, R_1 is the linearized flow resistance $(2(\bar{p}_1 - \bar{p}_2)/\bar{q})$ at the mean values of pressures and flow in the operating range, and K_1 is a constant of integration arising from the linearization.

Similarly the flow through the regulating valve will be given by the usual equation which may be linearized to

$$q = (p_2 - p_3)/R_2 + K_v m + K_2 \qquad (2\text{--}22)$$

where p_2 and p_3 are the upstream and downstream pressures, R_2 is the linearized flow resistance at the mean values (i.e. $2(\bar{p}_2 - \bar{p}_3)/\bar{q}$), K_v is the linearized valve characteristic $(\bar{q}/\bar{m})$, and K_2 is a further constant of integration. The 'system equation' can be derived by eliminating the intermediate pressure p_2 from the two equations, giving

$$q = K_a m + K_b(p_1 - p_3) + K_3$$

where $K_a = R_2 K_v/(R_1 + R_2) = (\bar{q}/\bar{m})(\bar{p}_2 - \bar{p}_3)/(\bar{p}_1 - \bar{p}_3)$
and $K_b = 1/(R_1 + R_2) = \bar{q}/[2(\bar{p}_1 - \bar{p}_3)]$

It will thus be seen that the transfer function relationship between the output variable q and the three input variables m, p_1, and p_3, is in each case the simple algebraic one of proportionality. The response of the system is immediate, any change in valve position or in the upstream or downstream pressure produces an immediate change in the rate of flow given by the change in the input variable modified by the appropriate gain (K_a or K_b).

Gas flow processes may be analysed by the same procedure, within the pressure drop limitation for negligible compressibility, but there is the additional complication that the weight flow rate of a gas depends on upstream temperature as well as pressure difference. The analysis is rather complex even when simplifying assumptions are made. The previous analysis may be used to write a process equation:

$$q = K_a m + K_b p_1 - K_c p_3 - K_d \theta + K$$

in which θ is the upstream temperature. The coefficients may now be evaluated from graphs of the flow rate against each of the four variables in turn. If the pressure ratio at the valve is greater than the critical value, the flow is not dependent on the downstream pressure and K_c will be zero.

Systems with Capacitance Only

If the flows of liquid into and out of a tank are both independent of the level in the tank, then the system does not possess a flow resistance since there is no relationship between flow and head. The system is therefore one with capacitance only, i.e. that of the tank. In the tank shown in Figure 2–23, the inflow is obviously independent of the level in the tank

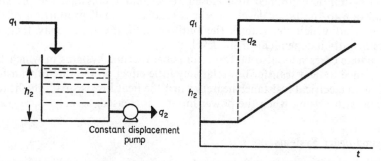

FIGURE 2–23. Tank with constant outflow (capacitance element only)

and, by using a constant displacement pump, so also is the outflow independent of level. The material balance for the tank is as previously:

$$A(dh_2/dt) = q_1 - q_2$$

but the outflow q_2 cannot now be related to the level h_2.

The response of such a system to a step change in either of the two input variables (q_1 and q_2), with the other remaining constant, will be a continuous change in level at a constant rate of change since dh_2/dt is constant, i.e.

$$dh_2/dt = q_1/A$$

when q_2 is constant, and

$$dh_2/dt = -q_2/A$$

when q_1 is constant.

The system thus does not exhibit the property of *self-regulation*, i.e. the tendency to regulate its own level which is shown by a system with both resistance and capacitance. In the present case a finite change in an input variable causes the level to change at a constant rate and a constant steady-state value can only be attained when the inflow and outflow are made equal; if resistance were present, a finite change in an input variable would ultimately lead to a constant steady-state value, as previously seen.

If the mass balance equation is transformed with the usual initial conditions and capacitance C written for area A, the following equation results:

$$Csh_2 = q_1 - q_2$$

and by re-arrangement the transfer function for either of the input variables is seen to be $(1/Cs)$:

$$h_2 = (1/Cs)(q_1 - q_2) \qquad (2\text{--}23)$$

Pure capacitance elements of this type are not uncommon in liquid level and pressure systems and the effect may also be observed when outflow resistance is also present under certain conditions, e.g. if a liquid level system is contained in a closed vessel under pressure, the pressure head due to the liquid level may be a relatively small proportion of the total head which determines the outflow, and the latter may then be significantly independent of the level.

Similar cases may also be found in some thermal systems in which the resistance to heat transfer has relatively little effect on the rate of transfer, e.g. with electrical resistance heaters when the heat loss is small, the heater temperature rising when the power input is increased.

Second-order Systems

The combination of two first-order systems in series yields a second-order response equation and therefore provides a second-order system. The transient response of such a system to a unit step change in input has already been shown (see Equation 2–20) to be of the form:

$$\theta_3 = 1 + B_1 \exp(s_1 t) + B_2 \exp(s_2 t)$$

where B_1 and B_2 are integration constants depending on the initial conditions, and s_1 and s_2 are the roots of the characteristic equation of the system obtained by equating the denominator of the transfer function of the system (Equation 2–18) to zero, i.e.:

$$(T_1 s + 1)(T_2 s + 1) = 0$$

In the present instance the roots are $s_1 = -1/T_1$ and $s_2 = -1/T_2$ and substitution of these values in Equation (2–20), followed by evaluation of B_1 and B_2 for the initial conditions will yield the response equation, Equation (2–19). As previously pointed out, the roots of the equation for the system formed of two first-order time-constant stages can only be real and negative, and the exponential terms in Equation (2–20) can only be non-oscillatory, and will decay to zero with increasing time.

Since the characteristic equation for a second-order system is, however, a quadratic, a more general form of the second-order characteristic equation may have roots which are conjugate imaginary or complex. This obviously cannot occur in the case of two first-order stages in series already discussed, since the quadratic is obtained from the product of two individual transfer functions and will always have real roots. However, if a mechanical system or the fluids in a process system are subject to accelerations, the behaviour of the system is described by a second-order equation which may have real or unreal roots, and an oscillatory response may be

obtainable. The mass suspended from a spring which was used as a preliminary illustration of process dynamic response is a classic example,
although in that example the system was simplified by the omission of
frictional damping; nevertheless the response equation (Equation 2–1) was
second-order due to the acceleration of the mass.

Second-order systems of this type, containing mass-spring-damping
elements or analogies thereto, differ fundamentally from the combination
of two first-order elements, and the term *second-order system* is often
restricted to this particular type of system. Such systems are of fundamental
importance in the control of mechanical devices with moving parts. In
process control such devices are usually found only in the measuring and
controlling instruments, and generally have an insignificant effect on the
dynamics of the system as a whole, the combination of first-order lags in
the process being greatly predominant. Yet a study of the behaviour of the
general second-order system is of very great value since it is found in
practice that, although a process system may require to be described by a
third- or still higher order of equation, the predominating mode of the
response is that of the second-order, and the transient response of a control
loop, as distinct from a series of process stages, may often be characterized
by that of a second-order system.

Derivation of the General Second-order Equation

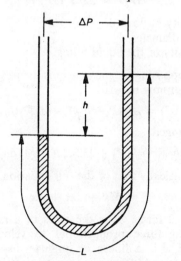

FIGURE 2–24. U-tube manometer

A typical general second-order system frequently encountered in process
control is the liquid filled manometer, as illustrated in Figure 2–24. As a
mechanical system the basic conservation equation is a force balance, the
inertial force accompanying the acceleration of the liquid being equal to

the applied force less the frictional resistance. The latter is assumed to be proportional to the liquid velocity and the acceleration is assumed to be uniform.

Consider that a pressure difference ΔP is applied to the manometer and that the difference in level between the two limbs at some time thereafter is h_2.

The inertial force is mass times acceleration, which can be written

$$(AL\rho/g)(\mathrm{d}^2 h_2/\mathrm{d}t^2)$$

where A is the cross-sectional area of the manometer tube,
   L is the total length of the liquid filling,
and ρ is the density of the liquid.

The applied force is that due to the pressure difference, less that due to the difference in levels between the two limbs, i.e. $A(\Delta P - h_2\rho)$. It is assumed that the density of any fluid over the liquid filling is negligible in comparison with ρ; this will be the case if the fluid is gaseous, if it is another liquid ρ is replaced in the last expression by the difference of the two liquid densities.

Since the flow is hardly likely to be turbulent within the relatively small range of movement, the frictional resistance is given by the Poiseuille-Hagen relationship which may be written,

$$\Delta P/V = 32\mu L/(D^2 g)$$

where V is the velocity of flow $(\mathrm{d}h_2/\mathrm{d}t)$,
   D is the tube diameter,
and μ is the viscosity of the liquid filling.

The frictional force opposing the movement is then $A[32\mu L/(D^2 g)](\mathrm{d}h_2/\mathrm{d}t)$. The force balance can now be written:

$$(AL\rho/g)(\mathrm{d}^2 h_2/\mathrm{d}t^2) = A(\Delta P - h_2\rho) - A[32\mu L/(D^2 g)](\mathrm{d}h_2/\mathrm{d}t)$$

which can be re-arranged

$$\Delta P = (L\rho/g)(\mathrm{d}^2 h_2/\mathrm{d}t^2) + [32\mu L/(D^2 g)](\mathrm{d}h_2/\mathrm{d}t) + h_2\rho \qquad (2\text{--}24)$$

This equation is basically that of the full equation of motion:

$$P = M\ddot{x} + D\dot{x} + Sx$$

where P is an applied force, M is the mass of a moving body, D is a frictional or damping force proportional to velocity, S is an elastic opposing force, and x is a linear or angular displacement. It is also analogous to the equation for the voltage in an electrical circuit containing inductance along with the more usual resistance and capacitance, i.e.

$$E = L(\mathrm{d}^2 q/\mathrm{d}t^2) + R(\mathrm{d}q/\mathrm{d}t) + (1/C)q$$

which states that the total applied voltage is equal to the sum of voltage across the inductance L (proportional to rate of change of current), that across the resistance R (proportional to current) and that across the condenser (charge divided by capacitance C). This then is a further process analogy whereby mass in a mechanical system is analogous to inductance in the electrical system.

Equation (2–24) may also be written, after dividing by the liquid density ρ, as

$$K_1(d^2h_2/dt^2) + K_2(dh_2/dt) + h_2 = \Delta P/\rho = h_1 \qquad (2\text{–}25)$$

in which the last term is the applied pressure divided by the liquid density and is thus the pressure head in terms of the manometer liquid. This is, of course, the steady-state difference in levels for a constant applied pressure difference, since when both derivatives are zero, $h_2 = \Delta P/\rho$. This term may then be written as h_1, the input variable, in the same units as the output variable, h_2.

It will be noted that there are three terms in the output variable in the system Equation (2–24) each with a constant coefficient. The number of parameters can be reduced to two by dividing two of the coefficients by the third (as was done in obtaining Equation (2–25). The type of parameter chosen is entirely a matter of convenience and the two following forms of second-order equation, using a generalized variable θ, are often encountered:

$$T^2(d^2\theta_2/dt^2) + 2\zeta T(d\theta_2/dt) + \theta_2 = \theta_1 \qquad (2\text{–}26)$$

and $\qquad (d^2\theta_2/dt^2) + 2\zeta\omega_n(d\theta_2/dt) + \omega_n^2\theta_2 = \omega_n^2\theta_1$

where T is a characteristic time (s/rad), ω_n is a natural frequency (rad/s), and ζ is a damping ratio. As can be seen, Equation (2–26) is of the same form as Equation (2–25) and the coefficients can be equated to find the values of T and ζ. In addition it will be noted that $T = 1/\omega_n$.

Step Response of the General Second-order System

Equation (2–26) can be transformed for the initial conditions of

$$\theta_1 = \theta_2 = (d\theta_2/dt) = 0$$

at $t = 0$, to give

$$T^2s^2\theta_2 + 2\zeta Ts\theta_2 + \theta_2 = \theta_1$$

and a transfer function obtained:

$$\theta_2/\theta_1 = 1/(T^2s^2 + 2\zeta Ts + 1)$$

To determine the response to a step function input of $\theta_1 = A$, the transform A/s is substituted for θ_1, giving

$$\theta_2(s) = A/[s(T^2s^2 + 2\zeta Ts + 1)]$$

and, as has been noted previously (see Equation 2–20), the response will be of the form:

$$\theta_2(t) = A[1 + B_1 \exp(s_1t) + B_2 \exp(s_2t)]$$

where s_1 and s_2 are the roots of the characteristic equation given by the denominator of the transfer function set equal to zero, i.e.

$$T^2s^2 + 2\zeta Ts + 1 = 0$$

Applying the usual quadratic formula, the roots are given by:

$$s_1, s_2 = [-\zeta \pm (\zeta^2 - 1)^{\frac{1}{2}}]/T \qquad (2\text{-}27)$$

In the generalized system there are four cases of interest:

(a) $\zeta > 1$, when the roots are real, unequal and negative,
(b) $\zeta = 1$, when the roots are real, equal and negative,
(c) $\zeta < 1$, when the roots are conjugate complex with negative real parts,
(d) $\zeta = 0$, when the roots are conjugate imaginary with no real parts.

Each of these cases will now be considered separately.

Over-damped Second-order Response ($\zeta > 1$)

When the damping ratio is greater than 1, the two roots of the characteristic equation are negative, real, and unequal, e.g. $s_1 = -a, s_2 = -b$, where a and b are real positive numbers.

Inverse transformation of the equation for $\theta_2(s)$ will thus yield

$$\theta_2(t) = (A/ab)\{1 + [b \exp(-at) - a \exp(-bt)]/(a - b)\}$$

which is the response equation for two non-interacting first-order stages with time constants, T_1 and T_2, in series, where a and b are respectively $1/T_1$ and $1/T_2$. Substitution of these values leads to Equation (2–19) and the response curve with respect to time will be similar to that of θ_3 in Figure 2–18. For reasons explained below this is referred to as the *over-damped* condition of a second-order system.

Under-damped Second-order Response ($\zeta < 1$)

When the damping ratio is less than one, $(\zeta^2 - 1)$ in Equation (2–27) is negative and the roots of the characteristic equation are then conjugate complex numbers which may be written

$$s_1, s_2 = (-\alpha \pm j\beta)$$

By reference to Equation (2–27), $\alpha = \zeta/T$ and $\beta = (1 - \zeta^2)^{\frac{1}{2}}/T$. The significance of these quantities will shortly become apparent.

The response defined by Equation (2–20) is now

$$\theta_2(t) = A[1 + B_1 \exp(-\alpha + j\beta)t + B_2 \exp(-\alpha - j\beta)t]$$
$$= A[1 + e^{-\alpha t}(B_1 e^{j\beta t} + B_2 e^{-j\beta t})]$$

The coefficient B_1 and B_2 can be found from the initial conditions, i.e. $\theta_2 = 0$ and $d\theta_2/dt = 0$ at $t = 0$, and are clearly also conjugate complex when the roots of the characteristic equation are complex. Thus in the present case at $t = 0$, from the above equation:

$$\theta_2 = A(1 + B_1 + B_2) = 0$$

and $$d\theta_2/dt = A[(-\alpha + j\beta)B_1 + (-\alpha - j\beta)B_2] = 0$$

Simultaneous solution of this last pair of equations gives

$$B_1 = (-\alpha - j\beta)/2j\beta; \qquad B_2 = (+\alpha - j\beta)/2j\beta$$

The step response is then

$$\theta_2 = A\{1 + e^{-\alpha t}[(-\alpha - j\beta)\, e^{j\beta t} + (\alpha - j\beta)\, e^{-j\beta t}]/(2j\beta)\}$$

which may be re-arranged to

$$\theta_2 = A\{1 + e^{-\alpha t}[(-\alpha/\beta)(e^{j\beta t} - e^{-j\beta t})/2j - (e^{j\beta t} + e^{-j\beta t})/2]\}$$
$$= A\{1 + e^{-\alpha t}[(-\alpha/\beta)\sin \beta t - \cos \beta t]\}$$

The sine and cosine of the angle βt can be combined by use of the identity for $\sin (A - B)$ to give

$$\theta_2 = A\{1 + e^{-\alpha t}[(\alpha^2 + \beta^2)^{\frac{1}{2}}/\beta]\sin (\beta t - \phi)\}$$

where $\phi = \tan^{-1}(-\beta/\alpha)$.

From the values of α and β as defined, $\alpha^2 + \beta^2 = 1/T^2$, hence:

$$\theta_2 = A[1 + e^{-\alpha t}(1/\beta T)\sin (\beta t - \phi)] \qquad (2\text{-}28)$$

Substituting for α and β the values defined for the original parameters, ζ and T, the response equation for the step function input of A is:

$$\theta_2 = A\{1 + [1/(1 - \zeta^2)^{\frac{1}{2}}]\, e^{-\zeta t/T} \sin [(1 - \zeta^2)^{\frac{1}{2}}t/T - \phi]\}$$

where $\phi = \tan^{-1}[(1 - \zeta^2)^{\frac{1}{2}}/-\zeta]$.

The appearance of the sine term in the response equation implies that the response will now contain an oscillatory component, but the negative exponential term in the amplitude of the sine term will make this oscillation *damped*, i.e. $\exp(-\zeta t/T)$ decreases to zero with increasing time, thus the amplitude of the oscillation also decreases to zero, and the sine term disappears from the steady-state part of the response which is obviously the magnitude of the input step A. The quantity defined by $\alpha(=\zeta/T)$ is the *damping factor* which determines the rate of decay of the transient; the quantity $\beta(=(1 - \zeta^2)^{\frac{1}{2}}/T)$ is the angular *frequency* of the damped oscillation.

The response of the second-order system for which $\zeta > 1$, considered in the previous section, can also be regarded as a damped oscillation which is, however, aperiodic, i.e. the degree of damping is sufficiently great to prevent any over-shoot by the transient part of the response beyond the final steady-state value. In the present case, for which $\zeta < 1$, the transient response does over-shoot the final value and then oscillates about the latter with decreasing amplitude until the transient has expired. The typical response of this *under-damped* condition is shown in Figure 2–25, the response becoming more oscillatory as the value of ζ approaches zero. An *over-damped* response ($\zeta > 1$) is also shown and should be compared with the response of two first-order stages (Figure 2–18).

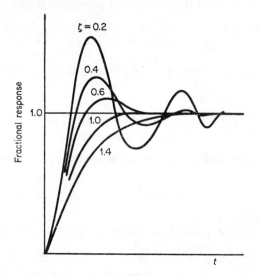

FIGURE 2–25. Step function response of second-order system with different degrees of damping

Critically-damped Second-order Response ($\zeta = 1$)

Figure 2–25 also illustrates the step function response of the second-order system when the damping ratio, ζ, is equal to 1. This is a special case in which the characteristic equation for the system has real and equal roots, since when $\zeta = 1$,

$$T^2s^2 + 2\zeta Ts + 1 = T^2s^2 + 2Ts + 1 = (Ts + 1)^2$$

The nature of the response can be deduced from the definitions of the damping factor α and frequency β in the preceding section. When $\zeta = 1$, $\alpha = 1/T$, and $\beta = 0$. Thus the response is damped since there will be a negative exponential term in the response, but the frequency of the oscillation is zero, i.e. the transient response will be of the aperiodic (non-oscillatory) type of the over-damped condition for which $\zeta > 1$. This is the special case of *critical damping*, which is the fastest response without overshooting of the final steady-state value by the transient.

Whilst the response equation for critical damping can be obtained from Equation (2–28) by considering the limiting case as $\zeta \to 1$, a simpler method is to use the Laplace transform inversion. The quadratic term of the transfer function can now be factorized into $(Ts + 1)^2$. Hence

$$\theta_2(s) = A/[s(Ts + 1)^2]$$

where the term in brackets is of the form, $1/[s(s + a)^2]$, where $a = 1/T$, and whose inversion is $(1/a^2)[1 - (1 + at)\exp(-at)]$. Hence

$$\theta_2(t) = A[1 - (1 + t/T)\exp(-t/T)] \qquad (2\text{–}29)$$

As already noted, the over-damped response is also obtained from two first-order stages in series with time constants T_1 and T_2 giving roots of $-1/T_1$ and $-1/T_2$ to the characteristic equation. It readily follows that the critically damped response will be obtained from two first-order stages whose time constants are equal.

Zero-damped Second-order Response ($\zeta = 0$)

When the damping ratio is zero, the damping factor $\alpha = 0$, and the frequency $\beta = 1/T$ or ω_n, i.e. the natural frequency of the system. Inserting these values in Equation (2–28), and noting that $\exp(0t) = 1$ and that $\phi = \pi/2$ for $\alpha = 0$, the response equation becomes

$$\theta_2 = A[1 + \sin(t/T - \pi/2)]$$

The zero-damped response to step forcing is thus a continuous sine wave of constant amplitude since the exponential term is now unity; the amplitude is the magnitude of the step function, A, i.e. the output variable cycles between the values of 0 and $2A$, and the frequency of the oscillation is the natural frequency of the system $(1/T)$. In practice this type of response is not physically realizable, since there is always some degree of natural damping in a real system which will make the response at least under-damped. As will be seen in due course, the situation is different in a closed feedback loop and zero-damping is then possible.

It is worthy of note that for zero-damping the characteristic equation of the system will have complex imaginary roots, since for $\zeta = 0$,

$$T^2s^2 + 2\zeta Ts + 1 = T^2s^2 + 1 = (1 + jTs)(1 - jTs)$$

and the roots are thus, $s_1, s_2 = \pm 1/jTs$.

Significance of the Second-order Parameters

The significance of the two parameters of a second-order system, ζ and T (or ω_n), will be apparent from the preceding discussion. The damping ratio ζ is a measure of the degree of damping of the transient response to a step input, the damping contribution arising from the negative real part of the roots of the characteristic equation which provides the term, $\exp(-\alpha t)$, in the response equation. The natural frequency, ω_n, is the reciprocal of the characteristic time, T, and is the frequency of oscillation of the zero-damped condition and is also the damping factor for critical damping.

To summarize, the responses of a second-order system are essentially determined by the value of ζ. When $\zeta = 0$, the response is zero-damped, and a step-forcing disturbance will set up a continuous oscillation at the natural frequency with a constant amplitude equal to the magnitude of the step. For $1 > \zeta > 0$, the system is under-damped and the response to a step-forcing is then a decaying oscillation with a frequency which decreases from the natural frequency to zero as ζ is increased from 0 to 1. At $\zeta = 1$, the system is critically damped and this is the fastest response without overshooting the final value. At values of $\zeta > 1$, the response is basically similar to critical damping but is slower and over-damped. In all cases except that of zero damping, the final steady-state value is the magnitude of the step function; for zero damping the steady state is a continuous oscillation.

It is interesting to note that negative values of ζ will yield a term with a positive exponent, i.e. $\exp(+\alpha t)$. This would produce an oscillatory response of increasing amplitude; this 'negative' damping is an unstable state which is not possible with the second-order or combined first-order systems so far considered but which is possible with most control loops.

Properties of the Under-damped Second-order Response

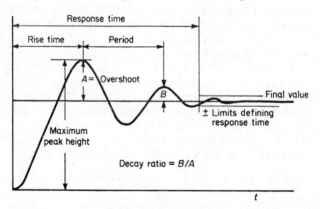

FIGURE 2–26. Terminology of the under-damped second-order response

In control system responses to step-function inputs the under-damped second-order type of response is of frequent occurrence, and it is useful at this point to consider some of the properties of this type of response and to define a number of terms which can be used for a quantitative description of a particular response. The terminology is illustrated in Figure (2–26). In general the quantitative relationships are functions of the damping ratio and/or the characteristic time (or natural frequency).

From the equation of the under-damped response (Equation 2–28), the *angular frequency* ω of the damped oscillation is the coefficient of t in the sine term, i.e. $\beta = (1 - \zeta^2)^{\frac{1}{2}}/T$, measured in radians per unit time. The *period* in units of time is given by $2\pi/\omega$ or $2\pi T/(1 - \zeta^2)^{\frac{1}{2}}$. The ratio of the actual (damped) frequency to the natural (undamped) frequency, ω/ω_n, is clearly $(1 - \zeta^2)^{\frac{1}{2}}$.

Overshoot is a measure of how far the response rises beyond the final value and is the difference between the maximum value of the output, which is the height of the first peak, and the final steady-state value. In control system responses the height of the first peak from the initial value is the *maximum deviation* or *peak error*. The *rise time* is the time necessary to reach the maximum value of the first peak from the zero time at which the step function input disturbance is imposed. These quantities can be obtained by differentiating the response Equation (2–28) and equating to zero, since the first peak occurs the first time the rate of change of the output variable becomes zero. The rise time is thus found to be $\pi T/(1 - \zeta^2)^{\frac{1}{2}}$, which is half the period. Substituting this value for t in the response equation gives the value of the variable at the first peak as

$$\{1 + \exp\left[-\pi\zeta/(1 - \zeta^2)^{\frac{1}{2}}\right]\}$$

for a unit step input ($A = 1$). Since the final value for a unit step input is also unity, the overshoot is $\exp\left[-\pi\zeta/(1 - \zeta^2)^{\frac{1}{2}}\right]$.

The *decay ratio* is the ratio of successive corresponding peak heights measured from the final value (i.e. B/A in Figure 2–26) and is a convenient way of expressing the degree of damping by measurement of peak heights from a response curve. The height of the first peak is the overshoot given above, the second peak occurs one cycle of oscillation of 2π radians later, the height will then be given by $\exp\left[-3\pi\zeta/(1 - \zeta^2)^{\frac{1}{2}}\right]$. The decay ratio is then the ratio of the two exponential terms, i.e. $\exp\left[-2\pi\zeta/(1 - \zeta^2)^{\frac{1}{2}}\right]$, which is the square of the overshoot. It can be seen that the larger the value of ζ, the greater is the degree of damping and the faster the decay. The degree of damping is often also defined by the *subsidence ratio* which is simply the reciprocal of the decay ratio (A/B in Figure 2–26), the subsidence ratio is thus $\exp\left[+2\pi\zeta/(1 - \zeta^2)^{\frac{1}{2}}\right]$.

All the above properties are dependent on the value of ζ and a plot of overshoot, decay ratio and frequency ratio *versus* ζ is given in Figure 2–27.

A further term used to described the under-damped response and which is of some importance in control systems is the *response* or *recovery time*. In principle this is the time required for the system to achieve the steady-state value or, in other words, the time for the transient part of the response to expire. Since the achievement of the steady-state is effectively determined in practice by the sensitivity of the instrument measuring the particular output variable, it is more usual to define the recovery time as the time required for the output response to come within certain specified limits of the final value and to remain within these limits, as shown in Figure 2–26. The limits are quite arbitrary and will often be determined by the accuracy with which the response variable is measured; values such as ±2 or ±5 per cent of the final value are typical. Owing to these arbitrary limits a mathematical definition is not possible, although the recovery time can be easily measured from the response curve of a recording instrument. The value is obviously determined by the magnitude of the input step and the decay ratio, hence the value of ζ which determines the decay ratio is a good indication for design purposes of the relative magnitude of the recovery time.

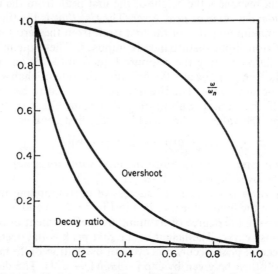

FIGURE 2–27. Relationship between damping ratio (ζ) and properties of the under-damped second-order response

Second-order Response to Sinusoidal Input

Since the over-damped and critically damped second-order systems have been seen to be equivalent to two first-order systems in series, the response to a sinusoidal input is the same as that of two first-order systems. The transient response is of little importance; the steady-state response will be an attenuated sine wave of the input frequency. There will be two stages of attenuation of the input amplitude and two stages of phase lag, each determined by the frequency of oscillation and the appropriate equivalent time constant.

The response of the under-damped second-order system to a sinusoidal input again shows the significance of the damping ratio and the characteristic time or natural frequency. The transient response is once more of much less interest than the steady-state response. By solution of

Equation (2–26), or the equivalent tranfser function, for an input of $\theta_1 = A \sin \omega t$, the steady-state sinusoid is given by .

$$\theta_2 = A \left[\frac{1}{(1 - \omega^2 T^2)^2 + (2\zeta\omega T)^2} \right]^{\frac{1}{2}} \sin (\omega t - \psi) \qquad (2\text{--}30)$$

where $\psi = \tan^{-1} [2\zeta\omega T/(1 - \omega^2 T^2)]$.

This solution is most easily obtained by substituting $j\omega$ for s in the transfer function and writing the resulting complex number in the polar form. As will be demonstrated later in the chapter on frequency response analysis (Chapter 7), the amplitude ratio of the output and input sinusoids is given by the magnitude of the complex number and the phase lag by the argument, for an oscillation of the particular frequency ω.

Comparing the input and output sinusoids, the output sine wave is of the same frequency and of fixed amplitude for a given frequency, but the amplitude and phase lag are determined by the values of ω, ζ, and T. At frequencies close to the natural frequency ($\omega_n = 1/T$), the amplitude ratio is greater than one at small values of ζ, i.e. the response wave is amplified. The phase lag is 90° at the natural frequency (at which $\omega T = 1$) and the limiting value at infinite frequency is 180° as in the over- and critically damped cases.

Estimation of System Parameters

Many process systems resemble second-order systems in their dynamic behaviour, in particular feedback control systems often exhibit responses corresponding to those of a second-order system when subjected to input disturbances. In the absence of the feedback of the closed control loop (i.e. on 'open' loop), most processes systems exhibit typical over-damped responses; when the loop is 'closed' all types of second-order response can usually be obtained by suitable adjustment of the control parameters. Most process systems are usually of an order higher than the second, but the transient behaviour of such systems will be determined, exactly as in the case of the second-order systems, by the roots of a characteristic equation for the system. The number of roots will be the same as the order of the system, but complex roots can only occur in conjugate pairs and each such pair will contribute an oscillatory mode to the overall response. In general the latter is dominated by the lowest frequency pair of complex roots and the response will thus show mainly the characteristics of the second-order response of this particular pair.

It is often difficult to carry out a detailed theoretical analysis of an existing process system, but it is often fairly simple to deduce from experimental data on the process the properties of a simple mathematical model which will provide a reasonably sound dynamic picture of the system. This does not imply that the experimental work to obtain the necessary data will be correspondingly simple; in fact considerable ingenuity may be required to obtain useful dynamic data. However, if reasonable transient responses to step function inputs can be obtained, the methods described in the following sections may be used to obtain estimates of the system parameters for a second-order model of the system.

Parameters for Over-damped Responses

For systems whose response approximates to that of an over-damped second-order system, the characteristic parameters are the effective time

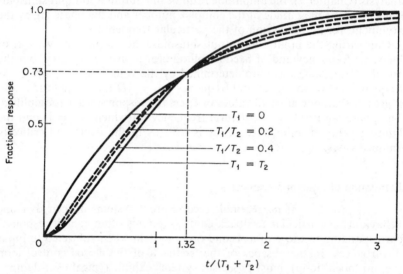

FIGURE 2–28. Normalized step function response of over-damped second-order system for different ratios of time constants

constants, T_1 and T_2. A number of methods are available for estimation of these parameters, of which that of Caldwell *et al.* [4] is to equate the time for the system to reach 73 per cent of the final steady-state value of the output after a step function input to 1.32 times the sum of the time constants. As shown in Figure 2–28, irrespective of the ratio of T_1 to T_2, when the step function response is plotted against the dimensionless ratio $t/(T_1 + T_2)$, all the response curves pass through a value of approximately 0.73 at $t/(T_1 + T_2) = 1.32$. The limiting cases of $T_1/T_2 = 0$ and 1 in Figure 2–28 correspond to the limiting cases of a first-order system with a single time constant, $T_2(T_1 = 0)$, and the critically damped response of equal time constants, $T_1 = T_2$. If a step function response up to the final steady-state value for the process is available, it is relatively simple to determine the time required for a 73 per cent response, and to divide this time by 1.32 to obtain the sum of the time constants. Then by interpolating the response at a shorter time between the curves of Figure 2–28, the ratio of the time constants can be determined and the individual values found. Alternatively, Figure 2–29 shows the fractional response for all possible ratios of T_1 to T_2 at a time equal to half that for the 73 per cent response, i.e. at $t/(T_1 + T_2) = 0.5$. For ease of use the abscissa of Figure 2–29 is plotted as $T_2/(T_1 + T_2)$.

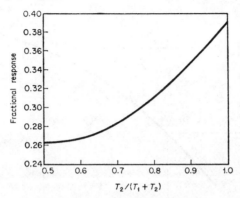

FIGURE 2–29. Fractional response of over-damped second-order system at $t = 0.5(T_1 + T_2)$

Example 2–5

A gas-fired furnace is at a steady temperature of 450°C when the gas supply to the burners is suddenly increased by 10 per cent. The resulting temperatures at 10 min intervals after the change are 499, 557, 601, 634, 659, and 678°C and the final steady-state temperature is 725°C. Determine the effective time constants if the system is regarded as second-order.

$$\text{Total temperature change} = 725 - 450$$
$$= 275°C$$
$$\text{Per cent change after first 10 min} = \left[(499 - 450)/275\right] \times 100$$
$$= 17.8 \text{ per cent}$$

Proceeding similarly, the percentage changes (i.e. the fractional responses) at 10 min intervals are

Time	10	20	30	40	50	60
Per cent response	17.8	38.9	54.9	66.7	76.0	82.9

The response curve is plotted in Figure 2–30 and from the curve, or by interpolation between the above values, the time to reach 73 per cent of the final value is 46 min. Hence the sum of the time constants, if the system is regarded as effectively second-order, is $46/1.32 = 34.8$ min.

At half this time, i.e. $0.5 \times 34.8 = 17.4$ min, the response is 35 per cent of the final value (from Figure 2–30 or by interpolation). From Figure

2–29, the value of $T_2/(T_1 + T_2)$ at 35 per cent fractional response is 0.9. Hence,

$$T_2 = 0.9 \times 34.8 = 31.3 \text{ min}$$

and

$$T_1 = 34.8 - 31.3 = 3.5 \text{ min}$$

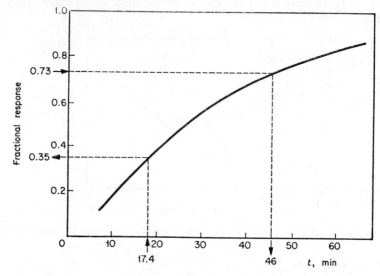

FIGURE 2–30. Step function response of temperature (Example 2–5)

Parameters for the Under-damped Response

For an under-damped oscillatory system the characteristic parameters are the damping ratio ζ and the characteristic time T. These can be estimated from a response curve without much difficulty, particularly if the system is appreciably under-damped with $\zeta < 0.5$. Figure 2–26 illustrates the relationship between ζ and the fractional overshoot and the decay ratio, hence a measurement of either of these quantities from the response curve can be used to estimate ζ. The characteristic time (or natural frequency) can also be found from the response curve, either by measuring the period of oscillation which has already been shown to be $2\pi T/(1 - \zeta^2)^{\frac{1}{2}}$, or by using the decay envelope of the response curve shown by the dotted lines on Figure 2–31. The decay envelope is a first-order decay curve defined by $[1 - \exp(-\zeta t/T)]$, hence the time constant is T/ζ and this is the time required to reach 63.2 per cent of the final value, as indicated in Figure 2–31.

Distributed Parameter Systems

In the preceding discussions of systems containing resistance and capacitance the systems considered have been analogous to linear electrical circuits in which all the resistance is concentrated in one location (the 'resistor') and all the capacitance in a different adjacent location (the 'capacitor' or condenser). Such circuits are, of course, ideal, and in practice components of one kind often have properties of the other, e.g. heat conducting materials also have thermal storage capacitance. In the majority of cases the additional property is often negligible compared to the major, or they can be assumed to be separate entities with little error. The system is accordingly regarded as a number of ideal elements located at adjacent points. The corresponding parameters are then *lumped* and, as already demonstrated, this assumption makes it possible to represent the system by an ordinary differential equation whose solution is not usually a matter of great difficulty.

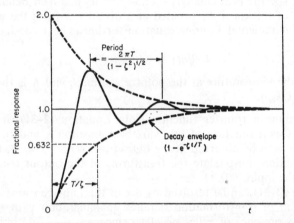

FIGURE 2-31. Step function response of under-damped second-order system

The wall of a vessel or a pipe has both thermal resistance and capacitance distributed throughout the thickness of the material, i.e. along the heat transfer path, but the walls are usually relatively thin with small capacitance compared to the contents and the resistance is also small compared to that of the fluid film on one or both sides. No great error is then involved in 'lumping' the resistance of the wall into the heat transfer coefficient on one side with the capacitance of the wall adjacent, or by using an overall heat transfer coefficient and assuming that the capacitance of the wall is negligible.

If these assumptions are not justifiable, as in the case of a relatively thick wall, or a more accurate analysis is required, the properties in question must be regarded as being distributed through the component. In place of the resistance and/or capacitance of the whole, an increment of resistance and capacitance must be allowed per unit of length or thickness. When these increments are constant, as is usually assumed to be the case, the system is then one with *distributed parameters*. The length or thickness of the transfer path must then be included in the system equation as a second independent variable in addition to time. Such a system may be regarded as an infinite number of identical inter-acting elements in series, each containing an increment of the distributed resistance and capacitance.

Response of a Distributed System

To illustrate the difference between a system with distributed parameters and a lumped parameter system, consider a slab of heat conducting material of thickness X. The input to the system is the temperature of one face where the distance into the slab is $x = 0$; the output is the temperature at the other face where $x = X$. Assuming constant thermal conductivity k, specific heat capacity c_p, and density ρ, a heat balance across an element of thickness dx situated at a distance x from the input face yields the fundamental Fourier equation for linear heat conduction in a solid, i.e.

$$(\partial\theta/\partial t) = K(\partial^2\theta/\partial x^2) \tag{2-31}$$

where θ is the temperature at the point x at time t, and K is the thermal diffusivity $(k/c_p\rho)$.

To determine a transfer function from Equation (2–31), it must be noted that θ is a function of two variables, position x and time t. The temperature is to be observed at the output face where $x = X$, hence the transfer function must relate the transforms of the output temperature, $\theta(X, t)$, to the input, $\theta(0, t)$.

Equation (2–31) can be resolved by use of the Laplace transformation; the operation of transformation requires an integration with respect to time but the presence of x has no effect since x is a second independent variable. Equation (2–31) can thus be transformed from the time-domain into that of the s-variable:

$$K(d^2\theta/dx^2) = s\theta(x, s) - \theta(x, 0)$$

where $\theta(x, 0)$ is the initial value of the temperature at point x at zero time. Following the usual convention this must be zero; hence the above equation simplifies to

$$K(d^2\theta/dx^2) = s\theta(x, s) \tag{2-32}$$

The partial differential equation in two variables x and t is thus reduced to an ordinary differential equation in one variable x; s is now merely a parameter in the equation since there are no derivatives with respect to s. As an ordinary second-order linear differential equation, Equation (2–32) can be solved without much difficulty. Thus a solution will be of the form

$$\theta = A_1 \exp(-ax) + A_2 \exp(ax)$$

By taking the second differential with respect to x, $(d^2\theta/dx^2) = a^2\theta$, and comparing with Equation (2–32), $a^2 = s/K$, hence $a = (s/K)^{\frac{1}{2}}$. The arbitrary coefficients A_1 and A_2 can be evaluated by considering the boundary conditions at $x = 0$ and X, i.e. at the input and output faces. The simplest case for consideration is that in which the distributed element does not interact with previous or succeeding elements; this requires that the input temperature be fixed and that there is no temperature gradient $(\partial\theta/\partial x)$ at the outer face, this latter condition being fulfilled if the outer face is insulated so that there is no further heat transfer. Assuming these conditions; at the input face, $x = 0$ and $\theta = \theta_1$, thus

$$\theta_1 = A_1 + A_2$$

At the output face, $x = X$, $\theta = \theta_2$, and $(\partial\theta_2/\partial x) = 0$. Hence

$$(\partial\theta_2/\partial x) = -aA_1 \exp{(-aX)} + aA_2 \exp{(aX)} = 0$$

Simultaneous solution of the last two equations yields

$$A_1 = \theta_1\, e^{aX}/(e^{aX} + e^{-aX})$$
$$A_2 = \theta_1\, e^{-aX}/(e^{aX} + e^{-aX})$$

Substituting these values in the equation for θ_2 at $x = X$,

$$\theta_2 = A_1\, e^{-aX} + A_2\, e^{aX}$$
$$= \theta_1[2/(e^{aX} + e^{-aX})]$$
$$= \theta_1[1/\cosh{(aX)}]$$

The required transfer function is thus

$$\theta_2/\theta_1 = 1/\cosh{[X(s/K)^{\frac{1}{2}}]} \qquad (2\text{--}33)$$

The inversion of this transform for a given input is difficult; Farrington [16] gives an approximate solution for a unit step input of

$$\theta_2(t) = [1 - (4/\pi)\exp{(-\pi^2 Kt/4X^2)}]$$

where the exponential term is the first term of a rapidly convergent series. This response function is plotted in Figure 2–32 against the dimensionless quantity (Kt/X^2). As will be shown later, X^2/K has the dimension of time and is effectively the time constant of a first-order system with the parameters lumped. Figure 2–32 also shows for comparison the response of a first-order system with this time constant.

Other boundary conditions for the same case of heat transfer yield similar transfer functions but with different hyperbolic terms.

The response of the distributed parameter system to a sinuoidal input will not be derived at this stage but will be considered later in the chapter on frequency response analysis (Chapter 7). It is sufficient to remark here of the distinctive property of the distributed system when subjected to sinusoidal forcing; this is that the phase lag between the input and output sinusoids increases without limit as the frequency of oscillation is increased. All lumped parameter systems show a limited value of phase angle (lag) as $\omega \to \infty$ of 90° for each time constant stage and 180° for a second-order element. This minimum phase angle is characteristic of lumped parameter systems; the *non-minimum* phase behaviour is characteristic of distributed systems.

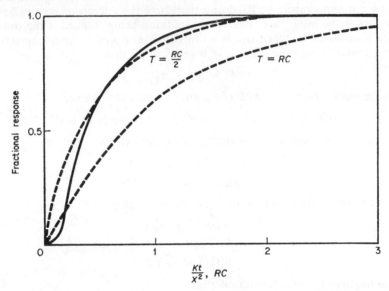

FIGURE 2–32. Step function response of distributed parameter system and first-order model with $T = RC$ and $\frac{1}{2}RC$

A further example of a distributed system often encountered in process control is the pneumatic transmission of signals, e.g. between a measuring element and a controller or recorder, between a controller and the control valve, etc. A pneumatic transmission line has resistance and capacitance and also some inductance distributed along its length, and the basic equation for signal transmission is completely analogous to that for an electrical transmission line. The inductance analogy arises from the inertia of the mass of air in the line, the pressure of which is the signal being transmitted. Inductance represents the force required to produce unit acceleration of the mass, and all three parameters can be expressed in terms of a unit length of pipe and so 'distributed'. The problem in applying the available transmission line theory to pneumatic transmission is to obtain the values of L, R, and C per unit length with proper allowance for radial gradients of velocity and temperature. Rigorous solutions are complex and not entirely in agreement with experimental measurements.

Fortunately some simplification is possible, e.g. it can be shown that inductance is negligible compared to the resistance over the range of frequencies normally encountered in process control, and the case can be approximated as one of resistance and capacitance only. For the usual situation where the volume of the receiving element at the end of the line is small compared to the volume of the line itself, application of the electrical transmission theory yields the same hyperbolic transfer function (Equation 2–33) as the heat transfer case already considered, with the parameter (X^2/K) replaced by RC, where R and C are the total resistance and capacitance of the line.

Lumped Parameter Models of Distributed Systems

In view of the complexity of the transfer functions for distributed systems, it is obviously useful to consider the possibility of a simpler mathematical model, such as a lumped parameter system, whose response behaviour is approximately the same as that of the distributed system. It will be seen from Figure 2–32 that the step function response of a distributed system is somewhat similar in general trend to that of a first-order lumped system, both show an exponential increase in the output and an asymptotic approach to the final value, although the slope of the curve for the distributed system at zero time is zero, which is more characteristic of a higher order time constant system. If the parameters of the distributed system are considered to be lumped, this is equivalent to regarding the system as consisting of a single resistance associated with a single capacitance; in the heat transfer case considered above, the lumped parameter model would have the resistance concentrated into a thin layer adjacent to a well-stirred mass of fluid providing the capacitance. Considering the same slab of heat conducting material of thickness X, the total capacitance is given by the usual relationship as $XA\rho c_p$, where A is the sectional area normal to the heat flow. Similarly the total resistance is taken as that offered to the flow of heat by conduction at steady-state between $x = 0$ and $x = X$. The flow of heat per unit time is given by $Q = kA\Delta\theta/X$, where k is the thermal conductivity (W/m K) and $\Delta\theta$ is the temperature difference between the two faces. Resistance is defined as potential per unit flow, i.e. $\Delta\theta/Q$, hence $R = X/kA$.

The time constant for this first-order lumped parameter model is thus given by

$$T = RC = (X/kA)(XA\rho c_p) = X^2(\rho c_p/k)$$

The quantity $(k/\rho c_p)$ is the thermal diffusivity K (m²/s), and thus the equivalent time constant becomes $T = X^2/K$, which will be recognized as the parameter with the dimension of time already identified in the solution of the transfer function of the distributed stage.

In the case of a pneumatic transmission line the lumped time constant is given directly by $T = RC$, where R and C are the total resistance and capacitance of the line.

The response of the first-order model is given by the usual response function for a unit step, i.e. $[1 - \exp(-t/T)]$, and this is the function plotted in Figure 2–32. The general trend of the response of the distributed and the lumped system is the same but there is considerable discrepancy between the two curves.

Other lumped parameter models may be constructed, e.g. further consideration of the distributed response of Figure 2–32 shows that the response is approximately 63 per cent completed at a value of the dimensionless group, Kt/X^2 of 0.5. The step response of the first order system reaches 63.2 per cent completion after the expiry of one time constant, and this suggests that the time constant of the lumped parameter first-order model should be half the time constant of the lumped parameters, i.e.

$$T = RC/2 = X^2/2K$$

This can be rationalized by regarding the average conduction distance to be half the transfer path with the total capacitance placed at the middle of the section. This response is also plotted in Figure 2–32; the agreement with the distributed response is somewhat better, and this lumped parameter model could be used with a reasonable degree of accuracy.

The slow initial response of the distributed system can only be simulated by use of a lumped parameter model of a higher order or by the addition of a time delay in the initial stages of the response. In the former case the total capacitance is split into a number of equal parts separated by an equal number of equal resistances, thus constructing a number of identical inter-acting stages with equal time constants. The true distributed system is, of course, an infinite number of such elements. There is, however, little to be gained in simplification if models of more than second- or third-order are employed. The alternative approach, the addition of a time delay to the model, is simpler in realization and will be considered further in a later chapter.

Time Delay

A phenomenon often encountered in process systems is a direct *time delay* or *dead time* between two related actions. In flow processes this is usually due to a distance-velocity effect, i.e. a finite time is required for the fluid conveying the signal to travel at a certain velocity between two points a given distance apart. A direct time delay is thus involved if, for example, the temperature of a flow stream leaving some item of plant is to be measured some distance downstream of the plant. With plug-flow reactors there may be a time delay equal to the residence time in the reactor before any particular property of the inflow can be measured at the outlet. Analytical instruments are often placed in separate sampling loops and there is a time delay before changes in the property measured can be passed from the plant to the instrument.

Time delays of this sort are often referred to as *distance-velocity* or *transportation lag*, but the use of the word 'lag' in this context is to be deprecated and the description of time delay or dead time is preferred. Although the terms are much confused in the literature and in general parlance, an essential and important difference should be drawn between a 'lag' and a 'delay'. In the case of a lag, such as that of a first-order system, there is some immediate response to an input forcing function and the output is a modified time function of the input. A first-order system shows an immediate response to a step input with a maximum rate of change; with second- or higher-order systems the response is less pronounced in that the initial rate of change of the response is zero, but nevertheless there is some immediate reaction. In contrast, a time delay gives no immediate response of any sort until the expiry of the delay period, hence the use of the term 'dead time'—a period during which nothing happens and there is no reaction whatsoever. The term 'lag' implies an element of resistance and it would seem preferable to reserve this term to the reaction of systems involving resistance and capacitance.

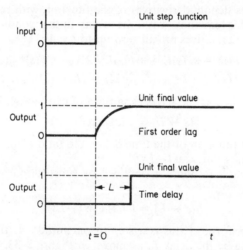

FIGURE 2-33. Comparison of step function response of first-order lag and time delay

The important practical difference between the effects of a lag and a delay in the process control loop is due entirely to this difference in the immediacy of the response. When a load disturbance occurs, the presence of a delay element in the loop allows the process to respond to the disturbance whilst the action of the controller is held back by the delay. The output variable may thus build up a deviation, sometimes of quite large magnitude, before any corrective action is applied by the controller. On the other hand, if only lag elements are present in the loop, some error signal, no matter how small, is passed to the controller almost immediately, and the corrective action then begins to be applied to oppose the effect of the disturbance on the output variable. As will be seen in due course, the effect of a time delay is to de-stabilize the control system, and a time delay is much more serious than a time constant lag of equal magnitude. The differences in responses to a step function input to a time delay and a first-order lag are illustrated in Figure 2–33.

Transfer Function of a Time Delay

Assuming that there is no inter-change of energy with the surroundings and that the input signal to a time delay element is a function of time, such as $f(t)$, the output signal will be identically the same function of time as the input but will be subjected to a delay in time of L, i.e. the response is displaced or *translated* along the time axis by the period of the delay. The output signal for an input, $f(t)$, will thus be $f(t - L)$. This latter term may be expanded by a Taylor series:

$$f(t - L) = f(t) - f'(t)L + f''(t)L^2/(2!) - f'''(t)L^3/(3!) + \ldots$$

where the primes denote derivatives of the function with respect to time. Transforming this equation with the usual initial conditions that the function and its derivatives are all zero up to $t = 0$:

$$F(t - L) = F(s) - sF(s)L + s^2F(s)L^2/(2!) - s^3F(s)L^3/(3!) + \ldots$$
$$= F(s)[1 - Ls + (Ls)^2/(2!) - (Ls)^3/(3!) + \ldots]$$

The term in brackets is the series expansion of $\exp(-Ls)$. Hence

$$F(s - L) = F(s)\exp(-Ls)$$

and the transfer function of the time delay, the ratio of the output to the input transform, is thus $\exp(-Ls)$.

The step function response is obtained by substitution of A/s for $F(s)$, giving the output function

$$F(s - L) = (A/s)\exp(-Ls)$$

The inverse transform of this is a step of magnitude A (the input step), displaced in time by the delay L, as shown in Figure 2–33.

For a sinusoidal input the amplitude of the oscillation will be unchanged but there will be a lag in phase due to the lag in time between the input and output sinusoids. The lag in phase can be determined by direct reasoning as follows: the time difference between corresponding peaks of the input and output sinusoids is the delay L, the phase difference is then that proportion of the complete cycle (2π radians) as the ratio of the delay time to the period of the complete cycle, i.e. if the phase lag is ψ,

$$\psi/2\pi = L/\text{Period}$$

Since the period is given by $2\pi/\omega$, where ω is the angular frequency in radians per unit time, the phase lag is then $\psi = \omega L$ radians, or $57.3\omega L°$. The phase lag thus has no limit but increases indefinitely with increasing frequency. The time delay element exhibits the non-minimum phase behaviour which was pointed out in the preceding section as characteristic of a system with distributed parameters.

Distance-velocity Delay as a Distributed System

That a distance-velocity delay is a type of distributed parameter system can be demonstrated without difficulty. Consider an incompressible fluid flowing at a linear velocity v with a flat velocity profile ('plug flow') through a pipe of length, X. Suppose the pipe is divided into n zones, each of length X/n. If each zone is now regarded as a well-mixed tank, the pipe is equivalent to n first-order stages each with a time constant of $T = (X/n)(1/v)$. The overall transfer function for n first-order stages in series for an input θ_1 and an output θ_2 is

$$\theta_2/\theta_1 = 1/(Ts + 1)^n = 1/[(X/n)(1/v)s + 1]^n$$

To distribute the parameters, n must be increased to infinity, hence:

$$\theta_2/\theta_1 = \lim_{n \to \infty} \{1/[(X/nv)s + 1]^n\}$$

which can be shown to be equal to $\exp[-(X/v)s]$; X/v is the length of the pipe divided by the velocity of flow and is thus the time delay L, and the transfer function is that previously derived, $\exp(-Ls)$.

Modified Distance-velocity Delay

In the discussion of time delays in the preceding sections, inter-change of energy with the surroundings has been specifically excluded; consequently the output function from a delay element is identically the same as the input but translated in time. Many distance-velocity effects in practice differ from the pure time delay in that some inter-change of energy from a fluid to the surroundings can occur, e.g. the loss of heat from a flowing fluid to the containing pipe walls. This will have an obvious effect on the magnitude of a temperature signal being conveyed by the fluid, in addition to the time delay; the output signal will not then be identically the same function of time as the input.

Consider then a fluid flowing with linear velocity v along a pipe of length L and diameter D. The resistance to heat loss through the pipe wall and the thermal capacitance of the fluid are now distributed along the pipe; consequently in any system equation the distance along the pipe from the entrance must be included as second independent variable with time.

A heat balance over an element of the pipe of length dx, situated at a distance x from the entrance, will as usual equate the accumulation of heat in the element dx with that entering and leaving the element per unit of time. If the fluid temperature at x is θ, that at $(x + dx)$ will be

$$(\theta + (\partial\theta/\partial x)\,dx)$$

and if the temperature of the pipe wall is θ_a, the heat balance is

$$Mc_p(\partial\theta/\partial t) = Wc_p[\theta - (\theta + (\partial\theta/\partial x)\,dx)] - UA(\theta - \theta_a)$$

where M is the mass of fluid in the element dx,

W is the mass rate of flow,

U is the overall heat transfer coefficient,

and A is the surface area of pipe in the element dx.

Expanding these properties in terms of the fluid density ρ, velocity v and diameter D;

$$[(\pi/4)D^2\,dx]\rho c_p(\partial\theta/\partial t)$$
$$= -[(\pi/4)D^2v]\rho c_p(\partial\theta/\partial x)\,dx - U(\pi D\,dx)(\theta - \theta_a)$$

whence

$$\partial\theta/\partial t = -v(\partial\theta/\partial x) - (4U/\rho c_p D)(\theta - \theta_a)$$

The coefficient of the last term, $4U/\rho c_p D$, has the dimension of reciprocal time and can thus be replaced by $1/T$. By using the Laplace transformation the equation is converted from a partial differential equation in x and t into an ordinary differential equation in x, thus:

$$s\theta = -v(d\theta/dx) - (1/T)(\theta - \theta_a)$$

Re-arranging,

$$-vT(d\theta/dx) = (Ts + 1)\theta - \theta_a$$

whence

$$\frac{d\theta}{(Ts + 1)\theta - \theta_a} = -\frac{dx}{vT}$$

Integrating from $\theta = \theta_1$ at $x = 0$, to $\theta = \theta_2$ at $x = X$,

$$\frac{(Ts + 1)\theta_2 - \theta_a}{(Ts + 1)\theta_1 - \theta_a} = \exp\left[-(X/v)(Ts + 1)/T\right] \quad (2\text{--}34)$$

The system thus has two input variables, the inlet temperature, θ_1, and the pipe wall temperature, θ_a, with the outlet temperature, θ_2, as the output variable. There will be two transfer functions relating the output to each individual input whilst the other input is constant. Thus, from Equation (2–34), for a constant pipe wall temperature, $\theta_a = 0$, the transfer function between the outlet and inlet temperatures of the fluid is

$$\theta_2/\theta_1 = \exp\left[-(X/v)(Ts + 1)/T\right]$$
$$= \exp\left[-(X/v)s - X/vT\right]$$
$$= \exp(-Ls)\exp(-L/T)$$

As can be seen, X/v is the time delay L, and the first exponential term is the transfer function of the time delay. The second exponential term is a constant whose magnitude is dependent on the time delay and also the 'time constant' of the heat loss. The result of the thermal leakage is thus a constant reduction in the magnitude of the output signal.

When the input function, θ_1, is kept constant and the forcing applied by a change in the wall temperature, θ_a, inasmuch as the variable is now distributed in space in the system, this will represent a *distributed parametric forcing*. For $\theta_1 = 0$, Equation (2–34) can be re-arranged to give the transfer function between θ_2 and θ_a:

$$\theta_2/\theta_a = [1/(Ts + 1)]\{1 - \exp[-(X/v)(Ts + 1)]\}$$
$$= [1/(Ts + 1)][1 - \exp(-Ls)\exp(-L/T)]$$

The expression in the first bracket is the transfer function of a simple first-order lag, that in the second is that of a displaced time delay with leakage.

In principle the response of the outlet temperature of the fluid can be found for any input function by introducing the appropriate transform into the transfer function above, but the resulting expression is complex and not easily inversely transformed into a time function. The frequency response characteristics can be obtained by substitution of $j\omega$ for s, as discussed in Chapter 7, and this analysis shows that for sinusoidal forcing both amplitude ratio and phase lag exhibit maxima and minima as the frequency is increased. This resonance effect has been demonstrated experimentally by Cohen and Johnson [8] on a simple steam-to-water heat exchanger.

The General Process System

Typical industrial process systems consist of a varying number of process stages containing resistance and capacitance, either singly or in combination, possibly with some distributed parameters and with some time delay. Allied to these sources of lag and delay in the process (plant) are other similar effects in the other parts of the control loop. These will include the regulating valve, the measurement transducer and the mechanism of the controller, and also the transmission lines which link the elements into a control loop. The nature of the lags and delays extraneous to the process itself are not fundamentally different in nature from those of the process, although usually of smaller and sometimes negligible magnitude. Second-order responses may be found in the mechanical elements of the measuring or controlling elements, but are not generally usual in the process unless this includes some mechanical elements or recycle streams.

Fortunately many of the lags and delays in the complete control loop are not significant, and the whole system is usually dominated by a relatively small number of significant lags and/or delays. The dynamic behaviour of the system can then be represented by an equation of relatively low order over a wide range of frequencies with reasonable accuracy, although the less significant lags or delays may become apparent if the frequency range is extended.

It is often desirable to represent an existing process system by an empirical equation, or by an empirical transfer function between the measured output variable b and the input manipulated variable m. The most general equation will represent the process by first- and second-order lag terms with a time delay. The first-order term provides for suitable attenuation of the input signal, the second-order term allows for a dominating oscillatory mode, and the time delay accounts for any distributed parameter effects or pure time delays. The complete empirical equation, including a steady-state gain, will be

$$b/m = K \exp(-Ls)/[(T_1 s + 1)(T^2 s^2 + 2\zeta Ts + 1)]$$

By adjustment of the five constants, this transfer function can be made to approximate to the dynamic response of almost any process system. The steady-state gain can be evaluated from the steady-state response following a step-function input, and the time delay is simply the displacement of the response along the time axis. The other constants can be found from the dynamic response by the methods already discussed.

In many cases simplification of this general system function is possible, e.g. if there is no tendency for the response to oscillate after a step input-forcing the under-damped second-order term can be replaced by an over-damped term and the first-order term omitted. Few process systems exhibit oscillatory tendencies in the absence of feedback unless there are internal recycle streams, and the step response usually approximates to that of the over-damped second-order system, the addition of a time delay often improving the representation. Note that the step response considered is that of the open loop, although the other elements in addition to the process are included in order to give the total lag and delay of the whole loop, the latter is not yet 'closed'. The input forcing is applied through the regulating valve which is disconnected from the controller, and the output of the latter is used as an indication of the measured value of the output variable.

It is interesting to note that the work of Ziegler and Nichols [30] and of later workers, in methods of assessing the type of control action required for existing processes and for determining optimum adjustments of control parameters, is based on a still simpler representation of the process control loop dynamic behaviour. This is the combination of a time delay with a first-order lag, i.e. $\exp(-Ls)/(Ts + 1)$ as the open-loop transfer function, which in many cases will give a satisfactory approximation. The methods referred to above are discussed further in Chapter 5.

PROBLEMS

2–1 A liquid storage tank is spherical in shape. Calculate the capacitance as a function of liquid level and linearize the relationship for small perturbations of the level about a normal value.

2–2 The flow of water over a V-notch (triangular) weir is related to the head by a relationship such as $q = C(2gh)^{5/2}$. Determine the linear flow resistance at a flow $\bar{q}$ and head $\bar{h}$.

2–3 Determine the transfer functions relating the input variable m and the output variable c and draw the appropriate block diagrams for the systems shown in Figure P2–1. Linear relationships between flow and head may be assumed.

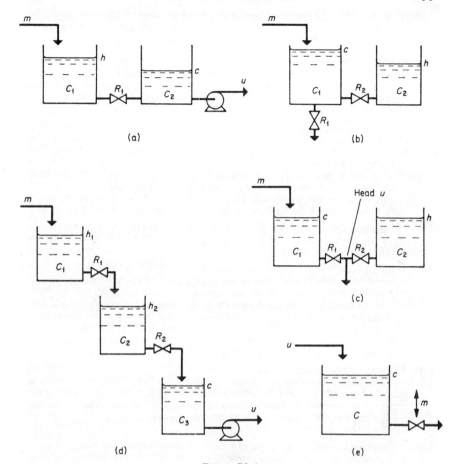

FIGURE P2-1

2-4 A process heating system consists of two insulated tanks in series, each of 2.5 m³ (100 ft³) hold-up volume, through which flows a liquid of density 800 kg/m³ (50 lb/ft³) and specific heat capacity equal to that of water, at a rate of 4 kg/s (500 lb/min). A heater in the first tank supplies heat at a rate of 10 kW (2500 Btu/min) per kN/m² (lbf/in²) air pressure applied to a control valve with a linear characteristic. The temperature is measured in the second tank by a thermometer with a time constant of 10 s (0.2 min) and which gives a linear pen displacement of 5 mm/°C (0.25 in/°F). Perfect mixing may be assumed and there is no dynamic lag in the valve. Draw a block diagram for the system with numerical values of the parameters, and evaluate the transfer function between pen displacement and a change in the inlet temperature.

2-5 A flow of 0.01 m³/s (5000 gal/h) of a weak salt solution of concentration (c_0) varying between 0.9 and 1.1 kg/m³ (gm/l) is fed to a stirred tank of 2.5 m³ (500 gal) hold-up volume. A strong salt solution of constant concentration (c_1) of 250 kg/m³ (gm/l) is added through a control valve whose flow varies linearly from zero to

50×10^{-6} m^3/s (0.6 gal/min) over the range of movement of the valve stem. The outlet concentration (c_2) is to be maintained at 2.0 kg/m^3 (gm/l) and is measured by an analysing meter with an input range of 1.8 to 2.2 kg/m^3 (gm/l) and a full-scale deflection of 0.25 m (10 in), situated at a point which the solution leaving the tank takes 100 s (2 min) to reach. Perfect mixing may be assumed and the amount of strong solution added is negligible compared to the total flow. Draw a block diagram of the system with numerical values of the parameters and determine the transfer function between the output concentration and movement of the valve stem.

2–6 A flow of 2 kg/s (250 lb/min) of an aqueous solution is fed to a system consisting of a heat exchanger of 0.1 m^3 (4 ft^3) volume, followed by a constant overflow tank of 0.25 m^3 (10 ft^3) volume. The temperature of the liquid is regulated by adding steam to the exchanger through a valve with a linear characteristic such that a 10 per cent movement of the valve stem produces a change of 10 kW (500 Btu/min) in the heat flow. There is no dynamic lag in the valve or thermometer. Draw a block diagram of the system with numerical values of the parameters and determine the overall transfer function relating outlet temperature to valve position.

2–7 Using the Laplace transformation, determine the response to a ramp forcing ($\theta_1 = At$) for a first-order system with time constant T, and for a second-order system with parameters ζ and T.

2–8 A stirred tank heater has a hold-up capacity of 0.075 m^3 (15 gal) with a flow of water of 5 kg/s (600 lb/min). The heated water passes into a well-insulated pipe of 0.02 m^2 (0.197 ft^2) cross-sectional area and the temperature is measured at a point 5 m (15 ft) downstream by a thermometer with negligible lag. An electrical heater provides a constant heat input of 100 kW. If the inlet temperature of the water is 50°C and cycles continuously with an amplitude of ± 2°C and a period of 50 s (1 min), determine the steady-state behaviour of the thermometer reading compared to the inlet temperature and to that in the tank.

2–9 An open cylindrical tank 1.5 m (5 ft) diameter is filled with water to a depth of 1 m (3 ft) when the inflow and outflow are 2.5×10^{-3} m/s (2000 gal/h). What is the new depth in the tank if the inflow is increased by 20 per cent, and what is the approximate time for the level to reach 95 per cent of the final value? Calculate the transfer functions relating level in the tank to the inflow and to the outflow valve position.

2–10 A step change in the desired value of a feedback control system operating on a low gain produces a non-oscillatory response with the following deviations measured at 5 min intervals after the step is imposed: 2.2, 4.8, 6.6, 7.7, 8.4 and the final steady-state value is 10.1. Determine the two effective time constants characterizing the response.

The controller gain is increased and the same step in desired value is applied in the reverse direction, producing a damped oscillation with a first peak deviation of -15.3 after 15 min and a second peak of -11.2 in 44.5 min. Determine the appropriate parameters for the oscillatory response.

Chapter 3: Elements of the Process Loop

As discussed in Chapter 1, the process control loop is a cyclic system of elements around which information, in the form of signals, is passed. The performance of such a cyclic system must necessarily depend upon the characteristics of every element in the loop; essentially the process control loop consists of three basic elements in addition to the process (or plant) itself. These are the elements which provide the functions of measurement (or feedback), control, and regulation. The present chapter is devoted to a discussion of these elements outside the process which comprise the essential 'hardware' necessary to exercise control over the process.

Measurement

The measurement of the instantaneous value of the controlled variable is necessary in order that it may be fed back to the controlling mechanism which must first determine the magnitude of the error or deviation from the desired value on which the generation of the control impulse is based. Primarily the function of the measuring element is to convert or *transduce* the controlled variable into a feedback variable which may be directly employed to actuate the control mechanisms. Thus, in Figure 3–1, the measuring means H converts the controlled variable c into the feedback variable b so that the controller may determine the actuating signal $e=(r-b)$ by direct comparison of the feedback variable with the reference input or desired value. The usual function of the process measuring instrument, i.e. display of the measured value by indication or recording is only incidental as far as the control of the particular variable is concerned.

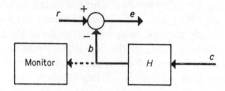

FIGURE 3–1. The measuring element

The feedback variable generated by the measuring element may be a mechanical displacement, a force or pressure, or an electrical signal such as voltage or current, depending essentially upon the design and operation of the measuring means. The latter, however, is necessarily inter-dependent with the means of operation of the controller since the feedback signal and the reference input must be of the same physical form so that subtraction of the one from the other to give the actuating signal can be carried out.

The measuring element will be considered here solely as it affects the performance of the control system; as will be seen in a later chapter, the dynamic characteristics of the measuring element, usually referred to as the *measuring lag*, directly influence the behaviour of the control system as well as affecting the indication and/or recording of the controlled variable. The latter, in process control, is naturally one of the common process variables which can be listed in an approximate descending order of frequency of occurrence as: temperature, pressure, rate of fluid flow, fluid stream composition, and liquid level. For a broader study of the principles of operation of the various measuring instruments used in process control the reader is referred to the numerous texts on this subject (see Bibliography).

Properties of the Measuring Element

The properties of the measuring element which must be briefly considered in addition to the dynamic characteristics are *accuracy*, *precision*, and *sensitivity*.

Accuracy and Precision

The accuracy of a measurement is the closeness with which the measured value approaches the true value; the precision is the reproducibility with which repeated measurements of the same magnitude can be made. For process control, precision of measurement is usually more important than a high degree of absolute accuracy.

Two differing types of accuracy may be distinguished, these are the *static*, or steady-state, accuracy which is the closeness of approach to the true value when the latter is constant in time, and the *dynamic* accuracy which is the closeness of approach when the true value is changing with time. The difference between the true and indicated values of the variable in the two cases are the *static* and *dynamic errors* respectively. Dynamic accuracy will obviously depend to some extent on the nature of the variation with time of the true value of the variable as well as on the properties of the measuring instrument. For process control systems, a practical specification of the time variation of the variable is the *linear* or *ramp* forcing, i.e. a change in value at a constant rate. Figure 3–2 shows the errors of a thermometer when forced at a constant rate of 10°/min. The thermometer shows a static error of 1° up to the start of the change, the dynamic error being zero up this point but this then increases up to a constant value by which the measurement lags behind the true value whilst the change in the latter continues. It can be shown by simple analysis that for a first-order transfer element the constant value of the dynamic error is the product of the rate of change of the variable and the time constant of the element.

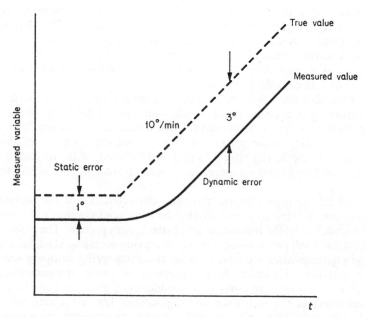

FIGURE 3–2. Static and dynamic error in the measuring element

Sensitivity

The sensitivity of a measuring element can be defined in two ways. The definition usually employed is the ratio of the magnitude of the output and input signals under steady-state conditions, in other words the steady-state gain of the instrument. Thus the greater the sensitivity, the greater is the change in output for a given change in input. As will be seen later, this gain sensitivity forms part of the overall gain of the process loop.

The alternative definition of sensitivity is the smallest change in the measured variable which will produce a change in the output signal, i.e. a *threshold sensitivity*. In most practical measuring instruments, input signals smaller than a minimum value do not produce any detectable change in output due to factors such as static friction and lost motion in moving parts; fortunately this threshold sensitivity is reasonably small in most cases of well-designed and constructed industrial instruments.

Speed and Nature of Response

As has already been seen in the discussions of the preceding chapter, most measuring instruments contain resistance and capacitance and sometimes also inductance when the moving parts are subject to accelerations. Thus all types of thermal elements similar to the bare thermometer immersed in a fluid, as considered on page 28, will show a typical first-order response with a transfer function such as $K_m/(Ts + 1)$, where K_m is the steady-state gain (sensitivity) and T is a time constant given by the product of the thermal resistance and capacitance. In practice it is unusual to expose a thermal element to the corrosive effects of process fluids, and the addition of a protective sheath produces a second-order over-damped system, as was considered on page 61.

In a similar way, fluid resistance and capacitance are the source of measuring lag in pressure elements when long fluid-filled lines connect the gauge to the point of measurement. If the inertia of the moving fluids and the moving parts of the gauge can be neglected the system is effectively first-order. If the inertia effects cannot be neglected, as in the case of the manometer considered on page 70, a second-order equation has to be employed.

It will be apparent that the dynamic characteristics of the measuring element are determined essentially by the same considerations as those of the process, i.e. by the resistances and capacitances present. The measuring element itself will generally contain some moving parts and elastic elements whose characteristics are described by the mass-spring-damping second-order equation. However, in the majority of cases, the second-order response of the moving parts is negligible when compared to the much greater effect of the resistance and capacitance lags associated with the sensing element. Thus in temperature measuring elements, the major lag is invariably that of the heat transfer to or from the sensing element and other sources of lag, such as in the deflecting milli-voltmeter with a thermo-couple, are not significant. The same remark applies to pressure gauges and flow meters when long fluid-filled lines are necessary to connect the point of measurement to the instrument. With short connecting lines or with moving parts of larger inertia, such as the liquid filling of a manometer, the response may be under-damped, showing that the second-order response predominates over the resistance-capacitance lags.

It must also be remembered that the dynamic characteristics of a process measuring element depend as much on the location of the sensing element with respect to the process as on the particular properties of the element itself. This is particularly evident with thermal elements since the heat transfer is determined by properties of the process fluid such as density, viscosity, etc. and by the velocity of flow over the element, all of which contribute to the heat transfer film coefficient. Thus the time constant of a thermometer, etc. in a gas stream will decrease as the velocity of flow increases; it will be very much less if the thermometer is immersed in a boiling liquid.

The effects of measuring lag on the performance of a control system will be considered in a later chapter.

Signal Transmission

An essential purpose of a process measuring instrument is the transmission of the indicated value from an often inaccessible point of measurement within the process to a more convenient point for indication or recording. This transmission of the signal is often an integral part of the measuring operation, e.g. the capillary tube connecting the bulb and Bourdon tube of a pressure-bulb thermometer is an essential part of the sealed pressure-filled system, but within limits can be made of any length to suit the particular application. In many control systems the point of measurement of the controlled variable is widely separated from the controller mechanism, and the feedback signal may have to be transmitted over a relatively long distance. In a similar way the output signal from the controller may have to be passed to a regulating unit on the plant some distance away. Whilst the controller output signal is usually suitable for direct transmission over suitable lines, there is a limit to the distance over which a conventional measuring instrument may transmit signals without excessive distortion or delay. In such cases it is necessary to convert the measuring instrument into an electrical or pneumatic force-balance design so that the output is a signal of the required type, or alternatively to add an electrical or pneumatic transmitter to convert the mechanical displacement of the instrument pointer into a signal suitable for transmission.

Electrical transmission is used whenever the process instrumentation is largely or wholly electronic or when it is necessary to transmit signals over distances longer than 50–60 m (200 ft). The dynamic response of an electrical signal is, of course, much too fast to contribute any detectable lag into the control system.

The limitation of pneumatic transmission is the lag in conveying the signal (change in air pressure) through a long connecting tube of relatively small diameter. This is effectively a resistance-capacitance effect but both parameters are distributed along the length of the tube; calculation of the lag is thus difficult and reliance is usually placed on experimental determination. The response may be characterized by a first-order lag coupled with a dead time, the latter arising from the distributed nature of the parameters. As an indication of the approximate magnitude of the effect, Eckmann [12] quotes the following values: over a length of 60 m (200 ft) of 5 mm ($\frac{3}{16}$ in) inner diameter tubing the time constant $T = \frac{1}{2}$ s and the dead time L is negligible; over 100 m (350 ft), $T = 1$ s and $L = \frac{1}{3}$ s; over 300 m (1000 ft), $T = 7$ s and $L = 1$ s.

The Controller

As illustrated in the block diagram of Figure 3–3, the control element of the process control loop comprises two essential parts, an error discriminator and an output element. The former compares the feedback signal b (the measured value of the controlled variable), with the reference input r, so producing an error or deviation variable e which is the actuating signal into the output element. This latter may also be referred to as the control action or function generator, since the output from this element is the controller output signal p which is applied to the regulating unit to manipulate a particular input variable to the process. Depending upon the design and method of operation of the controller, an input element may also be required to convert the desired value v (measured in units of the controlled variable), into the reference input r; in principle the input element is a signal transducer converting the mechanical setting of the desired value (usually by movement of a pointer over a suitable scale of units of the controlled variable) into a signal of the same physical nature as the feedback signal provided by the measuring element.

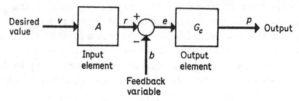

FIGURE 3–3. The control element

The practical design of process controllers varies considerably in the provision of these basic requirements. The source of power used to generate the control action may be mechanical (i.e. forces derived from the measuring element), pneumatic, hydraulic, electrical, or electronic; the method of operation may be based on displacement or force-balance techniques. For convenience of assembly the measuring element with an associated indicating or recording monitor may be placed in the same instrument case; alternatively the measuring and control elements may be completely separate with a signal transmitter linking the two. The design of commercial instruments is considered in detail by several authors, notably A. J. Young [29].

The method of operation of the controller is not, however, of direct consequence to system design, except insofar as some practical instruments do not conform exactly to theory, the control action developed departing to a greater or less extent from the theoretical definitions under certain conditions.

Control action is the functional relationship between the manipulated variable m and the controller actuating signal e, the error or deviation of the feedback variable from the desired value. Being respectively the output and input of the control action generator, this relationship can conveniently be expressed as a transfer function in the usual way so long as there is a continuous mathematical relationship between the two. As noted in the previous section, it is more usual in process control for the manipulated variable to be regulated by a unit additional to the controller, i.e. a control valve whose output is the manipulated variable and whose input is the controller output signal p. The control-action transfer function will then be defined between the controller output p and the error signal e. If the regulating unit is a valve with a linear characteristic, i.e. a direct proportionality between output and input, and without lag or delay, the control action defined by either m or p will differ only in the term representing the gain of the system.

Discontinuous Control Action

The simplest type of control action is a discontinuous relationship in which one of two possible values of controller output or manipulated variable is selected by the sign (positive or negative) of the error or deviation. This *two-position* action is undoubtedly the most widely used type of control in both industrial and domestic service, instruments of this type being relatively cheap, virtually fool-proof, and simple and rugged in construction. The action is conveniently generated by an electrical switch which has only two positions, open and closed, or on and off, and this action is thus the basis of most types of electrical thermostats and pressure-stats. Owing to the use of a switch, the action is often termed *on-off* action although strictly this term should be applied to the rather special application in which one of the switch positions corresponds to a zero input to the process. In the more general case the two positions may be regarded as corresponding to maximum and minimum values of input to the process; this can be accomplished by the use of double-throw switches, by-pass resistors, a by-pass flow around a solenoid valve, etc.

The action is expressed mathematically by two statements:

$$m = M_1 \quad \text{when } e > 0$$

$$m = M_2 \quad \text{when } e < 0$$

where M_1 and M_2 are the pre-determined values of the manipulated input variable. The action is thus described by two equations each applying in a certain region of deviation, or as can readily be seen, when the controlled

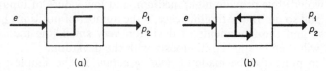

(a)   (b)

FIGURE 3–4. Block diagram representation of two-position action: (a) without differential, (b) with differential

variable is greater or less than the desired value. It is not possible to write a simple transfer function in this case, and in signal flow diagrams the graphical representation of Figure 3–4 is used, where the 'step' illustrates the instantaneous change from one position of input to the other.

In practice it is not usually possible to arrange for the change of values to occur exactly and instantaneously at the point when the deviation is zero, nor is it usually desirable for this to occur. A very sensitive controller responding to the change in sign of very small deviations would be subjected to excessive wear of the electrical contacts and the moving parts. A two-position controller will generally exhibit a *differential*, i.e. an overlap of the two zones of operation, so that as the deviation passes through zero the manipulated variable retains its value and does not change until a certain deviation of opposite sign has built up. This occurs naturally in practical instruments due to static friction and lost motion in instrument linkages but is often deliberately increased by damping and similar devices to prevent over-frequent operation of switches, relays, and solenoid valves, and to reduce arcing at contacts.

For systems with a large capacitance and subject to small disturbances, two-position action can provide efficient control at the minimum capital outlay, but as may readily be visualized the response of the process system is inherently oscillatory in nature. In thermal systems the two values of the manipulated input variable, i.e. two heat flow rates into the process, will correspond to two potential values of the controlled variable (temperature); one must be greater than the desired value and the other less so that the temperature can be driven in the appropriate direction to reduce the deviation to zero. Neither, however, can be equal to the desired value and the temperature must then always be changing in the direction of one or other of the two potential values and will thus cycle continuously about the desired value at all times. If the capacitance of the system is sufficiently large and the magnitudes of the potential corrections wisely chosen, the cycling may be slow and of quite small amplitude.

A mathematical analysis of two-position control presents some difficulty but many of the properties may be deduced from a qualitative study as developed in the next chapter.

The concept of two-position action may be widened to a *multi-position* action by the provision of additional operating points, i.e. subsidiary desired values at certain fixed increments of deviation, to provide additional incremental values of the manipulated variable. A three-position control would thus provide high, medium, and low values of input to the process, and so on. In the limiting case, if the number of positions becomes very large and the increments of deviation very small, the manipulated variable will then change continuously with the deviation.

Mention must also be made of *floating action*, in the simplest case of which the manipulated variable changes at a constant rate (thus *single-speed* floating action), and the *direction* of the change is determined by the direction (positive or negative) of the deviation. This action may be developed quite simply by use of an electrical three-position instrument supplying current to a reversing motor, the third position being an intermediate neutral or dead zone where the motor may come to rest when the deviation is relatively small. This type of action is not used very often and will not be discussed further, but it is interesting to note that a multiple-speed floating action can be developed which in the limit will correspond to a continuous relationship between the rate of change of the manipulated variable and the deviation.

Continuous Control Actions

In this case the relationship between controller output (or manipulated variable) and the error signal is a continuous mathematical function, i.e. p (or m) $= f(e)$. A controller transfer function can thus be defined:

$$G_c(s) = p/e$$

Whilst theoretically the continuous control function can take almost any form, in practice the function must be linear, and for other practical reasons the functions used in process control are limited to proportional, integral, and derivative functions, and various combinations of these.

Proportional Action

The simplest continuous linear function is that of proportionality. In proportional control action, the changes in the controlled variable with a constant desired value are repeated and amplified in the output of the controller, i.e.

$$p = K_c e + p_0$$

where K_c is the proportional sensitivity or *proportional gain*, and p_0 is a datum relationship which determines the normal value of the controller output at zero deviation ($e = 0$). For flexibility in operation, both K_c and p_0 must be capable of adjustment within certain limits in a practical instrument.

A plot of p *versus* e (Figure 3–5) is a straight line of slope K_c passing through the value of p_0 at $e = 0$. Since e is the difference between the measured and desired values of the controlled variable, if the desired value is constant, the deviation scale can be replaced by a scale of measured values on which $e = 0$ is the desired value. It can then be seen that there is a range of measured values of the controlled variable within which the controller output is proportional to the deviation, the boundaries of this range corresponding to the zero and maximum values of the controller output. This range of measured values over which the controller output changes from zero to maximum, expressed as a percentage of the instrument scale, is the *proportional band*, a term which is still commonly used to define the sensitivity (K_c) on practical instruments, although there is an increasing tendency to replace this by either 'sensitivity' or 'gain'. The width of the proportional band is inversely proportional to the sensitivity but is tied to the range of a particular instrument by the definition as a percentage of the scale. High sensitivity is equivalent to a narrow proportional band, a wide band to low sensitivity. The following example may clarify these points.

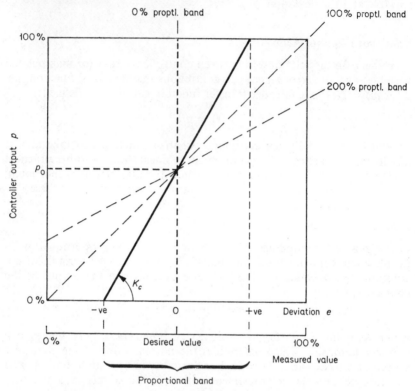

FIGURE 3–5. Proportional band and sensitivity

Example 3–1A

Determine the sensitivity and proportional band of a pneumatic proportional controller with a scale range of 0 to 120°C when the output changes from 20 to 100 kN/m² as the temperature rises from 95° to 110°.

$$\text{Proportional band} = \frac{\text{Temperature change}}{\text{Scale range}} \times 100$$

$$= (110 - 95)/(120 - 0) \times 100$$

$$= 12.5 \text{ per cent}$$

$$\text{Sensitivity} = \frac{\text{Output change}}{\text{Input change}}$$

$$= (100 - 20)/(110 - 95)$$

$$= 5.33 \text{ kN/m}^2 \text{ °C}$$

If the proportional band is changed to 50 per cent, determine the sensitivity and the temperature change required for the full change in output.

$$\text{Temperature change} = \text{Proportional band} \times \text{Scale range}$$
$$= (50/100) \times 120$$
$$= 60°C$$

$$\text{Sensitivity} = 80 \text{ kN/m}^2/60°C$$
$$= 1.33 \text{ kN/m}^2 \text{ °C}$$

Example 3–1B

The full range of output of pneumatic instruments of the British and American units is from 3 to 15 lbf/in^2. Using this range in the above example, the proportional band is obviously the same (12.5 per cent) if the same scale range and temperature change is defined.

$$\text{Sensitivity} = 12 \text{ lbf/in}^2/15°C$$
$$= 0.8 \text{ lbf/in}^2/°C$$

Increasing the proportional band to 50 per cent again requires a 60° change in temperature to achieve full change in output.

$$\text{Sensitivity} = 12 \text{ lbf/in}^2/60°C$$
$$= 0.2 \text{ lbf/in}^2/°C$$

Adjustment of the sensitivity varies the width of the proportional band but does not affect the value of p_0, thus maintaining the datum relationship between output and deviation. The band width is theoretically variable between zero and infinity; most practical instruments normally have band widths variable between 2 and 150 per cent. At values over 100 per cent, the output change cannot cover the full range since part of the band width lies outside the range of the instrument scale of measured values. At zero width the output changes from zero to maximum at the desired value, corresponding to an on-off action which may thus be regarded as a limiting case of proportional action. In practice a very narrow proportional band (*circa* 2 per cent) produces effectively on-off control.

The constant, p_0, is the value of the controller output required to maintain the controlled variable at the desired value under a particular load condition. Changing the value of p_0 should not affect the sensitivity in a well-designed instrument; the slope (K_c) of the operating line relating p to e stays the same but the intercept at zero deviation is altered to a new value of p_0 (Figure 3–6). This adjustment effectively shifts the position of the proportional band with respect to the desired value. It is a manual adjustment and not automatic, hence the term *manual re-set* is applied (based on the re-setting of the proportional band).

Proportional action is the basic type of continuous function control action, but a disadvantage results from the action being such as to produce a particular output for a particular deviation (since the two are related by a strict proportionality). This disadvantage is that proportional control can be subject to a steady-state error or *offset*. A mathematical explanation will be given later but a qualitative explanation can be given at this point.

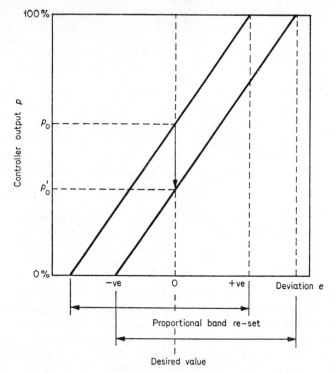

FIGURE 3–6. Manual re-set

Consider the liquid level control system of Figure 3–7 in which the level is followed by the float which is directly connected to the stem of the inflow valve by the lever linkage. There is thus a proportional relationship between the float movement and the change in valve position and it may be assumed further that flow through the valve will be linear with valve position over small ranges of movement. For a given outflow from the tank a condition of equilibrium will be established, i.e. the inflow will equal the outflow and the level will become stationary. Suppose now the load on the system is increased by opening the outflow valve to give an increased value of the outflow. To obtain a new steady-state condition after the inevitable disturbance, the inflow must be made equal to the new outflow but, owing to the fixed relationship between the float and the inflow valve position, this can only be obtained by the level falling to a new value to allow the valve to reach the new position. The system thus stabilizes at a different (lower) value of the level. If the level under the first condition is regarded as the desired value, that in the second condition is offset from the desired value. The situation can be corrected by adjusting the datum relationship between the float and the valve position in order to maintain the new inflow at the original value of the level, but this is, of course, a manual adjustment—it is, in fact, the manual reset discussed above. As will be seen later, the offset will be small if the sensitivity is high, and *vice versa*.

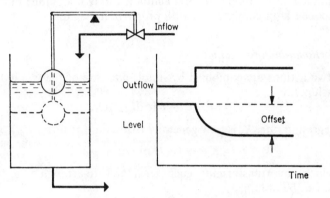

FIGURE 3–7. Offset in proportional control

Reverting to the proportional action equation, p_0 is constant whilst the control is in operation (changing the value of p_0 creates a new situation), and p may then be measured from the datum value of p_0. The action equation then reduces to

$$p = K_c e$$

where p is now the *change* in controller output from that at $e = 0$. Transforming this equation gives the transfer function for the proportional action:

$$G_c(s) = p/e = K_c$$

Integral Action

Integral action is defined by a change in the controller output proportional to the time integral of the deviation, i.e.

$$p = K \int e \, dt$$

Differentiating this relationship gives $dp/dt = Ke$, showing that the rate of change of the output is proportional to deviation; this is thus a continuous function floating action.

The particular feature of integral action is that the action recognizes not only the magnitude of the deviation but also the time during which the deviation exists. For a deviation which is constant in time the controller output will change at a constant rate, theoretically without limit. It will be seen that the dimensions of the constant K will contain reciprocal time (T^{-1}) so that K represents a rate of operation which may be termed the *floating rate* by analogy with the other discontinuous forms of floating action. The transfer function for integral action is obviously K/s.

Integral action is limited in application to processes of small capacitance and fast response which are subject to large changes in load, such as are found in some flow control systems. Consequently, apart from these rather limited applications, integral action is rarely used alone but is used more frequently in combination with proportional action.

Proportional-integral Action

The two actions are combined by simple addition of the two individual relationships:

$$p = K_c e + K \int e \, dt + p_0$$

The integrating rate, K, is now generally written in the form, K_c/T_i. Hence

$$p = K_c e + (K_c/T_i) \int e \, dt + p_0$$

from which, by transforming each term and re-arranging, a transfer function can be obtained:

$$G_c(s) = \frac{p}{e} = K_c \left(1 + \frac{1}{T_i s}\right)$$

The proportional-integral action thus has two adjustment parameters, the proportional sensitivity (K_c) as defined for the proportional action and the parameter, T_i, which is the integral action time, or more simply the *integral time*. The latter is defined by comparison with the proportional action, being the time required for the integral action to generate a change in output equal to that of the proportional action when the deviation is constant. Thus, in Figure 3–8, a constant deviation $(e = A)$ is applied at zero time; the proportional action produces an immediate step response of AK_c which is proportional to the applied deviation. The integral action begins to integrate the deviation with respect to time, so causing the output to increase at a constant rate whilst the deviation remains constant; the rate of change is $AK_c t/T_i$. It will be seen that after a certain time has elapsed the amount of the integral action response will be equal to that of the initial proportional response, i.e.

$$AK_c = AK_c t/T_i$$

at time t, from which the time required is the integral time, T_i.

The distinct advantage gained by the addition of integral action to proportional can be illustrated by assuming the constant deviation of Figure 3–8 to be removed so returning to the original condition of zero deviation. The initial step of the proportional action is removed and the integrating action ceases since the deviation is now zero. The integral action does not reverse its direction since to do so a deviation of opposite sign is required. Effectively, then, a new condition for zero deviation has been set up, the value of the controller output p_0 corresponding to zero deviation at the zero time has been changed to a new value, p_0', for the condition of zero deviation at the later time. As seen previously, with the proportional action alone, the alteration of p_0 is a manual adjustment (manual reset) which can be used to correct for the offset following a change in load on the process. With the combined actions, this adjustment is made by the controller whilst overcoming the deviation caused by the change in load; hence the term *automatic reset* which is applied to integral action when used in combination with proportional action. The integral action will thus eliminate the offset associated with proportional action when the latter is used alone; it can be demonstrated that the automatic reset of the integral action causes a shift of the proportional band with respect to the desired value, as does the manual reset.

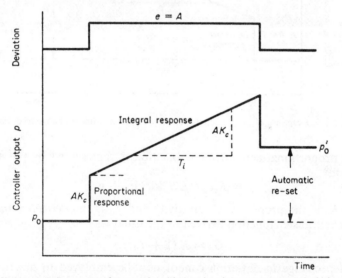

FIGURE 3–8. Response of proportional-integral action to a constant deviation

It will be noted that the reciprocal of the integral time is a rate of operation; this is the *reset rate* which is used as an alternative to integral time in calibration of the controller adjustment, particularly in instruments of American origin.

Proportional-derivative Action

The derivative control action is defined by a change in controller output proportional to the rate of change of deviation, i.e. $p = K(de/dt)$. A little consideration shows that this action alone is incapable of exercising control since it cannot recognize a constant deviation (since $de/dt = 0$). It must therefore be used in combination with proportional action or with the proportional-integral combination.

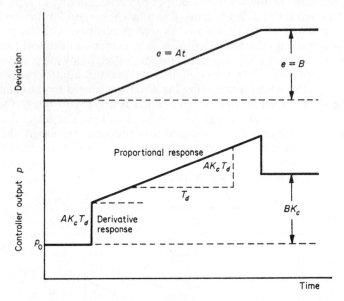

FIGURE 3–9. Response of proportional-derivative action to a linear change of deviation

The proportional-derivative combination is represented by the action equation

$$p = K_c e + K_c T_d (de/dt) + p_0$$

where K_c is the proportional sensitivity and T_d is the *derivative* (*action*) *time*. The transfer function derived in the usual way is

$$G_c = K_c (1 + T_d s)$$

A step change in deviation cannot now be employed to describe the combined response since the time derivative of a step is infinite at the time of the change. Instead a linear (ramp) change of deviation ($e = At$) must be employed as in Figure 3–9. The response is now a step of $AK_c T_d$ from the derivative action (since $de/dt = A$) followed by a linear rate of change of $AK_c t$ due to the proportional action. The effect of the derivative action is to anticipate the proportional action by an immediate step before the proportional response begins. It can readily be seen that the derivative time is the time taken by the proportional action to reproduce the initial step of the derivative action.

The effect on the constant p_0 (the output for the condition of zero deviation) can be deduced by considering that the deviation becomes constant for a short period and then returns to zero at a constant rate. It will be found that the output returns to its original value of p_0; there is thus no re-setting action with derivative and the offset of proportional control will not be eliminated. Nevertheless, the use of derivative action is generally beneficial in stabilizing a control system and allowing the proportional sensitivity to be increased. This will lead to a reduction in the offset following a load change and an increase in the speed of the system response, i.e. a shorter recovery time following a disturbance.

Three-term Control Action

For control systems in which offset must be eliminated with a short recovery time, it is often necessary to combine all three continuous control actions into a three-term combination, i.e. a proportional-integral-derivative (PID) action. The reasoning is that the integral action will provide the automatic reset to eliminate offset following a load change and the derivative action permits a reduction in proportional sensitivity, which will not now affect the offset, but will reduce the peak deviation as well as providing a faster recovery. In practice the combination achieves these objectives, but some compromise is necessary between the rate of recovery of offset and the overall recovery time.

The control action equation is now

$$p = K_c e + K_c T_d (de/dt) + (K_c/T_i)\int e\ dt + p_0$$

There are thus three adjustment parameters, K_c, T_d, and T_i, and the transfer function is

$$G_c = K_c(1 + T_d s + 1/T_i s)$$

Other Continuous Control Actions

It will be seen that the three-term combination produces a symmetrical action equation, particularly if written as follows:

$$p = K_c \left(T_d \frac{de}{dt} + e + \frac{1}{T_i} \int e\ dt \right) + p_0$$

The question arises as to the possible logical extension of the term in the brackets to include still higher successive integral and differential terms beyond those used. As a practical matter, there is little advantage at this stage of the process control art of doing so in present applications; there might even be some severe disadvantages, as the higher order differentiations are extremely sensitive to modest changes in the controlled variable and to signal 'noise'. In other fields of automatic control, the situation is rather different and it is not unusual to find position controls employed with velocity (dx/dt) and acceleration (d^2x/dt^2) errors.

Non-ideal Responses

The transfer functions for the continuous control actions derived in the preceding sections are based on the theoretical definitions of the control actions. As already mentioned, some practical instruments do not entirely conform to these ideal equations. This occurs particularly with pneumatic instruments, and is due to the finite gain of the pneumatic amplifier and to inter-action between the integral and derivative generators. In general the response departs from the ideal equations at very low and very high frequencies, causing a limitation of the gain and phase relationships. Fortunately the effect is not serious over quite wide operating ranges, apart from the integral-derivative interaction effect but this can also be allowed for. These departures from ideality are dependent on the mechanical design of the instrument, and further discussion is deferred to Chapter 8.

The Final Control Element

In process control the final control element is the mechanism by which one of the input process variables is adjusted in response to the controller output signal, and is usually physically separate from the controller mechanism. In most process applications this element is a valve regulating the flow of a fluid stream which conveys material or energy into the process. Many heating processes employ electrical energy, and apart from the straightforward on-off switching technique, modulating voltage regulators such as rheostats, variable transformers and saturable core reactors, are used. In the following discussions, however, attention is directed entirely to regulating valves.

Valve Actuators

Since the operation of a valve is mechanical in nature, apart from the relatively few examples of mechanical (self-actuating) controllers such as float-operated level control valves, etc., the first essential is to convert the pneumatic or electrical signal from the controller into a linear or rotary motion to operate the valve. The most commonly used type of valve actuator or 'motor' is the pneumatic diaphragm, illustrated in Figure 3–10. This is basically a chamber to which the pneumatic output pressure of a controller is applied, one wall of which is formed by a limp (i.e. non-elastic) diaphragm, of a material such as neoprene, supported by a pressure plate. The diaphragm is spring-loaded in opposition to the air pressure and the mechanical displacement of the diaphragm is directly proportional to the pressure applied, the range of movement being in the range of 1–10 cm ($\frac{1}{4}$–4 in). The standard operating pressure range for pneumatic instruments at present defined is from 3 to 15 lbf/in^2, a standard range in SI units has not yet been defined but will presumably be of corresponding order of magnitude, possibly 20 to 100 kN/m^2. As diaphragm sizes vary from some 15–60 cm (6–24 in) in diameter, considerable forces can be developed at the maximum air pressure, but it must be remembered that the actuator has to overcome static friction and any unbalance forces in the valve. These latter cause a hysteresis effect in the pressure-valve stroke relationship which must generally be less than 1 per cent of the full stroke. For this reason, and particularly when wide proportional bands are used and very small valve movements are required, the actuator may be provided with a supplementary servo-mechanism known as a *valve positioner*.

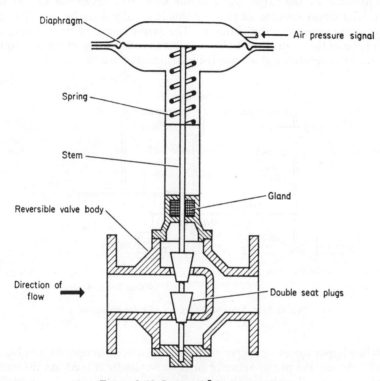

Diaphragm

Air pressure signal

Spring

Stem

Gland

Reversible valve body

Direction of flow

Double seat plugs

FIGURE 3–10. Pneumatic diaphragm valve

This provides, through a relay amplifier, sufficient pressure to ensure that the valve takes up the correct position as required by the controller output pressure. As seen in the schematic diagram of Figure 3–11, the valve position is compared directly with a measurement of the output pressure and any deviation operates the relay accordingly. A further advantage of the use of the positioner is that much higher operating pressures than the usual instrument supply pressure (about 20 lbf/in² or 150 kN/m²) can be used to operate the valve.

It should be noted that in the event of an air supply failure or leakage, the spring-loaded diaphragm must return to its zero position. This means that the attached valve will, under such circumstances, take up one or other of its extreme positions depending upon the direction of operation of the actuator, i.e. whether the arrangement is 'air to open' or 'air to close'. It is conventional industrial practice to arrange for installations to 'fail safe', i.e. in the event of any such failure the valve must take up the safer of the two alternative positions. In general this will require raw-material feed and heating supply valves to close on failure and product take-off, and cooling supply valves to open. This requires that the direction of operation of the valve and actuator should be reversible to suit the particular circumstances. In practice this is usually done by inverting the valve with respect to the actuator. The requirement of reversibility of operation of the control valve also requires that the action of the controller should be reversible and all control instruments have this facility.

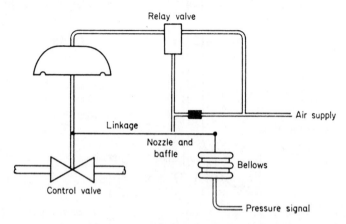

FIGURE 3–11. Schematic diagram of valve positioner

When larger forces or longer strokes are required to operate a valve or similar device, the piston-actuator or power-cylinder is used. As the name implies this is a double-acting hydraulic or pneumatic ram, the operating fluid being supplied to the appropriate side of the piston by relay valves operated by an integral servo-positioner.

Electrical actuation of control valves is somewhat limited in application. Reversing motors and solenoids are often used for two-position controls, but for the more refined control actions where precise valve positioning is required, an electrical actuator is usually too costly and often relatively unsatisfactory in performance. It is, in fact, the common practice in systems using electrical or electronic controllers to convert the electrical output signal of the controller into an air pressure to operate a conventional diaphragm actuator. This electro-pneumatic transducer is virtually a valve positioner with an electrical input.

Control Valves

Valves for use as regulating units fall into two main categories, sliding-stem and rotary-shaft valves. The former requires a linear displacement, and the stem carrying the valve plug can be positioned directly by the diaphragm actuator (Figure 3–10). The stem must pass through some type of packing gland, and may carry a single- or double-seated plug.

The single-seat plug is less costly and has the advantage of a tight shut-off, but is subject to large unbalance forces due to the pressure drop across the opening between the plug and the seat. The double-seated plug avoids this by balancing the two pressure drops against each other so that a minimum force is required to move the stem. Double-seated valves are somewhat limited for small rates of flow and cannot give a completely tight shut-off. A large variety of valve plug shapes is available (Figure 3–12) to give almost any lift-flow characteristic over a wide range of flow rates.

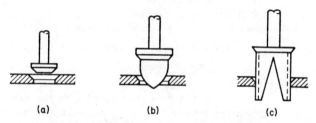

FIGURE 3–12. Valve plugs: (a) bevel plug, (b) parabolic plug, (c) V-port

Also included in the sliding-stem category are pinch valves in which the linear movement compresses a flexible section of pipe or forces a flexible diaphragm on to a weir, as in the well-known Saunders valve.

Rotary valves require a rotational movement of the shaft of at least 90° and include rotary plug valves (cocks), and butterfly and louvre valves. The linear motion of the diaphragm actuator can be converted to a rotary motion by means of lever linkages. The rotary valves are much less flexible than the sliding-stem pattern as far as the flow-lift relationship is concerned.

Valve Characteristics

Control valves are characterized by *rangeability, turndown ratio,* and the *flow-lift relationship.*

Rangeability is the ratio of the maximum to minimum controllable rates of flow. The later is not normally zero but may be 2 to 5 per cent of the maximum depending upon the type and size of valve. The rangeability of a typical sliding-stem valve will be between 20 and 70.

Turndown ratio is really a property of the process which affects the selection of the valve, being the ratio of the normal maximum operating flow to the minimum controllable flow. A practical rule of thumb is that the valve should be of sufficient size for the normal maximum flow required by the process to be about 70 per cent of the maximum possible flow. The turndown ratio should then be about 70 per cent of the rangeability.

The flow-lift characteristic is the relationship between the rate of flow through the valve and the amount of opening measured by the lift, i.e. the vertical displacement of the valve plug. The movement of the valve plug changes the area of opening of the port, or the area through which the fluid flowing is constrained to pass. The valve may be regarded in principle as a variable orifice and, to a sufficiently close approximation, the orifice flow equation can be written

$$q = KA(\Delta p/\rho)^{\frac{1}{2}}$$

where q is the rate of flow,
$\quad K$ is a combined flow coefficient,
$\quad A$ is the area of opening through which the flow passes,
$\quad \Delta p$ is the pressure difference across the valve,
and $\quad \rho$ is the density of the fluid.

The equation shows that the rate of flow is proportional to the area of the port opening if the pressure difference and flow coefficient are constant. Under these ideal conditions the flow characteristic of the valve is determined by the relationship between the area exposed and the lift of the stem, and by suitable design of the shape of the plug this area characteristic may be made to any desired relationship. Apart from the quick-opening valves used for two-position control, the two commonly-used area characteristics are the *linear*, which requires no explanation, and the *equal percentage*. The latter name arises from the fact that the flow increments resulting from a given change in lift are a constant proportion of the actual flow regardless of the valve position; the area thus increases more rapidly with the lift as the valve opens. For an ideal equal percentage valve the sensitivity or gain would be directly proportional to the flow ($dq/dx = kq$) and the characteristic would be a straight line on semi-logarithmic coordinates.

Unfortunately the condition of a constant pressure difference and constant flow coefficient rarely obtains in practice, and the actual relationship between flow and lift may change radically from that predicted by the area characteristic.

A control valve is usually installed in a pipeline between two points of more or less constant pressure so that a constant pressure difference exists along the length of the pipe. The latter, with fittings such as orifice plates for flow metering, shut-off valves, bends, tees, etc. and process plant which may be included, forms a resistance to flow in series with that of the control valve. When the fluid flows through the pipe and fittings, pressure is consumed by frictional effects, the loss in pressure thereby being deter-mined by the square of the fluid velocity. Thus as the valve is opened and the flow rate increased, the loss in pressure due to frictional effects in-creases rapidly at the expense of the pressure difference available across the valve, this latter decreasing progressively as the flow increases. A valve with a linear area characteristic then shows a markedly non-linear flow characteristic if there is appreciable pressure loss in the lines. The maximum rate of flow is also reduced below that obtaining at the maximum pressure drop at the valve, as shown in Figure 3–13. With a linear valve, this effect

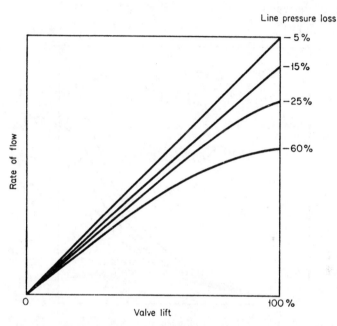

FIGURE 3–13. Effect of line pressure drop at constant pressure difference on flow characteristics of a linear characterized valve

is basically a decrease in the valve gain as the flow increases, and this may make the process difficult to control if there are large changes in load requiring large changes in the rate of flow through the valve. In such cases the equal percentage valve is more frequently used. The effect of line pressure loss on such a valve is shown in Figure 3–14, where, although the valve gain is unstable at low and high flows, it is almost constant over the intermediate range. The changes in the valve gain are undesirable when the process gain is constant, but for many systems the process gain tends to decrease with increasing load, and an equal percentage valve can make the overall gain almost constant over wide ranges of flow.

Since both the valve position and the pressure difference across the valve may both change in normal conditions of operation, the actual flow-lift characteristics are rather more complicated than that given by either the flow equation or plots of the flow-lift relationship. At a particular load condition, the behaviour of the valve may be approximated linearly by using the slope of the flow-lift curve for the actual pressure difference at the mean value of the lift. This slope is effectively the value of the steady-state gain K_v, and is the effective transfer function for the valve since the dynamic response is generally too rapid to require any further elaboration of the transfer function. Any lag which is involved is usually due to transmission of the signal from the controller.

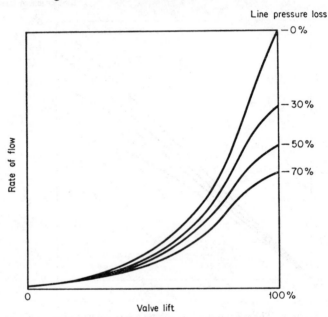

FIGURE 3–14. Effect of line pressure drop at constant pressure difference on flow characteristics of an equal percentage valve

Selection and Sizing of Control Valves

The factors to be considered in selection of a control valve are:

(1) the normal operating loads of the process,
(2) rangeability,
(3) the pressure drop available at the valve at the maximum and minimum operating flows,
(4) the nature and properties of the fluid handled.

The control valve must be suitably sized to dominate the process, the rangeability of the valve must exceed that of the process by at least 20 per

cent. For a turndown ratio of 70 per cent of the valve rangeability suggested above, a safety margin of some 40 per cent in rangeability is provided.

The pressure drop may be a design variable or it may be fixed by other
process requirements. A high pressure drop increases the range of controllable flow and makes it easier to obtain a desired characteristic, but it
also represents an excess use of power.

In many process plants the range of operating loads is deliberately
narrow, the plant being operated for long periods at an optimum throughput. In such cases the control valves maintain almost constant rates of
flow, and linear or even uncharacterized valves operating at 60–70 per cent
of the maximum flow would be adequate. In cases where plants have to
operate on widely fluctuating loads, equal percentage valves are almost
essential to provide a reasonably constant sensitivity at the different
levels.

The nature of the process fluids obviously determine the materials of
construction of the valve, and may also to some extent determine the type
of valve required.

The size of a control valve is determined by consideration of the required
rate of flow and the variable pressure drop. Most manufacturers have now
standardized on this, and quote in their literature a valve flow coefficient
C_v, which is defined as the flow rate of water in gallons per minute passed
by the fully open valve at a pressure difference of 1 lbf/in^2. The value of
C_v for a particular installation is calculated by the use of flow equations
such as, for liquids,

$$q = C_v(\Delta p/G)^{\frac{1}{2}}$$

where q is the liquid flow rate (gal/min),

Δp is the pressure drop (lbf/in^2),

and G is the specific gravity of the liquid (water $= 1$).

Similar equations are available for gas and steam flows and are often
embodied in nomograms and slide rules for ease of calculation. The flow
coefficient of a given design and size of valve is determined by actual
testing by the manufacturer, values range from 4–60 for a 2 in (nominal
pipe size) valve and vary roughly with the square of the nominal pipe
size.

Transmission Lines

The control valve is often situated some distance from the controller
and the pressure signal is conveyed by transmission lines, usually 5–10 mm
($\frac{3}{16}-\frac{3}{8}$ in) bore. Reference has already been made to the problems of
signal transmission with regard to measurement, and the same remarks
apply here. The exact response is often difficult to obtain but the lag is
often small compared to that of the process, and approximate methods of
obtaining the lag, such as that of using a lumped parameter model, are

usually satisfactory. If such rough calculations show the lag to be critical, a more rigorous treatment can be used, or, preferably, experimental tests can be made on the actual installation.

PROBLEMS

3–1 A thermometer with a time constant of 5 s and a static error of 1°C is placed in a bath of liquid whose temperature is increasing at a constant rate of 1°C/s. Determine the error in the measured value.

3–2 Determine the transfer function of a thermocouple brazed to the inside tip of a metal pocket, 3 mm (0.125 in) outer diameter and 2 mm (0.075 in) inner diameter, which is inserted to a depth of 50 mm (2 in) into a gas line. The heat transfer coefficient to the pocket is 25 W/m² K (5 Btu/h ft² °F) and the properties of the metal are: density, 7.9×10^3 kg/m³ (specific gravity, 7.9); thermal conductivity, 27.7 W/m² K (16 Btu/h ft °F); specific heat capacity, 0.35 kJ/kg K (0.15 Btu/lb °F).

3–3 A thermocouple and pocket each have a simple time constant. Show that there is little difference in the response to a linearly changing input by neglecting the smaller time constant when the ratio of the time constants is greater than 10 to 1.

3–4 Compare the response of proportional, integral, and derivative control actions to a sinusoidal input, $e = A \sin \omega t$.

3–5 A unit step change in the error signal is introduced into a proportional-integral-derivative controller for which $K_c = 10$, $T_i = 1$, and $T_d = 0.5$. Plot the response of the output p.

3–6 Show that the response of a proportional-derivative controller does not reset the proportional band when the error signal rises to a maximum value and returns to zero.

3–7 A pneumatic temperature controller has an output range of 3 to 15 lbf/in² g and a chart calibration of 0–200°C. If the output is 6 lbf/in² at 70°C, determine the proportional sensitivity when the proportional band is 25 per cent and the output at 100° and 125°C.

3–8 An electronic temperature controller has an output range of 0 to 15 mA and a chart calibration of 50–200°C. If the output is 5 mA at 80°C when the sensitivity is 0.5 mA/°C, determine the proportional band and the output at 90° and 110°C.

3–9 Determine the time constant of a spring and diaphragm actuator of a control valve connected through a transmission line of resistance R and negligible capacitance, when the capacitance of the diaphram chamber is C, the area of the diaphragm is A and the spring constant is K.

3–10 The area of opening of a valve plug is given by $A = a + bx^2$, where x is the valve lift. Derive the flow-lift characteristic for two different pressure drops of Δp and $0.75\Delta p$, assuming a constant flow coefficient.

Chapter 4: Two-position Control

Although simple in operation and widely used in practice, two position control is difficult to analyse mathematically owing to the discontinuous nature of the changes applied by the controller to the manipulated variable. Nevertheless, a qualitative consideration can lead to a better understanding of the principles of application. It is possible to consider a generalized case without specifying a particular controlled variable, but this action is limited in practical application to relatively few of the usual process variables. For example, fluid flow control does not lend itself to two-position action owing to the lack of capacitance, and neither does pressure when controlled on a flow basis. Liquid level is infrequently controlled by two-position action when the requirement is for a constant pressure head, since the self-actuated float-operated inlet valves are essentially simple devices giving effectively proportional action. Two-position control of level is widely employed when used with a wide differential zone for the emptying of sump tanks and the like, this being a rather similar application in principle to the on-off control of compressors where a large storage capacitance is available. The widest application of two-position action is undoubtedly in control of temperature where a considerable range of thermostatic devices is available. Consideration, however, will be given first to a liquid-level control system since this can represent the most elementary case and the reactions can be most easily visualized.

Two-position Control of a Liquid Level System

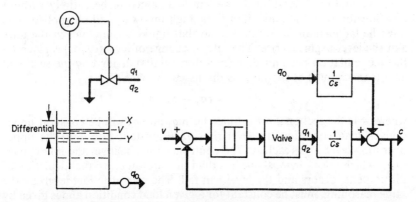

FIGURE 4–1. Two-position control of single capacitance level control system without self-regulation

Consider the system illustrated in Figure 4–1 where the tank provides a single capacitance and the outflow q_o is by means of a constant-displacement pump. The system is not therefore self-regulating and the outflow is not affected by any changes in the level providing a pressure head. The level controller LC measures the level by any suitable means and, via the control valve, selects the inflow. Since the control is to be two-position only, two values of inflow will be made available; these will be a 'high' value q_1 and a 'low' value q_2, the selection (by the controller) depending upon whether the level is below or above the desired value. In practice all two-position controllers have a differential zone situated about the desired value; thus if the latter is represented by the level V in the tank, the operating levels at which the inflows are changed over will be X for a rising level and Y for a falling level.

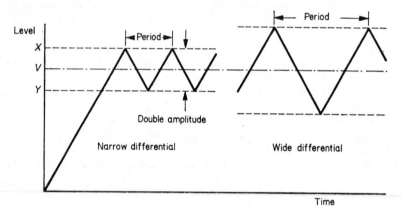

FIGURE 4–2. Response of level in ideal case of Figure 4–1

In the first consideration the system is assumed to be entirely without lag. Starting with the tank empty, the high inflow q_1 will be applied; this must be larger than the outflow q_o so that liquid accumulates in the tank and the level begins to rise. The rate of change of level (dh/dt) is given by the accumulation per unit time (in volume units) divided by the sectional area (capacitance) of the tank as discussed on page 67, i.e.

$$dh/dt = (q_1 - q_o)/A$$

where A is the sectional area. Since both q_1 and q_o will be constant under a given load condition the level rises at a constant rate until the upper operating level X is reached. The controller then changes the inflow over to q_2, which must obviously be less than the outflow q_o, so that depletion of the contents occurs and the level will fall. The rate of falling level, by the same reasoning, must be constant for a given load condition and is given by

$$dh/dt = (q_2 - q_o)/A$$

When the falling level reaches the lower operating level Y, the high inflow q_1 is again applied to make the level rise again. Thus a continuous 'saw-tooth' oscillation of the level, Figure 4–2, is set up about the desired value, and in this ideal case without lags the double amplitude of the oscillation is the width of the differential zone, which then represents the 'tightness' of the control, i.e. the range of variation of the level in this particular application. The period of the oscillation is determined by the rates of change of the level and the width of the differential zone; if the latter is increased, so also is the period. With two-position action a large but not excessive period is desirable since this reduces the frequency of operation of the moving parts, arcing at contacts, etc. and so prolongs the life of the apparatus. There is, in fact, an optimum frequency of operation commensurate with a reasonable working life of the mechanism; this optimum period effectively determines the width of the differential and hence, in this idealized case, the closeness of control to the desired value.

Effect of Magnitude of Correction

The rates at which the level rises and falls are determined by the magnitudes of the flow differences, i.e. $(q_1 - q_o)$ and $(q_2 - q_o)$; these may be termed the corrections applied by the controller to cause the level to change in one direction or the other. Initially these corrections must be opposite in sign if control is to be exercised at all, i.e. $q_1 > q_o > q_2$. If the differences are equal, then the rates of rise and fall of the level will be equal, and a symmetrical saw-tooth pattern will be set up. If the values are large then the rates of change will be fast and the period of oscillation short; if

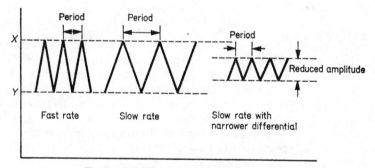

FIGURE 4–3. Effect of magnitude of correction

the values are small the rates will be slow and the period longer (Figure 4–3). With slower rates of change the differential zone can be made narrower so that by reducing the magnitudes of the corrections the control may be made tighter, i.e. a lower amplitude of cycling about the desired value without unduly shortening the period and thus the life of the mechanism.

To obtain a slow cycle of small amplitude, the two values of the inflow should then be only slightly larger and smaller than the outflow. This would be permissible in the present case if there were no load changes in the system, but the effect of such changes is to alter the absolute values of the flows. There is thus a danger that if the corrections are made too small, the order of inequality of the flows may be changed and the tank will then either fill (if $q_2 > q_o$) or empty (if $q_o > q_1$).

Effect of Load Changes

Two types of load changes in the system may be distinguished; a supply change, such as a change in pressure upstream of the control valve, which will change the values of the inflows q_1 and q_2, not necessarily by an equal amount; and a demand change, which will change the value of the outflow q_o. Since the outflow is by means of a constant displacement pump in this instance, such a change could be accomplished by a change in the pump speed to satisfy the requirements of a succeeding process.

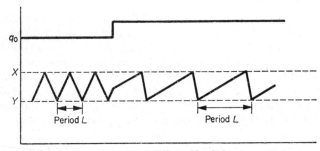

FIGURE 4–4. Effect of load change

It can readily be seen that in spite of these possible changes in the flow rates, the level will continue to oscillate between the limits of the differential zone so long as the inequality, $q_1 > q_o > q_2$, is observed, the main effect being to alter the rates of change of the rising and falling level. This is illustrated in Figure 4–4 for a demand load change, an increase in the outflow, which produces a slower rise and a faster fall in level, and also some change in the frequency of the operation. Control is, however, still retained and the effects of the load change are negligible owing to the absence of lags.

Effect of System Capacitance

The effect of changing the capacitance of the system (the sectional area A) is also readily seen. Increasing the value of A will slow down the rates of both rise and fall of the level with results exactly the same as considered in Figure 4–3. A large capacitance is thus beneficial in obtaining a smaller amplitude of cycling and tighter control.

Effect of Lags

The simplest type of lag to introduce into the idealized system is a time delay in the operation of the controller. This often occurs in practice due to static friction and lost motion in moving parts producing a hysteresis effect in the operation of the mechanism. The effect is simply to delay the operation of the controller in reversing the inflows by a short period. Thus if a delay time of L is introduced, the level will rise to the upper operating level and will continue to rise for the period L before the controller operates to change the inflow. The level then falls and similarly continues to fall below the lower operating level for the period L before the controller again operates. As a result (Figure 4–5) the level overshoots the differential zone on either side, and the amplitude and period of the cycling are increased. If the rates of rise and fall are equal, the amount of overshoot is the same on either side and the cycling is symmetrical about the desired value. If the rates are not equal, the cyling is unsymmetrical and the mid-point of the oscillation is 'offset' above or below the desired value. As seen above, these represent different load conditions and the effect of load changes when a delay is present is thus to cause a shift in the mean value of the cycling; there will also be changes in the value of the amplitude and period but these are not likely to be significant.

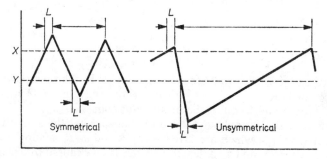

FIGURE 4–5. Effect of dead time

Effect of Self-regulation

If the system is made self-regulating by passing the outflow through the resistance provided by a valve instead of the constant displacement pump, the system becomes a single time-constant transfer stage and the outflow now is dependent on the pressure head provided by the level in the tank (see page 26). The inflows q_1 and q_2 will now correspond to two potential values of the level h_1 and h_2, and without the action of the controller the level would theoretically stabilize at one or other of these values as the outflow changes with the pressure head to become equal with the inflow. These potential values of the level may not be physically attainable as one or both may lie outside the confines of the tank, but the concept is still valid.

The response of the level in each direction is now that of a single time-constant stage defined by the exponential term, $[1 - \exp(-t/T)]$, where T is the time constant RC. The change of level is not now linear but the rates of change full off as the potential value is approached and the oscillation is now formed of segments of the exponential curves (Figure 4–6). In general the response of the level is now slower than the case without self-regulation, the period of cycling will be extended and the differential zone can be narrowed. The effect of self-regulation is thus similar to that of increased capacitance, and is generally beneficial in all applications. The period of oscillation now depends on the time constant as well as on the width of the differential zone; increasing the time constant slows down the response and further increases the period.

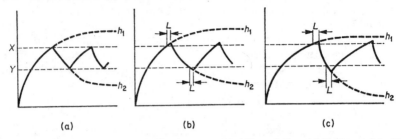

FIGURE 4–6. Two-position control of level system with self-regulation: (a) ideal case, (b) with dead time, (c) after load change

Adding a time delay produces overshooting as in the previous system. The effect of load changes is again to distort the cycling from a symmetrical pattern and to offset the mean value, but the reasoning is rather more complex. Both supply and demand changes alter the magnitudes of the potential values, which are given by q_1/R and q_2/R; a supply change causes a change in q_1 and q_2, a demand change is now caused by a change in the outflow resistance R, and this changes the potential values but also changes the time constant.

With self-regulation present, the effect of the magnitude of the corrections is much easier to visualize since the two potential values can be assessed directly against the desired value.

Additional Capacitance

Two-position control of systems with more than one storage capacitance does not differ essentially in general respects from the single capacitance case considered. The effect of additional storage capacitances in the system is to 'round off' the peaks of the oscillation and to produce a cycling which is more or less sinusoidal in general character.

Two-position Control of Temperature

A thermal system will contain at least two energy-storage capacitances, at least one each in the process and the measuring element, and very often more. The response of the temperature as measured in such a system to a step change in the heat input is the typical response of a multi-capacitance system (see Figure 2–18), i.e. there is initially a very slow response, building up to a maximum rate of change and then decreasing to zero as the potential value is approached. Thermal systems are always self-regulating, and the two heat inputs of two-position control will then always correspond to two potential values.

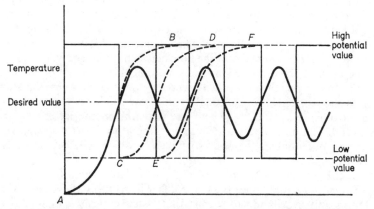

FIGURE 4–7. Temperature response of multi-capacitance system with two-position control

The response of the temperature in a multiple-capacitance system under two-position control is developed in Figure 4–7. The temperature rises from its zero time value according to the exponential curve AB, under the driving force of the higher potential value, until the desired value (or the upper edge of a differential zone) is reached. The controller will then apply the alternative input corresponding to the lower potential value, and the temperature must ultimately fall since the heat input is not now sufficient to maintain the temperature at the desired value.

The further development of the response now employs the principle of *super-position*, i.e. that when a number of forces or potentials act simultaneously on a system, each can be considered as producing its own effect independently of the others, and the total effect is equivalent to the algebraic sum of the several independent effects. This concept applies strictly to linear systems with constant coefficients (which is the case here) but the principle can be applied to most control problems as a reasonable approximation. In the present example it is assumed that the original potential value is maintained, causing the temperature to continue to rise along the curve AB, and that the new potential is presented by a change of opposite sign. The change in potential which is applied is thus the difference between the two potential values, which when applied in the positive direction would produce a response from the lower potential value given by the curve CD. This potential is, however, applied negatively to reduce the potential value from the high level to the low, and the total response is then given by subtracting curve CD from curve AB. It will be seen that during the initial period after the change of potential values, the original curve AB is rising faster than the additional curve CD; hence the overall response will continue to rise even though the high potential value has been removed. Ultimately the curve CD will begin to rise faster than curve AB; the overall response will then be reversed and the temperature will begin to fall. The physical explanation of this apparent anomaly of the temperature continuing to rise for a period after the heat input has been reduced is the 'transfer lag' between the capacitances. Considering only a two-stage system, the potential in the first stage may fall due to a reduction in the input, but for some period the flow between the stages will be greater than the outflow from the second stage so that the potential of the latter will continue to increase. This will decrease the flow between the stages owing to the self-regulation; ultimately this flow will become less than the outflow from the second stage and the potential in the latter will then start to fall. The measured variable in a temperature control system is, of course, the potential in the last capacitance, which will be the sensing element such as the thermometer bulb.

When the falling temperature reaches the desired value (or the lower edge of the differential zone) the controller will re-apply the higher of the two inputs and so will add to the existing system a positive change equal to the difference between the two potential values. The response is now the sum of the three individual responses, i.e. curves AB, CD (negative), and EF. The temperature will continue to fall below the desired value but will ultimately turn back as the rate of change of the new input (EF) exceeds that of the previous one ($AB - CD$), and will rise again to the desired value. A continuous cycle is thus set-up; the response of the temperature is more or less a sinusoidal curve which is exactly one half-cycle out of phase with the control action, i.e. when the temperature wave shows a positive half-cycle above the desired value, the correction wave is negative at the lower potential value, and *vice versa*.

In this argument and in Figure 4–7 the differential zone has been ignored. For temperature control, owing to the 'overshooting' effect of the transfer lag between capacitances, the effect of a differential zone is relatively small. The discussion can, however, be applied with operating points above and below the desired value as indicated in the text, and the same result will be obtained but with an additional spread of the amplitude of cycling due to the differential zone.

From the previous discussion on page 125, if the positive and negative corrections (i.e. the differences between the desired value and each of the potential values) are equal, then the wave will be symmetrical about the desired value and the times of application of the two potential values will also be equal. If the corrections are not equal, then there is a greater energy potiental available during one half-cycle and there will be a quicker return to the desired value than during the other half-cycle. The times of application of the two potential values will not then be equal and the temperature wave becomes unsymmetrical with respect to the desired value, i.e. the mean value of the cycling is 'offset' from the desired value. This is the effect observed after a load change which effectively changes the magnitudes of the two potential values and so also the positive and negative corrections, as shown for an increase in load in Figure 4–8. The overshoot of the temperature beyond the normal amplitude of the cycling immediately after the load change is because the average heat input is now inadequate to meet the new demand.

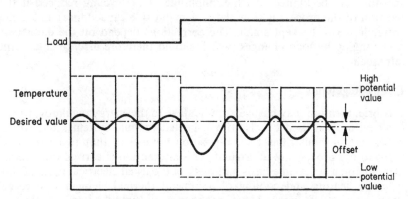

FIGURE 4–8. Effect of load change on two-position control of temperature

On-off Control

On-off control is the special case of two-position control in which the lower of the two positions represents an 'off' condition, i.e. a zero energy input into the system. In a self-regulating system the lower potential value will thus be a 'zero' value of the controlled variable; in a thermal system this will be the prevailing ambient temperature to which the system would cool if the 'off' condition were maintained for a long enough period.

From the preceding discussion it is apparent that on-off control will be most suitable and effective in systems of relatively low potential level. If this is the case the positive and negative corrections can be reasonably small and approximately equal, so giving a temperature cycle of low amplitude and low frequency. If, on the other hand, the potential level of the system is high, the negative correction (desired value to zero) is bound to be relatively large, and the cycling will thus be of greater amplitude and higher frequency. It is also likely in such circumstances that the negative correction will be larger than the positive correction, so giving a cycling which will be offset below the desired value.

Summary

It will be seen that there are some limitations in the applications of two-position control. Initially the response is inherently cyclical, and this will immediately rule against two-position control in some applications. In many other applications the cycling will be immaterial so long as the amplitude is reasonably small. Assuming that measuring lag is not negligible, as is often the case in temperature control, the process must have suitable characteristics if this type of control is to be satisfactory.

The most important requirement for obtaining a small amplitude of cycling is that transfer lag between capacitances, and any time delay, should be as small as possible, since these are the primary causes of over-shooting of the cycle outside the controller differential zone. The over-shooting will be limited and the amplitude of the cycling reduced if the reaction of the process is slow (large demand side capacitance) and if the corrections can be kept small. The corrections depend on the danger of over-ranging by load changes, which should therefore also be small and infrequent.

Energy Regulators

A brief mention may be given at this stage to the energy regulator, which is a device often used on electrically-heated thermal systems, and which bears certain similarities to two-position control in that two values of input to the process are alternated. This alternation is carried out fundamentally on a strict time basis, although in the usual electrical units of this type the heating and cooling of a bimetal thermal element is used to operate an on-off switch. There is, however, in these devices no reference to the measured value of the controlled variable, i.e. there is no feedback, and the energy regulator is therefore an open-loop type of control.

The behaviour may be deduced by considering the response of a thermal system with temperature as the variable so that the two energy inputs correspond to two potential values. By the same discussion used in the section on Two-position Control of Temperature (page 129), a continued sequence of alternations of the potential values will produce a uniform and out-of-phase oscillation of the temperature essentially similar to that of two-position control. The difference is the important one that the changes of potential value are not determined by the temperature in relation to the desired value but the alternation is purely arbitrary. The alternation is usually rather slow and the amplitude of the cycling is consequently somewhat larger than with two-position control, but the effect of load changes is more serious. Electrical units of the type mentioned above are stabilized to some extent against supply voltage changes by the bimetal element which is heated by the supply voltage, but demand changes have the same result as with two-position control in changing the magnitudes of the potential values. With two-position control in this circumstance, the average energy input is adjusted by the automatic change in the ratio of the times of application of the two values which arises naturally from the control action (Figure 4–8). This is not possible with the energy regulator which must continue to alternate the two values on the same time cycle. The average energy input thus remains the same even though the demand load has changed. The temperature cycling must then be displaced considerably more than the offset which occurs with two-position control (Figure 4–9).

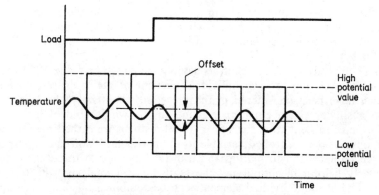

FIGURE 4–9. Energy regulator response to load change

It must be appreciated that the energy regulator is simply an input regulating device and not a controller in the usual sense. So long as this limitation is recognized, the device is capable of wide application in situations where close limits of control are not required.

PROBLEMS

4–1 Discuss the application of on-off control to (a) a laboratory thermostat bath operating at 25°C, (b) a drying oven at 100°C, and (c) an electric muffle furnace at 1000°C.

4–2 The boiler of a domestic central-heating system is subject to a constant and continuous demand for long periods and is operated by an on-off controller with a differential zone which supplies 30 kW (100 000 Btu/h) when 'on'. At a demand rate of 10 kW (30 000 Btu/h) the heating supply is on for 5 m in each cycle. Determine the period of the complete cycle for demand rates of 5 and 25 kW (10 000 and 75 000 Btu/h) and find the demand rate giving the minimum period of oscillation. Sketch the type of temperature response for each demand load.

4–3 The level in a tank of 0.5 m² (4 ft²) sectional area is controlled by a two-position controller with a differential zone of 100 mm (3 in) at a desired value of 1.5 m (5 ft). The tank has a constant outflow of 1.5×10^{-3} m³/s (3 ft³/min) and the inflows are 2.5×10^{-3} and 1×10^{-3} m³/s (4.5 and 2 ft³/min). Plot the response of the level against time and determine the effect of (a) a delay time of 3 s and (b) a 20 per cent decrease in the two inflows.

Chapter 5: Transient Response of Linear Control Systems

In the present chapter consideration will be given to the mathematical analysis of the control loop as a whole, and to the determination of the dynamic response of the controlled variable following an input disturbance to the system.

The analysis of the system behaviour requires first the differential equations relating to each individual element of the loop, relationships which have been considered in the preceding chapters, to be combined into a single system equation relating to the whole control loop. The solution of this equation for a given input disturbance is the time function response of the controlled variable to the particular disturbance. The use of the Laplace transformation and transfer functions at this stage is particularly helpful in simplifying the mathematical analysis. The transfer functions of the individual elements can be combined by simple multiplication to give an overall transfer function of the open loop and, by simple algebraic manipulation, to give the transfer function for the closed loop. Substitution into the latter of an input variable transform in the usual way provides the output variable transform, which need only be inversely transformed to yield the required temporal response.

First, consideration will be given to the combination of transfer functions of the individual elements of the control loop into the open and closed-loop functions.

Block Diagram Algebra

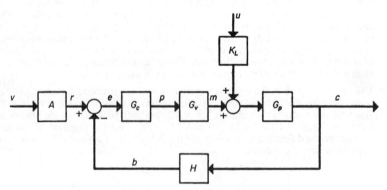

FIGURE 5–1. Process control loop block diagram

Consider the block diagram of the control loop shown in Figure 5-1, which shows the standard symbols for the variables and the transfer functions relating the output and input variable of each element of the loop. The variables are: a desired value v; reference input r; controlled variable c; feedback variable b; one load variable u; controller output p; manipulated variable m; and an error or deviation signal e. The transfer functions shown in the appropriate blocks for the separate elements are: input element A; control function generator G_c; regulating unit (valve) G_v; process G_p; and measurement feedback H. The series of blocks between the error signal e and the controlled variable c is the *forward path*, and that between the controlled variable c and the feedback variable b is the *feedback path*. The use of G for transfer functions in the forward path and H for those in the feedback path is a commonly used convention. The load variable transfer function N is also used in the forward path since the gain (K_L) of the load function is often different from that of the process transfer functions (such as G_p) through which the load variable passes.

In more complex systems the block diagram may contain additional feedback paths and internal loops, as will be seen in later chapters, but many systems may be represented by the simple block diagram of Figure 5-1, although there will often be more than one load variable entering at different points in the loop.

The next step is to determine the overall transfer functions relating the controlled variable to the reference input or desired value and to the load variables. These provide considerable information about the control system and are used to determine the response of the controlled variable to changes in the other (input) variables. Analysis of the system must always be partial in the sense that changes in only one variable at a time can be considered and all other input variables must be held constant. Since all variables shown on the block diagram are transformed, i.e. are functions of s and not of t, any changes of these variables are perturbations from the normal values which are conventionally zero at zero time. Thus to obtain the overall transfer function for a change in desired value, the load variable u will be zero, and the block diagram can be simplified by the omission of the load variable, as shown in Figure 5-2(a). The diagram can then be further simplified, or *reduced*, by use of the simple rule that over a series of transfer stages the overall function is the product of the individual functions. This follows from the definition of the transfer functions as the ratio of the output and input transforms over the particular element. The following equations can then be written by inspection:

$$c = G_p m \qquad m = G_v p \qquad p = G_c e$$

and by combining:

$$c = G_c G_v G_p e = Ge \tag{5-1}$$

The three blocks of Figure 5-2(a), with functions G_c, G_v, and G_p, may then be replaced by a single block with a function $G = G_c G_v G_p$, as in Figure 5-2(b).

The closed-loop transfer function can then be obtained by writing the further relationships

$$b = Hc \qquad e = r - b \qquad r = Av$$

which, by eliminating e from Equation (5–1), gives the overall function

$$\frac{c}{v} = \frac{AG}{1 + GH} \tag{5–2}$$

The product GH is the *open-loop* transfer function relating the feedback variable b to the reference input r when the feedback link is disconnected from the error discriminator, i.e. when the loop is opened. $G/(1 + GH)$ is the *closed-loop* transfer function relating the controlled variable c, to the reference input when the loop is closed; since $b = Hc$, multiplication by H gives $GH/(1 + GH)$ as the closed-loop function relating the feedback variable b to the reference input. Figure 5–2(b) may be reduced still further to Figure 5–2(c) by replacing the forward and feedback paths by the single forward element with the closed loop function relating c and r. Multiplication by A extends the relationship to the desired value giving the transfer function of Equation (5–2).

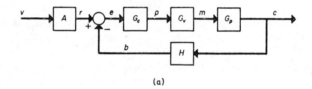

(a)

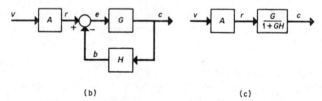

(b)   (c)

FIGURE 5–2. Reduction of block diagram, constant load

For a change in a load variable such as u, the desired value and reference input are both constant, i.e. $v = r = 0$, and Figure 5–1 can be immediately reduced to Figure 5–3(a) by omitting these variables and the input element. From the diagram the following equations can be written

$$c = G_p(m + K_L u)$$

$$m = G_v p \qquad p = G_c e$$

$$b = Hc \qquad e = -b$$

Combining to eliminate e,

$$\frac{c}{u} = \frac{N}{1 + GH} \qquad (5\text{–}3)$$

where $G = G_c G_v G_p$ as above, and $N = K_L G_p$, i.e. the load gain times the process transfer functions through which the load variable passes. Figure 5–3(a) can thus be reduced to the single block of Figure 5–3(b).

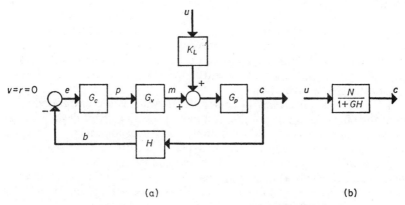

(a)   (b)

FIGURE 5–3. Reduction of block diagram, constant desired value

It should be noted that the denominator of the closed loop transfer function is always the same regardless of the pair of variables to which the function applies, i.e. $(1 + GH)$ in the case of the two examples (Equations 5–2 and 5–3) derived above.

These results can be generalized for a single loop system into a simple rule that the closed-loop transfer function relating any pair of variables, say x and y, is given by

$$y/x = \Pi_f/(1 + \Pi_l)$$

for negative feedback, where Π_f is the product of the transfer functions of the individual elements in the forward path between the input signal x and the output signal y, and Π_l is the product of all the transfer functions of the elements forming the closed loop. It can also be shown by a similar argument that in the case of a loop exhibiting a positive feedback, the corresponding function will be

$$\Pi_f/(1 - \Pi_l)$$

Whilst a positive feedback loop in itself is unstable and not capable of control of a process, such loops are often found as inner loops in multiple-loop systems.

The above procedure may be extended to multiple-loop systems by reducing the innermost loops first to their single block equivalents.

Equations (5–2) and (5–3) define the theoretically ideal conditions for the two main categories of automatic control to which reference has earlier been made. Equation (5–2) defines the response of the controlled variable to changes in desired value in the absence of load changes, and is therefore characteristic of a servo-operation where the objective is for the output of the system to follow as closely as possible a changing input. Equation (5–3) defines the response of the controlled variable to fluctuations in a load variable at a constant desired value, and is thus typical of process control or regulator operation. The designer must be aware of the general purpose of the control operation, since in general terms a system giving optimum performance as a servo-control will not be the best for a process control. A large capacitance or inertia, for example, will help to minimize deviation in process control but will tend to make a servo-system sluggish in operation.

In practical applications, of course, servo-systems are not entirely free from the effects of load changes, just as in process control desired values are not necessarily completely constant, and it may be necessary in some cases to consider the effects of both types of disturbance. In the present study, however, attention will be restricted almost entirely to the effects of load changes which are the principal source of disturbance in process control, and the effects of changes in desired value will be considered only in the simpler cases.

Analysis of Transient Response

The term 'transient response' strictly means the response of a system following any type of change in an input variable from the time the change is imposed up to the imposition of a new steady-state condition. For systems analysis, as considered in this chapter, a rather narrower definition is employed by which the transient response is taken to mean the response of the output to a step change in an input variable. The step change is used mainly for convenience; mathematical solutions are somewhat easier to obtain than for other types of input changes, and the effect can be produced without difficulty in existing systems for practical experiment and confirmation. For realism in design, the step change is the most severe disturbance which can be introduced into a system, and it will show the maximum response which can arise for a given change in input. It is therefore useful for comparisons between different systems, or different values of system or control parameters in the one system, since the system showing the best response to a step change should also show the best response to the random fluctuations in a particular variable likely to be encountered in practice. As far as the stability of a system is concerned, the type of change imposed is immaterial so long as the system is linear, since if such a system is unstable for one type of change it will be unstable for all changes. With servo operations the probable types of input changes can usually be predicted in advance, but with process control the fluctuations in load are completely random and are hardly ever specified at the design stage. Without this information the use of the step input response is the best method of evaluation of the system performance, either with or without the additional evaluation of frequency response analysis.

The response to a step change in either desired value or load variable is obtained by replacing the appropriate variable in Equation (5–2) or (5–3), as applicable, by the Laplace transform of a step change (i.e. $1/s$ for a unit step), and then using the inverse transformation to obtain the response of the controlled variable as a function of time. In some cases it will not be necessary to complete the inverse transformation in detail since the properties of the response, in particular the stability, can be deduced from the characteristic denominator of the closed-loop function.

In order to simplify the analysis in the following sections, some simplifying assumptions will be made. The controller input element usually provides only a constant proportionality between the desired value and the reference input, and by putting $A = 1$ so that $v = r$, the function A can be omitted. Lags in the controller will be neglected and the ideal transfer functions of Chapter 3 will be used for the controller function G_c. In addition the measurement feedback will also be assumed to be unity ($H = 1$) initially, although the effect of a measurement lag or time delay will be considered in due course.

Proportional Control of a First-order Process

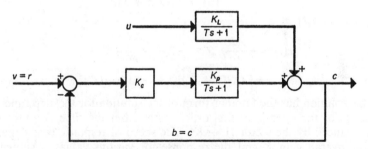

FIGURE 5–4. Proportional control of first-order process

The simplest case for consideration is that of a first-order process with proportional control, the final control element (the valve) having no dynamic lag and a constant gain (K_v) which, for simplicity, may be included in the process gain (K_p). As was seen previously, the time constant for load changes before the capacitance is the same as that for changes in the manipulated variable, but the load gain will usually be different. Hence the transfer function for a change in load applied before the capacitance will be $K_L/(Ts + 1)$. For a proportional controller the transfer function is the proportional sensitivity, K_c. The system is shown in the block diagram of Figure 5–4.

For a change in desired value, Equation (5–2) will apply with A and H both equal to 1, i.e.

$$c/v = G/(1 + G)$$

$$G = K_c K_p/(Ts + 1)$$

$$1 + G = (Ts + 1 + K_c K_p)/(Ts + 1)$$

$K_c K_p$ is the product of the gains of the transfer functions in the loop and may be termed the *overall gain* of the loop. Let $K_c K_p = K$. Then

$$\frac{c}{v} = \frac{K}{Ts + 1} \left(\frac{Ts + 1}{Ts + 1 + K} \right)$$

$$= \frac{K}{Ts + 1 + K}$$

If the desired value is now subjected to a step change of magnitude A, the Laplace transform of v is A/s. Hence

$$c(s) = (A/s)[K/(Ts + 1 + K)]$$

$$= AK/[s(Ts + 1 + K)]$$

To put into a standard form for inversely transforming, the numerator and denominator are both divided by $(1 + K)$, giving

$$c(s) = \frac{AK}{1 + K} \left(\frac{1}{s(T's + 1)} \right)$$

where $T' = T/(1 + K)$. Hence

$$c(t) = \frac{AK}{1 + K} \mathscr{L}^{-1} \frac{1}{s(T's + 1)}$$

$$= [AK/(1 + K)][1 - \exp(-t/T')] \qquad (5\text{--}4)$$

This solution has the familiar form of the solution for the step function response of the open-loop first-order system, but the time constant has been reduced by the factor $(1 + K)$. The speed of response thus depends on the overall gain K and the response for various values is shown in Figure 5–5.

For a change in the load variable, Equation (5–3) is used, simplified to

$$c/u = N/(1 + G)$$

$$N = K_L/(Ts + 1)$$

$$1 + G = (Ts + 1 + K)/(Ts + 1)$$

where $K = K_c K_p$, as above. For a step change in u of magnitude A, $u(s) = A/s$, and

$$c(s) = (A/s)[K_L/(Ts + 1 + K)]$$

$$= [AK_L/(1 + K)]\{1/[s(T's + 1)]\}$$

Inversely transforming,

$$c(t) = [AK_L/(1 + K)][1 - \exp(-]/T')]\qquad(5\text{--}5)$$

where $T' = T/(1 + K)$.

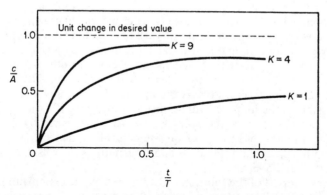

FIGURE 5–5. Response to change in desired value for proportional control of first-order process

The response of Equation (5–5) is shown in Figure 5–6 for different values of K. It will be seen that for load changes the initial value of the slope of the response curve is not dependent on the value of K; this is because the controller does not begin to act until the load change has started to take effect. The initial rate of change (dc/dt at zero time) is

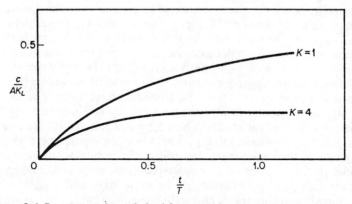

FIGURE 5–6. Response to change in load for proportional control of first-order process

given by AK_L/T and is therefore determined by the magnitude of the load change and the load gain and time constant. In the preceding case the change of desired value is introduced through the controller; the response of the variable is due entirely to the control action and the initial slope is dependent on the value of K.

From Equations (5–4) and (5–5), and as seen in Figures 5–5 and 5–6, the response in each case approaches a steady-state value as the exponential term decays. For a change in desired value the steady-state value of the variable is $AK/(1 + K)$; hence the final value is offset from the new desired value A by an amount $A - AK/(1 + K)$ i.e. $A/(1 + K)$. For a change in load the final value is offset from zero (the desired value) by $AK_L/(1 + K)$. Offset following a change in an input variable is a fundamental property of proportional control, and a physical explanation was given in Chapter 3. It can readily be seen that the magnitude of the offset is inversely proportional to $(1 + K)$ and is thus inversely dependent on the magnitude of the overall gain K. The latter is the product of the individual transfer function gains in the loop, including the controller gain K_c. All except the latter are constants; hence the offset is inversely determined by the value of the proportional sensitivity set on the controller; increasing the controller gain increases the value of K and thus reduces the offset, as shown in Figure 5–5 and 5–6.

The transient part of the solution is determined by the exponential term, $[1 - \exp(-t/T')]$. This factor determines the amount of approach to the final steady-state condition and, as seen on page 54, after the expiry of four time constants the change is over 98 per cent complete. The time for the system to reach the steady-state is thus determined by the effective time constant T', which is given by $T/(1 + K)$. As with the offset, the effective time constant is inversely dependent on the overall gain K, and so on the proportional sensitivity K_c. The larger the controller gain, the smaller is the effective time constant of the system and so the faster is the time for the system to reach the steady state and so recover from the disturbance.

A high value of the controller gain thus reduces the offset and speeds up the response. Theoretically, in this particular case a very high gain could be used since the equations indicate no change in the type of the response as the overall gain approaches infinity. In any practical case, there would be an upper limit to the proportional sensitivity at which oscillations would be set up in the response, these being due to other small lags such as those in measurement and in the controller which are always present in a real system. These may be neglected in any practical case at low values of gain, but they would become apparent at high gains, the system is then no longer first-order and oscillatory responses are possible. In simple cases of first-order processes, such as a level control, the maximum gain would probably be quite large and a narrow-band proportional controller (i.e. a high gain controller) could give satisfactory results.

Proportional Control of a Second-order Process

A rather more practical example, from the type of responses obtainable, is provided by proportional control of a second-order process, i.e. a process with two time-constant stages, the time constants being T_1 and T_2, and the process stage gains, K_{p1} and K_{p2}. It is now possible to introduce load changes before the first stage of the process or between the first and second stages, as shown in Figure 5–7. As will be demonstrated, the point of imposition of the load change makes no fundamental difference to the type of response obtained but alters only the magnitude of the deviation; as would be expected, the further the load change is away from the controller, the greater is its effect on the maximum deviation.

For the present analysis, a load change before the first stage of the process (u_1 in Figure 5–7) will be considered, the load change thus has to pass through both time constant stages and the load transfer function is given by

$$N = K_L/[(T_1s + 1)(T_2s + 1)]$$

In this case, $K_L = K_{p1}K_{p2}$, since Figure 5–7 shows the load change entering the system with a unit gain.

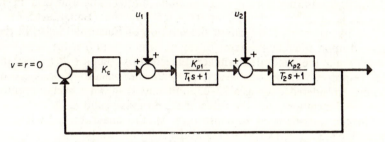

FIGURE 5–7. Proportional control of two-stage process

As before, Equation (5–3) can be written

$$c/u = N/(1 + G)$$

$$G = K_cK_{p1}K_{p2}/[(T_1s + 1)(T_2s + 1)]$$

$$1 + G = [(T_1s + 1)(T_2s + 1) + K_cK_{p1}K_{p2}]/[(T_1s + 1)(T_2s + 1)]$$

whence

$$\frac{c}{u} = \frac{K_L}{(T_1s + 1)(T_2s + 1)} \frac{(T_1s + 1)(T_2s + 1)}{(T_1s + 1)(T_2s + 1) + K}$$

where $K_cK_{p1}K_{p2} = K$, the overall gain.

The above equation reduces to

$$c/u = K_L/[T_1T_2s^2 + (T_1 + T_2)s + 1 + K]$$

To reduce to a standard form for subsequent inversion, both numerator and denominator are divided by $(1 + K)$, giving

$$\frac{u}{c} = \frac{K_L}{1 + K}\left[\left(\frac{T_1T_2}{1 + K}\right)s^2 + \left(\frac{T_1 + T_2}{1 + K}\right)s + 1\right]^{-1}$$

The quadratic term can then be written in the standard form

$$\frac{c}{u} = \frac{K_L}{1 + K}\left[\frac{1}{T^2s^2 + 2\zeta Ts + 1}\right] \tag{5-6}$$

where

$$\left.\begin{array}{l} T = [T_1T_2/(1 + K)]^{\frac{1}{2}} \\ \zeta = \frac{1}{2}(T_1 + T_2)/[T_1T_2(1 + K)]^{\frac{1}{2}} \end{array}\right\} \tag{5-7}$$

For a step change in u of magnitude A, $u(s) = A/s$. Hence

$$c(s) = \frac{A}{s}\left(\frac{K_L}{1 + K}\right)\left(\frac{1}{T^2s^2 + 2\zeta Ts + 1}\right)$$

whence

$$c(t) = \frac{AK_L}{1 + K}\mathscr{L}^{-1}\frac{1}{s(T^2s^2 + 2\zeta Ts + 1)} \tag{5-8}$$

The second-order system thus exhibits a quadratic function in the denominator of the closed-loop transfer function (Equation 5-6), as would be expected. The form of the solution of Equation (5-8) will then depend upon whether the zeros of the quadratic term (i.e. the values of s which make the quadratic term zero) are real or complex. It is clear that the response will be oscillatory, as with the open-loop second-order system considered on pages 72–75, and that it may be over- or under-damped depending upon the value of the damping ratio ζ.

For the over-damped response ($\zeta > 1$), the quadratic term has real factors, e.g. $(T_as + 1)(T_bs + 1)$, and the solution will be

$$c(t) = [AK_L/(1 + K)]\{1 - [T_a \exp(-t/T_a) \\ - T_b \exp(-t/T_b)]/(T_a - T_b)\} \tag{5-9}$$

The form of the solution (Equation 5–9) may be compared with that for the step response of over-damped open-loop second-order system (Equation 2–19).

For the critically damped response ($\zeta = 1$) the quadratic term has equal factors and becomes $(Ts + 1)^2$, and the solution is

$$c(t) = [AK_L/(1 + K)][1 - (1 + t/T)\exp(-t/T)] \tag{5-10}$$

and this solution may be compared with that for the open-loop system (Equation 2–30).

For the under-damped response ($\zeta < 1$) the zeros of the quadratic term are now conjugate complex and the solution is

$$c(t) = [AK_L/(1 + K)] \\ \times \{1 + [1/(1 - \zeta^2)^{\frac{1}{2}}]e^{-\zeta t/T}\sin[(1 - \zeta^2)^{\frac{1}{2}}t/T - \phi]\} \tag{5-11}$$

where

$$\phi = \tan^{-1}[(1 - \zeta^2)^{\frac{1}{2}}/-\zeta]$$

and this solution may be compared to that of the under-damped open loop system (Equation 2–29).

Considering the three possible forms of solution, it will be seen that each contains a steady-state term, $AK_L/(1 + K)$; thus in each case there is a steady-state error whose magnitude is inversely proportional to $(1 + K)$ and so is inversely dependent on the proportional sensitivity K_c, which is the only variable parameter in the overall gain K.

The effect of the overall gain on the damping may be assessed by considering the definition of ζ given above (Equation 5–7), i.e.

$$\zeta = \tfrac{1}{2}(T_1 + T_2)/[T_1 T_2(1 + K)]^{\frac{1}{2}}$$

It will be seen that when $K = 0$ (i.e. when $K_c = 0$), ζ must be greater than one if T_1 is not equal to T_2. Thus in the absence of control ($K_c = 0$) the response of the system must be over-damped, as was seen previously for two first-order stages in series. Increasing the value of K_c from zero will increase the value of K and decrease the value of ζ, so that eventually a value of K will be found at which $\zeta = 1$; the response is then critically damped. Any further increase in K produces values of ζ less than one, and the response is under-damped. Typical responses are plotted in Figure 5–8.

In the rather special case of equal time constants ($T_1 = T_2$), the system is critically damped in the absence of control, and any value of K will produce an under-damped response.

The maximum deviation shown by the response is the height of the first peak which is given by the product of the residual offset, $AK_L/(1 + K)$, and the expression given on page 77 for the height of the first peak following a unit step change with an under-damped second-order system, i.e. $[1 + \exp(-\pi\zeta/(1 - \zeta^2)^{\frac{1}{2}}]$. It will be seen from Figure 5–8 that, although the relative amount of overshoot beyond the final value increases with decreased damping (lower values of ζ), the absolute value of the maximum deviation decreases owing to the larger decrease in the residual offset. In the present case of a load change at the start of the process the peak deviation is always less than twice the final offset.

The case of an intermediate load change (such as u_2 in Figure 5–7) can be handled in a similar way; the only difference is that the load transfer function N contains only one of the process time constants (T_2) and the transform of the response then has the other time constant term ($T_1 s + 1$) in the numerator, i.e.

$$N = K_L/(T_2 s + 1)$$

where $K_L = K_{p2}$, if the load change has a unit gain at the point of entry as shown in Figure 5–7.

$$\frac{c}{u} = \frac{N}{1 + G}$$
$$= \frac{K_L}{T_2 s + 1} \frac{(T_1 s + 1)(T_2 s + 1)}{(T_1 s + 1)(T_2 s + 1) + K}$$
$$= \frac{K_L}{1 + K} \frac{T_1 s + 1}{T^2 s^2 + 2\zeta T s + 1}$$

where T and ζ have the same values as defined by Equation (5–7).

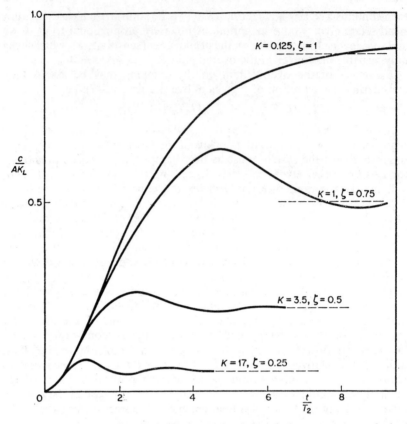

FIGURE 5–8. Response to load change for proportional control of two-stage process for load change before first capacitance

For a step change of magnitude A,

$$c(t) = \left(\frac{AK_L}{1 + K}\right) \mathscr{L}^{-1} \frac{(T_1 s + 1)}{s(T^2 s^2 + 2\zeta T s + 1)} \tag{5–12}$$

The solution of the inverse transform can be obtained from tables for the different values of ζ; only that for the under-damped response need be given here, i.e.

$$c(t) = \left(\frac{AK_L}{1 + K}\right) \left\{ 1 + \left[\frac{1 - 2\zeta T_1/T + (T_1/T)^2}{(1 - \zeta^2)}\right]^{\frac{1}{2}} \right.$$

$$\left. \times e^{-\zeta t/T} \sin\left[(1 - \zeta^2)^{\frac{1}{2}}\right]/T + \phi\right\} \tag{5–13}$$

where

$$\phi = \tan^{-1}\left[\frac{(1 - \zeta^2)^{\frac{1}{2}} T_1/T}{1 - \zeta T_1/T}\right] - \tan^{-1}\left[\frac{(1 - \zeta^2)^{\frac{1}{2}}}{-\zeta}\right]$$

Comparison of this equation with that for the previous case (Equation 5–11) shows that the damping factor, frequency and decay ratio are the same in each case; these properties depend on the elements in the closed loop which are identical in the two cases and which contribute to the denominator of the closed loop transfer function i.e. $(1 + G)$. Typical

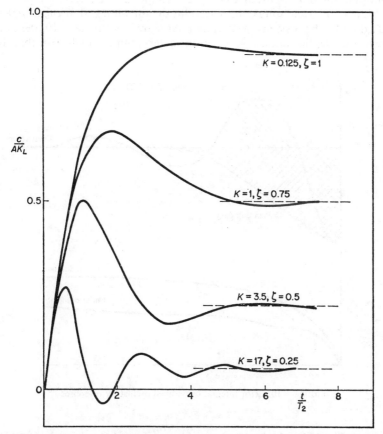

FIGURE 5–9. Response to load change for proportional control of two-stage process with load change between capacitances

response curves for this second case are shown in Figure 5–9, from which it can be seen that for the same value of the load gain (K_L) as in the previous case of Figure 5–8 (by assuming unit process gains for both cases), the residual offsets and damping ratios are the same irrespective of the values of K, but that the maximum peak deviations are much greater. This is particularly so if the second time constant is relatively small compared to the first, since the corrective action of the controller is delayed by the longer first time constant.

To summarize, increasing the controller gain for this system with two time-constant stages will decrease the offset and the peak deviation, but the larger values of K_c will produce under-damping and a recovery time greater than that of critical damping, which is the fastest response without overshooting. In practice the absolute magnitude of the peak deviation is not of paramount importance compared to the time the deviation persists, and it is more usual to take into account the time-integral of the deviation (based on the final steady-state value). As shown in Figure 5–10, the time-integral of the deviation is less for an under-damped oscillation than for

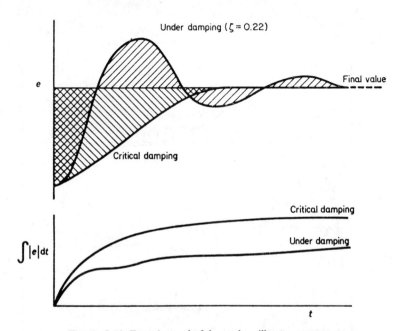

FIGURE 5–10. Error integral of damped oscillatory responses

the critically damped, the increased frequency and lower peak deviation more than compensating for the effect of the greater number of oscillations and the longer time for the oscillations to die out. Thus, as in the case of a first-order process with proportional control, the highest possible controller gain would seem to be indicated to give the best response to a step change. A real system, however, would again have a number of other small lags and time delays which would be negligible at low gains but would be significant and tend to produce instability at high gains. In practice the controller gains are chosen to give a reasonable degree of under-damping with ζ in the range of 0.2 to 0.3, corresponding to a subsidence ratio of about 4:1.

If the time constants of the system are known, the required value of K_c can be calculated from the relationship between ζ and K (Equation (5–7)), which can be re-arranged to give

$$K = (1/4\zeta^2)(T_1/T_2 + T_2/T_1 + 2) - 1$$

It is interesting to note that there is no difference in the stability of the resulting oscillations whether the first or second time constant is the larger; the only effect of inter-changing the time constants is on the peak height of the deviation following a load change introduced between the two time-constant stages.

Stability

In the discussion of damped second-order responses in Chapter 2, it was noted that the theoretical limit of damping is the zero-damped oscillation ($\zeta = 0$), which is a continuous oscillation of constant amplitude. Control systems up to second-order are inherently stable since, from the definition of ζ by Equation (5–7), it is not possible for ζ to be made equal to zero. In control systems of higher order it is possible for $\zeta < 0$, and the system becomes unstable.

In classical terms a stable oscillation is one of limited amplitude, i.e. the amplitude is not necessarily constant but cannot increase. Thus damped and continuous oscillations of fixed amplitude are stable, the only unstable condition is that with an increasing amplitude, as shown in Figure 5–11. All these cases are described by the simplified equation

$$\theta = A\, e^{-\alpha t} \sin \beta t$$

in which α and β are real numbers. In the case of damped oscillations, α is positive so that $\exp(-\alpha t)$ decreases with time; for zero damping $\alpha = 0$ so that the amplitude is constant; for instability α is negative so that $\exp(-\alpha t)$ increases with time.

It is a particular property of all feedback systems higher than second order that the responses can take up any of these forms by suitable adjustment of the control parameters. It must be accepted that a controller should impose a non-oscillatory steady-state after a disturbance, and it is necessary to consider the conditions which produce the zero-damped oscillation which is the limiting condition of stability. A physical explanation will be considered later; at this stage a mathematical analysis will be given.

The transfer function for closed-loop responses for changes in either the desired value or a load variable have already been derived as Equations (5–2), (5–3):

$$c/v = AG/(1 + GH) \qquad c/u = N/(1 + GH)$$

For any particular change in the input variables, v or u, the transform response of the controlled variable can be written in general terms as a fraction whose numerator and denominator are polynomials in s, i.e.

$$c(s) = \frac{N}{D} = \frac{F_1(s)}{F_2(s)}$$

$$= \frac{A_m s^m + A_{m-1} s^{m-1} + \ldots + A_2 s^2 + A_1 s + A_0}{B_n s^n + B_{n-1} s^{n-1} + \ldots + B_2 s^2 + B_1 s + B_0}$$

where n and m are integers and the As and Bs are constants.

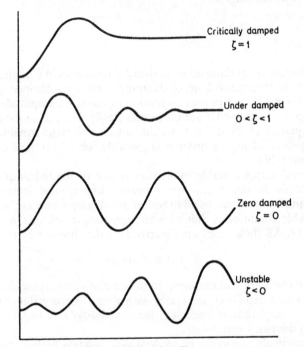

FIGURE 5–11. Stability of oscillatory responses

To solve the equation to find the time-function response of c, the fraction must be inversely transformed. Without the use of tables of transform pairs, it is necessary to divide the expression into simpler standard transforms by means of a partial fraction expansion. This requires the denominator to be factorized, thus:

$$c(s) = \frac{F_1(s)}{(s - r_1)(s - r_2) \ldots (s - r_n)}$$

whence

$$c(t) = \mathscr{L}^{-1} \left[\frac{C_1}{s - r_1} + \frac{C_2}{s - r_2} + \ldots + \frac{C_n}{s - r_n} \right] \qquad (5\text{–}14)$$

where $C_1, C_2, \ldots, C_n$ are constants arising from the partial fraction expansion, and $r_1, r_2, \ldots, r_n$ are roots of the polynomial denominator set equal to zero, i.e.

$$F_2(s) = B_n s^n + B_{n-1} s^{n-1} + \ldots + B_2 s^2 + B_1 s + B_0 = 0 \quad (5\text{–}15)$$

which is thus an equation characteristic of the system.

The polynomial in s of Equation (5–15) arises from the expansion of the denominator of the closed-loop transfer function, $(1 + GH)$, multiplied by the denominator of the transform of the change in v or u imposed on the system, e.g. s from A/s for a step change of magnitude A. To test the stability of the system, the change must itself be stable, so the stability of the system is determined only by the denominator of the transfer function of the closed loop, and the effective characteristic equation is

$$1 + GH(s) = 0 \qquad\qquad (5\text{–}16)$$

Carrying out the inverse transformation of Equation (5–14), each term is of the form, $1/(s + a)$, which is the transform of $\exp(-at)$. Hence

$$c(t) = C_1 \exp(r_1 t) + C_2 \exp(r_2 t) + \ldots + C_n \exp(r_n t) \quad (5\text{–}17)$$

The roots r_1, r_2, etc. of the characteristic equation may be real, imaginary or complex numbers. If, for example, the root r_1 is real, it may be either positive or negative; if negative, the term $\exp(r_1 t)$ will decrease with time to zero, and the term containing r_1 is thus a transient which makes no contribution to the final response; if r_1 is positive, $\exp(r_1 t)$ will increase with time. In the latter case the system must be unstable, for if only one root is positive the total response must show a continuous increase with time since all the negative roots must ultimately produce a zero contribution to the total response. The system must be unstable since there is no final steady state.

If r_1 is imaginary, say $j\beta$, then there must also be a conjugate root, $-j\beta$, as the polynomial contains only real terms. The response will then contain a group $[\exp(j\beta t) + \exp(-j\beta t)]$, which will yield an oscillatory term such as $\sin \beta t$ of constant amplitude.

Finally r_1 may be a complex number with a real and an imaginary part, e.g. $\alpha + j\beta$. Again there must be a conjugate, $\alpha - j\beta$, and these now produce in the response the group

$$[\exp(\alpha + j\beta)t + \exp(\alpha - j\beta)t]$$

or

$$\exp(\alpha t)[\exp(j\beta t) + \exp(-j\beta t)]$$

This now yields an oscillatory term but with an exponential term in the amplitude, i.e. $e^{\alpha t} \sin \beta t$. If the real part of the complex roots is positive, the amplitude will increase with time and the oscillation is unstable; if the real part is negative, the amplitude decays with time and the oscillation is damped and therefore stable.

A control system will thus be stable if the denominator of the closed-loop transfer function has no positive zero or no complex zeros with positive real parts. The maximum gain at which the system becomes unstable can thus be found by equating the denominator of the closed-loop function to zero (thus forming the characteristic equation 5–16) for various values of the gain and determining the value that just makes one root of the equation positive, or take on a positive real part in the case of a complex root. Alternatively, since a pair of imaginary roots produce a continuous oscillation of constant amplitude (which is the condition of limiting stability), the gain for this condition can be determined by finding the value necessary to produce one pair of imaginary roots without any other root being positive or having a positive real part. A graphical method of investigating the behaviour of the roots of the characteristic equation is considered in the next chapter; a mathematical test for stability follows in the next section.

Example 5–1

For the system of Figure 5–7, the transfer functions are $G_c = 5$, $G_{p1} = 2/(2s + 1)$, $G_{p2} = 0.1/(s + 1)$ and $H = 1$. Find the characteristic equation and determine if the system is stable.

The open-loop transfer function is

$$GH = 5 \times \frac{2}{2s + 1} \times \frac{0.1}{s + 1} \times 1$$

$$= \frac{1}{(2s + 1)(s + 1)}$$

The characteristic Equation (5–16) is thus

$$1 + \frac{1}{(2s + 1)(s + 1)} = 0$$

or
$$2s^2 + 3s + 2 = 0$$

Using the quadratic formula, the roots are

$$s_1, s_2 = [-3 \pm \sqrt{(9 - 16)}]/4$$
$$= -\tfrac{3}{4} \pm j\sqrt{(\tfrac{7}{4})}$$

Since the real parts of the roots are negative $(-\tfrac{3}{4})$, the system is stable (being second-order, it cannot theoretically be unstable).

The Routh Test

The main difficulty in assessment of the stability of a system from its characteristic equation is that it is necessary to find roots of a polynomial of third or higher order. Whilst not impossible this becomes increasingly difficult as the order of the equation increases. The Routh test [22] is an exact and purely algebraic method of determining if any of the roots of a polynomial equation are positive or have positive real parts, which can be used to test whether the characteristic equation of a system has such roots and is thus a criterion as to whether or not the closed loop system is stable. The method does, not, however, give any indication as to how close a system may be to stability or instability.

The method is first to write the characteristic equation in descending order of the powers of s, i.e. in the form

$$a_0 s^n + a_1 s^{n-1} + a_2 s^{n-2} + \ldots + a_{n-1} s + a_n = 0$$

where a_0 is positive. If any coefficient is negative then the system is immediately known to be unstable, but the test may still be used to determine the number of roots with positive real parts. An array is next constructed by arranging the coefficients alternately in the first two rows, as shown:

Row				
1	a_0	a_2	a_4	a_6
2	a_1	a_3	a_5	a_7
3	b_1	b_2	b_3	
4	c_1	c_2	c_3	
5	d_1	d_2		
6	e_1	e_2		
7	f_1			
8	g_1			

The elements in the remaining rows are found from the following formulae, the elements in any row after the second being determined from the elements in the preceding two rows:

$$b_1 = \frac{a_1 a_2 - a_0 a_3}{a_1} \qquad b_2 = \frac{a_1 a_4 - a_0 a_5}{a_1} \qquad \text{etc.}$$

$$c_1 = \frac{b_1 a_3 - a_1 b_2}{b_1} \qquad c_2 = \frac{b_1 a_5 - a_1 b_3}{b_1} \qquad \text{etc.}$$

The construction of the array terminates when only zeros are obtained for the additional elements; in general there will be $(n + 1)$ rows, where n

is the degree of the polynomial equation. During computation any row can be divided throughout by a constant without changing the result.

Routh's criterion states that the number of changes of algebraic sign in the first column of the array is equal to the number of roots having positive real parts. As a corollary, for all roots to have negative real parts, as required for a stable system, all elements of the first column must be positive and non-zero.

A further theorem of the Routh test is that if *one* pair of roots is purely imaginary ($\pm jx$) and all other roots have negative real parts, all the elements of the nth row will vanish and none of the elements of the preceding row will vanish. The values of the pair of imaginary roots can then be found by solving the equation $Ax^2 + B = 0$, where A and B are the first and second elements respectively in the $(n - 1)$th row. This particular theorem is very useful in finding values of the gain in the root-locus method of analysis considered in Chapter 6.

Example 5–2

Consider the stability of a system with closed-loop transfer function of $10/(5s^3 + 6s^2 + 3s + 15)$.

The characteristic equation is

$$5s^3 + 6s^2 + 3s + 15 = 0$$

The Routh array is

Row		
1	5	3
2	6	15
3	−9.5	
4	15	

Since there are two changes of sign (+ve to −ve and −ve to +ve) in the first column, two of the three roots of the cubic equation have positive real parts and the system is thus unstable.

Example 5–3

A four-stage process has time constants of 5, 5, 3, and 1 min with a gain of 2.

(a) Is the system stable with proportional control if $K_c = 2.5$?

(b) If so, will the system still be stable if integral action with $T_i = 4$ min is added?

(a) $G = (2.5 \times 2)/[(5s + 1)^2(3s + 1)(s + 1)]$

The denominator of the closed-loop function $(1 + G)$ is then

$(5s + 1)^2(3s + 1)(s + 1) + 5 = 75s^4 + 130s^3 + 68s^2 + 14s + 6$

The Routh array is

$$\begin{array}{ccc} 75 & 68 & 6 \\[1em] 130 & 14 & (0) \\[1em] 59.9 & 6 & \\[1em] 0.98 & & \\[1em] 6 & & \end{array}$$

Since there is no change in sign in the first column, the system is stable.

(b) $G = (2.5)(1 + 1/4s)(2)/[(5s + 1)^2(3s + 1)(s + 1)]$

$$= \frac{5(4s + 1)}{4s(75s^4 + 130s^3 + 68s^2 + 14s + 1)}$$

The characteristic equation, $1 + G = 0$, is

$4s(75s^4 + 130s^3 + 68s^2 + 14s + 1) + 5(4s + 1)$
$$= 300s^5 + 520s^4 + 272s^3 + 56s^2 + 24s + 5 = 0$$

The first two rows of the Routh array are

$$\begin{array}{ccc} 300 & 272 & 24 \\[1em] 520 & 56 & 5 \end{array}$$

but the first row may be divided throughout by 24 and the second row by 5 to simplify computation, whence the full array is constructed as follows:

$$\begin{array}{ccc} 12.5 & 11.33 & 1 \\[1em] 104 & 11.2 & 1 \\[1em] 9.99 & 0.88 & \\[1em] 2.04 & 1 & \\[1em] -4.89 & & \end{array}$$

Since there is one change in sign in the first column, the system is now unstable.

Example 5–4

Determine the limiting gain and the values of the imaginary roots for a system whose characteristic equation is the cubic

$$s^3 + 7s^2 + 14s + 8 + K = 0$$

The Routh array is

$$
\begin{array}{cc}
1 & 14 \\
7 & (8 + K) \\
b_1 &
\end{array}
$$

The Routh theorem states that if one pair of roots is imaginary (the condition of limiting stability) and all others have negative real parts, all the elements of the nth row must be zero. Applying the formula for the first element in the third row (b_1) and equating to zero,

$$b_1 = [7(14) - 1(8 + K)]/7 = 0$$

Solving for K,

$$K = 90$$

The array is then

$$
\begin{array}{cc}
1 & 14 \\
7 & 98 \\
0 &
\end{array}
$$

There is thus no change of sign in the first column to indicate instability, but the last element is zero which indicates the condition of limiting stability.

The value of the imaginary roots is given by the equation $As^2 + B = 0$, where A and B are the elements of the $(n - 1)$th row taken in order from left to right, i.e. the elements of the second row, $A = 7$ and $B = 98$. Thus

$$7s^2 + 98 = 0$$

$$s = \pm j\sqrt{14}$$

The Routh test is relatively simple to apply but is restricted to systems described by a polynomial equation; it cannot strictly be used if there is a time delay in the system since this would introduce an exponential term into the equation. Whilst this may be expanded into the usual series, it then provides an infinite number of terms; taking only the first few terms does permit an approximate solution but this may be subject to error.

As indicated above, the Routh test shows only whether or not a system is stable; it cannot show how close an unstable system is to stability. Alternative methods of investigating stability which do show how close a system is to stability or instability have the further advantage of being graphical and somewhat easier to apply than a purely algebraic method. These include the root-locus technique, frequency response analysis, and the Nyquist criterion, which are discussed in Chapters 6 and 7.

Effect of Controller Gain on Stability

The effect of the controller gain on the stability of the system can be demonstrated by considering proportional control of a third-order process, i.e. three process stages in series with transfer functions of $K_{p1}/(T_1s + 1)$, $K_{p2}/(T_2s + 1)$, and $K_{p3}/(T_3s + 1)$. Assuming unity feedback ($H = 1$), the system is as shown in Figure 5–12.

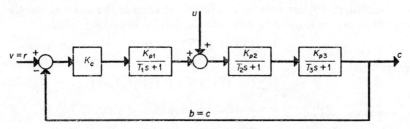

FIGURE 5–12. Proportional control of three-stage process

The open loop function GH is now given by
$$GH = K_cK_{p1}K_{p2}K_{p3}/[(T_1s + 1)(T_2s + 1)(T_3s + 1)]$$
The product $K_cK_{p1}K_{p2}K_{p3}$ is equal to K, the overall gain. Hence
$$1 + GH = 1 + K/[(T_1s + 1)(T_2s + 1)(T_3s + 1)]$$
and the characteristic equation obtained by equating $(1 + GH)$ to zero will be
$$(T_1s + 1)(T_2s + 1)(T_3s + 1) + K = 0$$
Expanding the term in brackets,
$$\sum_3 Ts^3 + \sum_2 Ts^2 + \sum_1 Ts + 1 + K = 0 \qquad (5\text{--}18)$$
where $\sum_n T$ = sum of all combinations of T taken n at a time.

Since the characteristic equation is a cubic there will be three roots, s_1, s_2, and s_3. For an oscillatory response, two of these roots will form a conjugate complex pair and the third must be real, since the equation contains only real terms. The roots may therefore be written
$$s_1 = -\alpha_1 + j\beta$$
$$s_2 = -\alpha_1 - j\beta$$
$$s_3 = -\alpha_2$$

where α_1, α_2, and β are real numbers. As seen in preceding discussions, the system will be stable only if the roots have no positive real parts; from the above definitions α_1 and α_2 must then be positive. The limiting condition for stability of the continuous zero-damped oscillation obtains when α_1 is zero and the conjugate pair are purely imaginary, i.e. $s_1, s_2 = \pm j\beta$. Substituting either of these values for s in Equation (5–18) gives an expression for the overall gain at the zero-damped condition, i.e.

$$\sum_3 T(j\beta)^3 + \sum_2 T(j\beta)^2 + \sum_1 T(j\beta) + 1 + K = 0$$

or

$$-j\sum_3 T\beta^3 - \sum_2 T\beta^2 + j\sum_1 T\beta + 1 + K = 0 \qquad (5\text{–}19)$$

Equation (5–19) can only be satisfied if the sums of the real and imaginary terms are each equal to zero. Hence, for the imaginary terms,

$$-j\sum_3 T\beta^3 + j\sum_1 T\beta = 0$$

from which

$$\beta^2 = \sum_1 T / \sum_3 T = (T_1 + T_2 + T_3)/(T_1 T_2 T_3) \qquad (5\text{–}20)$$

From the real terms of Equation (5–19),

$$-\sum_2 T\beta^2 + 1 + K = 0$$

and substituting the value for β^2 from Equation (5–20),

$$K = (\sum_1 T \sum_2 T)/\sum_3 T - 1$$

or

$$K = \frac{(T_1 + T_2 + T_3)(T_1 T_2 + T_2 T_3 + T_3 T_1)}{T_1 T_2 T_3} - 1 \qquad (5\text{–}21)$$

Equations (5–20) and (5–21) give the limiting values of the overall gain K and the frequency β which produce the continuous zero-damped oscillation. Since these are critical values for the limiting condition of stability the symbols K_{max} and β_0 will be used. Since K is the product of the process stage gains and the controller gain K_c, K_{max} corresponds to the maximum controller gain $K_{c\,max}$ at which the system reaches the limit of stability. That the limiting value is a maximum is shown in the previous discussion for a second-order system (pages 143–5); increasing the value of K_c from zero reduces the damping and gives more oscillation and so less stability.

The expression for the maximum overall gain (Equation 5–21) can be re-arranged to give

$$K_{max} = \left(1 + \frac{T_2}{T_1} + \frac{T_3}{T_1}\right)\left(1 + \frac{T_1}{T_2} + \frac{T_1}{T_3}\right) - 1$$

which indicates that the maximum gain is determined by the *ratios* of the time constants and not the absolute magnitudes. The maximum gain becomes very large if one time constant is significantly larger or smaller than the others.

To determine the effect of the controller gain on the relative stability of the system, the case of complex roots must now be considered. If the complex root $s_1 = -\alpha_1 + j\beta$ is substituted for s in Equation (5–18), the following equation results:

$$\sum_3 T(-\alpha_1 + j\beta)^3 + \sum_2 T(-\alpha_1 + j\beta)^2$$
$$+ \sum_1 T(-\alpha_1 + j\beta) + 1 + K = 0 \quad (5\text{–}22)$$

Equation (5–22) must have real and imaginary parts which are both equal to zero; after expanding the terms, for the imaginary parts

$$\beta[\sum_3 T(3\alpha_1^2 - \beta^2) - 2\sum_2 T\alpha_1 + \sum_1 T] = 0$$

whence

$$\beta^2 = [3\sum_3 T\alpha_1^2 - 2\sum_2 T\alpha_1 + \sum_1 T]/\sum_3 T \quad (5\text{–}23)$$

For the real parts of Equation (5–22),

$$\sum_3 T(-\alpha_1^3 + 3\alpha_1\beta^2) + \sum_2 T(\alpha_1^2 - \beta^2) - \sum_1 T\alpha_1 + 1 + K = 0$$

Substituting for β^2 from Equation (5–23) and collecting like terms,

$$8\sum_3 T\alpha_1^3 - 8\sum_2 T\alpha_1^2 + 2[\sum_1 T + (\sum_2 T)^2/\sum_3 T]\alpha_1$$
$$- (\sum_1 T\sum_2 T)/\sum_3 T + 1 + K = 0$$

From Equation (5–21),

$$(\sum_1 T\sum_2 T)/\sum_3 T - 1 = K_{max}$$

Hence

$$K = K_{max} - 8\sum_3 T\alpha_1^3 + 8\sum_2 T\alpha_1^2 - 2[\sum_1 T + (\sum_2 T)^2/\sum_3 T] \quad (5\text{–}24)$$

The damping constant α_1 and the frequency β, given by Equations (5–23) and (5–24), are plotted against K in Figure 5–13 for a three-stage process with time constants in the ratio $1:2:5$. It can be seen that the critical value of K (i.e. K_{max}) is 12.6, at which value the damping constant becomes zero and any further increase in K would give a negative value of α_1. If K is reduced below the critical value, the damping constant increases and the frequency decreases until a point is reached at which the frequency becomes zero and the response is then critically damped. The diagram may thus be divided into zones of over-damped, under-damped, and unstable responses. The dotted line on the diagram indicates the relative magnitude of the offset which, as previously seen, is proportional to $1/(1 + K)$, and it is apparent that the offset is not appreciably reduced by increasing K beyond about 50 per cent of the maximum value (i.e. about $K = 6$ in the present case). It is a common recommendation by controller manufacturers, based on practical experience, that the proportional sensitivity should be reduced to the order of 50 per cent of the value which just produces a continuous oscillation after a disturbance, the latter value being obviously $K_{c\,max}$. This is equivalent to making $K = 0.5K_{max}$, and this is found to give a subsidence ratio of the under-damped oscillation which results of about 4:1.

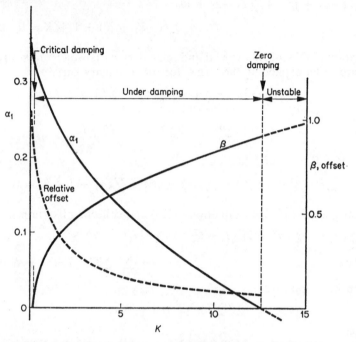

FIGURE 5-13. Damping factor, frequency and offset *versus* overall gain for proportional control of three-stage process

Integral Control

With integral control the output of the controller is proportional to the time integral of the error, and the controller transfer function may be written $G_c = 1/T_i s$, where T_i is the integral time.

Since integral control is generally used only in combination with other control modes, only the response of a simple first-order process need be considered for purely integral control in order to illustrate the particular features of the action. The following equations give the response of integral control to a change in load on a first-order process, as shown in the block diagram of Figure 5-14, and these should be compared to the equations for proportional control of a first-order (pages 139–142) and a second-order process (pages 143–5).

From Figure 5–14,

$$N = K_L/(T_1s + 1)$$

$$1 + G = 1 + K_p/[T_is(T_1s + 1)]$$

$$\frac{c}{u} = \frac{K_L}{T_1s + 1} \cdot \frac{T_is(T_1s + 1)}{T_is(T_1s + 1) + K_p}$$

$$= \frac{K_L T_is}{T_iT_1s^2 + T_is + K_p}$$

$$= (K_L T_i/K_p)\left[\frac{s}{(T_iT_i/K_p)s^2 + (T_i/K_p)s + 1}\right]$$

For a step change in load of magnitude A, $u(s) = A/s$. Hence

$$c(t) = AK_L T_i/K_p \mathscr{L}^{-1} \frac{1}{T^2s^2 + 2\zeta Ts + 1}$$

where

$$T = (T_iT_1/K_p)^{\frac{1}{2}} \quad \text{and} \quad \zeta = \tfrac{1}{2}(T_i/K_pT_1)^{\frac{1}{2}} \tag{5–25}$$

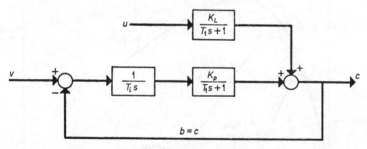

FIGURE 5–14. Integral control of first-order process

A first-order process with integral control thus yields a second-order equation, and damped oscillatory responses will be obtained in the simplest case; if the order of the process is increased to second or higher order then instability will be possible. The most significant difference between the equations for integral and proportional control, however, is that the denominator of the response transform for integral control does not contain the factor s which is present in the proportional-control response transforms. The reason is that this factor cancels with the s term in the numerator introduced by the integral control transfer function. As a result the characteristic equation will not have a zero root, and the time function of the response, given by the inverse transformation of Equation (5–25), will not contain a term invariant with time. Consequently, if the system is stable and the other roots produce damped transient terms, the steady-state value of the response will be zero, i.e. the controlled variable will return to the desired value after the disturbance and there is no residual offset. For example, the solution for the under-damped case of Equation (5–25) is

$$c(t) = (AK_L T_i/K_p)[1/T(1 - \zeta^2)^{\frac{1}{2}}] \, e^{-\zeta t/T} \sin [(1 - \zeta^2)^{\frac{1}{2}} t/T]$$

The effect of the damping coefficient on the response is shown in Figure 5–15, the curves being similar to those for proportional control of a second-order process except that there is no offset. As seen from the definition of ζ (Equation 5–25), reducing the integral time T_i decreases the damping ratio and makes the response less stable, prolonging the oscillation. Based on a criterion of minimum error-integral the lowest integral time would theoretically be the best, but again small additional lags would become apparent and produce instability in such circumstances. It is usual, therefore, to choose an integral time to give a reasonable subsidence ratio of about 4:1 and the corresponding value of ζ (about 0.22) can be used to calculate the integral time from Equation (5–25).

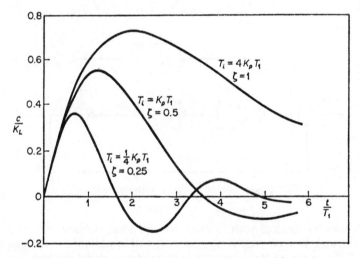

FIGURE 5–15. Response to load change for integral control of first-order process

Proportional-integral Control

The reasons for combining integral control with proportional will now be apparent. Proportional control can be adjusted to give a stable response, but suffers from the residual offset from the desired value following a load disturbance. An added integral action (which may now be legitimately termed a reset action) will eliminate the offset. In practice the offset would be reduced to zero with most electronic controllers and a few pneumatic controllers which have an ideal integral action; most types of the latter do not develop a completely ideal integral action and a residual offset of about 1 per cent of the potential value of the load change will remain, but this will be considerably less than the offset if the integral action were not used.

In most practical controllers, the integral action varies with the controller gain as well as with the integral time and the ideal transfer function is

$$G_c = K_c(1 + 1/T_i s)$$

The case of proportional-integral control of a first-order process will be considered since this is the only case for which direct solutions can be written without having to take factors of a cubic or higher order equation. The block diagram is given in Figure 5–16, and the derivation of the system response can be compared with that for proportional control of a second-order process (pages 143–5).

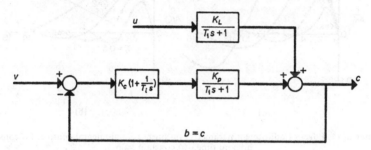

FIGURE 5–16. Proportional-integral control of first-order process

From Figure 5–16,

$$N = K_L/(T_1 s + 1)$$

$$1 + G = 1 + K_c(1 + 1/T_i s)K_p/(T_1 s + 1)$$

$$\frac{c}{u} = \frac{K_L}{T_1 s + 1} \frac{T_i s(T_1 s + 1)}{T_i s(T_1 s + 1) + K(T_i s + 1)}$$

where $K = K_c K_p$, the overall gain.

$$\frac{c}{u} = (K_L T_i/K)\left[\frac{s}{(T_i T_1/K)s^2 + (1 + K)(T_i/K)s + 1}\right] \qquad (5\text{–}26)$$

Equation (5–26) is very similar to that derived for integral control alone, but when the quadratic term is written in the standard form the damping ratio is increased by the factor $(1 + K)$, i.e. from Equation (5–26),

$$T = (T_i T_1/K)^{\frac{1}{2}} \qquad \text{and} \qquad \zeta = \frac{1}{2}(1 + K)(T_i/T_1 K)^{\frac{1}{2}}$$

The system now has two adjustable parameters in K_c and T_i, both of which contribute to the value of ζ. Figure 5–17 shows the effect of changing each parameter independently of the other. In Figure 5–17(a) increasing K_c for a fixed value of T_i improves the response by decreasing the maximum (peak) deviation and also by damping the response by the increased value of ζ. For a fixed value of K_c (Figure 5–17b), a decrease in T_i decreases the maximum deviation and period, but makes the response more oscillatory as ζ is now decreased. The overall effect is relatively small in view of the wide range of variation of the parameters.

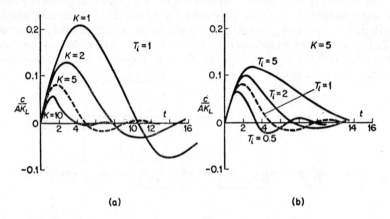

FIGURE 5–17. Effect of adjustment parameters on proportional-integral control: (a) constant integral time, (b) constant gain

For systems of a higher order, the equation for the transient response becomes increasingly complex as the order increases; the second-order process will yield a cubic equation and the response will be similar to that of proportional control of a third-order process. The general effect of changing K_c and T_i on a higher order process is shown in Figure 5–18. K_{max} is the maximum overall gain for the condition of limiting stability (continuous oscillation) for proportional control alone, and it can be seen that the system with proportional-integral control becomes unstable at a lower gain than with proportional control alone, the stability decreasing as the integral time is reduced. It is often necessary in practice to reduce the proportional gain by a small amount when integral action is added. A moderate amount of integral action is usually required to overcome the offset in a reasonable time, and this amount of integral action may often lead to relative instability if K_c is adjusted on the basis of proportional control alone. It should also be noticed that the frequency of oscillation is reduced as the integral component is increased (reduced integral time) and the proportional-integral controller will have a slower response than a pure proportional controller for the same degree of damping.

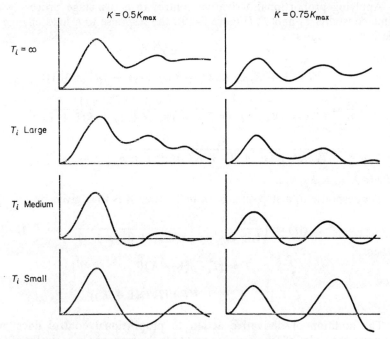

FIGURE 5–18. Response of proportional-integral control for a multi-stage process

Proportional-derivative Control

Derivative action is generally called for when the process has a large number of storage elements and thus considerable transfer lag. It cannot be used alone but must be combined with either proportional or proportional-integral action. Like integral action, the derivative action in most proportional-derivative controllers varies with the controller gain, and the ideal transfer function is

$$G_c = K_c(1 + T_d s)$$

where T_d is the derivative time.

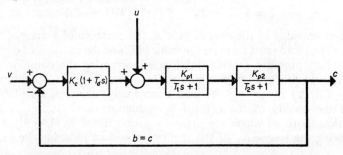

FIGURE 5–19. Proportional-derivative control of two-stage process

Applying proportional-derivative control to a two-stage process with time constants T_1 and T_2 (Figure 5–19), the response to a load change is derived as follows:

$$N = K_L/[(T_1s + 1)(T_2s + 1)]$$

$$1 + G = 1 + K_cK_{p1}K_{p2}(1 + T_ds)/[(T_1s + 1)(T_2s + 1)]$$

$$\frac{c}{u} = \frac{K_L}{(T_1s + 1)(T_2s + 1)} \frac{(T_1s + 1)(T_2s + 1)}{(T_1s + 1)(T_2s + 1) + K(1 + T_ds)}$$

$$= \frac{K_L}{T_1T_2s^2 + (T_1 + T_2 + KT_d)s + 1 + K}$$

where $K = K_cK_{p1}K_{p2}$.

The response to a step change of magnitude A is thus given by

$$c(t) = \left(\frac{AK_L}{1 + K}\right) \mathscr{L}^{-1} \frac{1}{s(T^2s^2 + 2\zeta Ts + 1)} \qquad (5\text{–}27)$$

where

$$T = [T_1T_2/(1 + K)]^{\frac{1}{2}}$$

and

$$\zeta = \tfrac{1}{2}(T_1 + T_2 + KT_d)/[T_1T_2(1 + K)]^{\frac{1}{2}}$$

The addition of derivative action to proportional control does not increase the order of the equation and thus the response is basically of the same form as that for proportional control of the same process (pages 143–5), the only essential difference being the changed value of ζ due to the derivative component. Offset is not eliminated since the first term of the partial fraction expansion of Equation (5–27) will contain s as the denominator which will yield a steady-state constant term in the inversion of the transform. The main effect of the derivative action is the increased value of ζ, which can be seen by comparing the value defined for proportional control in Equation (5–7),

$$\zeta = \tfrac{1}{2}(T_1 + T_2)/[T_1T_2(1 + K)]^{\frac{1}{2}}$$

with that given above in Equation (5–27) for proportional-derivative control,

$$\zeta = \tfrac{1}{2}(T_1 + T_2 + KT_d)/[T_1T_2(1 + K)]^{\frac{1}{2}}$$

For a given value of the overall gain K, the presence of a derivative time greater than zero must increase the value of ζ, and the addition of derivative action thus provides greater stability by increasing the damping of the response. In practice the damping ratio is restored to its optimum value $(0.2 - 0.3)$ by increasing the gain accordingly; this will inevitably improve the quality of the control by reduction of the offset and peak deviation, both of which are inversely proportional to $(1 + K)$, and by increasing the frequency of the oscillation and so reducing the recovery time.

The major advantage of adding derivative action to proportional control is the increase in gain which is possible without producing excessive oscillation; it is often possible that the addition of derivative action to a proportional control will permit a sufficient increase in gain to reduce the offset to such an extent that the use of integral action to eliminate the offset is unnecessary.

Example 5-5

Consider the effect of adding derivative action to proportional control of a two-stage process with both time constants equal to 10 min with an overall gain of 24, and for an increased gain of 48.

(a) *Proportional control.* $T_1 = T_2 = 10$ min, $K = 24$. From Equation (5-7), $T = 2$ min, $\zeta = 0.2$.

For a unit load change before the first stage of the process the response is given by Equation (5-11), which can be written in the more general form

$$c = [K_L/(1 + K)][1 + A\, e^{-\alpha t} \sin(\beta t - \phi)]$$

where $1/(1 + K) = 0.04$,

$$A = 1/(1 - \zeta^2)^{\frac{1}{2}} = 1.02,$$
$$\alpha = \zeta/T = 0.1,$$
$$\beta = (1 - \zeta^2)^{\frac{1}{2}}/T = 0.49 \text{ rad/min} = 28°/\text{min},$$
$$\phi = \tan^{-1}[(1 - \zeta^2)^{\frac{1}{2}}/-\zeta] = 101.5°$$

and the response is thus given by

$$c/K_L = 0.04[1 + 1.02\, e^{-0.1t} \sin(28°t - 101.5°)]$$

(b) *Proportional-derivative control.* $T_1 = T_2 = 10$ min, $K = 24$. Let $T_d = 0.25$ min; from Equation (5-27), $T = 2$ min, $\zeta = 0.26$.

The response to a unit load change is given by the above equation but the constants are changed owing to the different value of ζ, i.e.

$$A = 1.04$$
$$\alpha = 0.13$$
$$\beta = 0.48 \text{ rad/min} = 27.7°/\text{min}$$
$$\phi = 106.5°$$

whence

$$c/K_L = 0.04[1 + 1.04\, e^{-0.13t} \sin(27.7°t - 106.5°)]$$

The response equations are plotted in Figure 5-20, along with those for larger derivative times of 1 and 2 min (corresponding to values of ζ of 0.44 and 0.68 respectively). It will be seen that all the curves stabilize at the same value of 0.04 since the same overall gain is used throughout.

(c) *Proportional-derivative control with increased gain.* Taking case (b) above, $T_1 = T_2 = 10$ min, $T_d = 0.25$, $K = 48$. From Equation (5–27), $T = 1.43$ min, $\zeta = 0.23$.

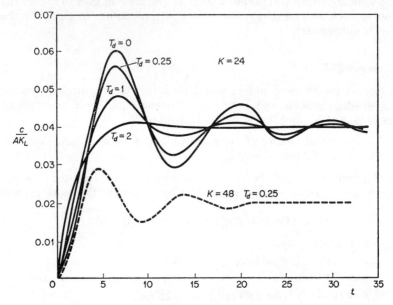

FIGURE 5–20. Effect of adjustment of derivative time with proportional-derivative control

The constants of Equation (5–11) are now given by

$$1/(1 + K) = 0.02$$
$$A = 1.03$$
$$\alpha = 0.16$$
$$\beta = 0.68 \text{ rad/min} = 39°/\text{min}$$
$$\phi = 103.2°$$

whence

$$c/K_L = 0.02[1 + 1.03 \, e^{-0.16t} \sin (39°t - 103.2°)]$$

This response is also plotted in Figure 5–20, and shows the great improvement resulting from the increased gain which has been made possible by the addition of the derivative action.

Proportional-integral-derivative Control

The theoretical transfer function for the ideal three-term controller is

$$G_c = K_c(1 + 1/T_i s + T_d s) \quad \text{or} \quad K_c(T_i T_d s^2 + T_i s + 1)/T_i s$$

The presence of the integral component raises the order of the system response equation by one degree; thus a first-order process will yield a second-order system equation. The nature of the overall response can be obtained by direct reasoning; the integral action will eliminate the offset, and the derivative action will increase the stability and speed of response and permit the use of a higher proportional gain and/or a lower integral time.

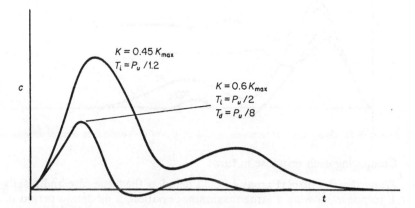

FIGURE 5–21. Effect of addition of derivative action to proportional-integral control (Ziegler-Nichols settings)

The theoretical response becomes more difficult to evaluate as the order of the equation is increased. Figure 5–21 shows the effect of adding derivative action to proportional-integral control of a four-stage process and the results are basically as noted above, the derivative action permitting an increased gain and reduced integral time and reducing the error integral by a factor of approximately four.

It should be noted that most pneumatic controllers employing three-term action do not develop the ideal transfer function, and in adjusting the three parameters, K_c, T_i, and T_d, attention must be given to the actual transfer function developed by the controller. This departure from the ideal equation is due to interaction between the integral and derivative function generators, and usually leads to a limitation of the amount of derivative action which can be developed and to an increase in the controller gain, determined by the ratio of the derivative and integral times. This effect and its consequences are discussed in Chapter 8.

Comparison of Responses of the Continuous Control Actions

To summarize the conclusions drawn from the preceding discussions of the continuous control actions, the effectiveness of the various control modes is illustrated in Figure 5–22, where the responses of a two-stage process with equal time constants to a unit step change in load are compared for the four control actions: proportional (P), proportional-integral (PI), proportional-derivative (PD), and proportional-integral-derivative (PID), each being adjusted to give the same degree of damping.

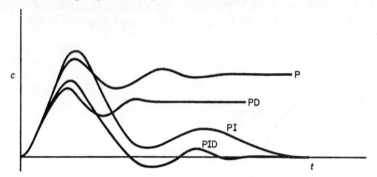

FIGURE 5–22. Comparison of response to load change of the continuous control actions

Comparing each response in turn:

Proportional control (P) may be considered as the basic case and results in a response showing a large maximum deviation, a moderate period of oscillation and the maximum offset.

Proportional-integral (PI) control shows no offset but the elimination of the offset is at the expense of a larger maximum deviation and a longer period of oscillation, due to the destabilizing effect of the integral action which necessitates a reduction of the proportional gain.

Proportional-derivative (PD) control brings the system to the final steady-state value in the shortest time with the smallest maximum deviation. The residual offset is smaller than for proportional control alone. These effects are due to the additional stability provided by the derivative action, which generally permits a considerable increase in the proportional gain.

Proportional-integral-derivative (PID) control is essentially a compromise between the advantages and disadvantages of PI control and the advantages of PD control. There is no offset owing to the integral action, and the stabilizing effect of the derivative allows the gain to be increased, so reducing the maximum deviation and increasing the speed of response compared to P and PI controls. The destabilizing effect of the integral action, however, will not permit such a large increase in gain as with PD control, thus the maximum deviation and recovery time are not quite as good as with PD control.

Integral control used alone has been omitted from these comparisons as this mode of control is best suited for processes with little or no capacitance. The performance of integral control with a process of the type considered above would be relatively poor in comparison (large maximum deviation and very slow recovery) and would not be truly representative of integral control when suitably applied.

It is important to appreciate that the above comments are generalized and exceptions may occur in some systems, but the basic characteristics of elimination of offset by integral action and an increased speed of response from derivative action are fundamental.

The question of which type or combination of actions to use in a particular application is not one which can generally be given a definite answer. Ideally, the simplest controller which will give adequate control is the essential requirement, but unfortunately it is not often possible to state in advance which will be the simplest controller for a given application. This can usually be done only if the application is relatively simple, or if a considerable background of information on similar applications is available.

The importance of designing the control system to suit the characteristics of the process must not be overlooked, but it must also be borne in mind that the selection of the controller must also depend on the operating requirements of the process and the tolerances permitted in the performance, e.g. the maximum offset, the minimum deviation and the maximum recovery time which can be allowed. If offset cannot be tolerated, then integral action must be included in the controller since this is the only way of eliminating offset or reducing it to a negligible amount in the presence of significant changes in load. The need for derivative action will be dictated by the maximum deviation and/or length of recovery time. If a small offset is not critical to the operation, then it may be possible to omit the integral action and the use of derivative will depend on the other factors and also on whether a small enough offset can be obtained from proportional control alone without the increased gain usually available with the added derivative action. It will be appreciated that some of these considerations may be relatively imponderable at the design stage and may be only determinable when the process is put into operation.

Effect of Time Delay on System Response

Many process systems exhibit a time delay or dead time between elements of the control loop, and systems with distributed parameters or a large number of time constant elements in series can often be approximated by a time delay with one or more time-constant elements. It is therefore necessary to consider the effect of the inclusion of a time delay into a process loop on the response to an input disturbance. Frequency-response analysis shows that such delays have a destabilizing effect on the system greater than that of a time constant of equal magnitude. However, the transient response in such cases is not easily obtained accurately, since the transfer function of the time delay is the exponential exp $(-Ls)$, where L is the length of the delay period. This introduces an exponential term into the system equation which, if expanded into the usual infinite series, provides an infinite number of roots for the characteristic equation. An approximate response can be obtained by using only the first few terms from the series expansion, but other approximations to the exponential term are probably easier to handle. For example, the Padé approximations are derived as follows:

$$\exp(-Ls) = \frac{\exp(-\tfrac{1}{2}Ls)}{\exp(+\tfrac{1}{2}Ls)}$$

$$= \frac{1 - \tfrac{1}{2}Ls + (\tfrac{1}{2}Ls)^2/(2!) - \ldots}{1 + \tfrac{1}{2}Ls + (\tfrac{1}{2}Ls)^2/(2!) + \ldots}$$

The first-order approximation is obtained by discarding all terms above the first power of s, giving

$$\exp(-Ls) \approx (1 - \tfrac{1}{2}Ls)/(1 + \tfrac{1}{2}Ls)$$

which is simple to use and reasonably accurate for many purposes. The second and higher order approximations can also be used but with an increasing degree of complexity.

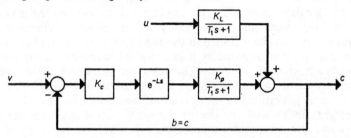

FIGURE 5–23. Proportional control of first-order process with time delay

The first-order approximation will now be used to compare proportional control of a two-stage process (pages 143–7) with that of a single-stage process with a time delay (Figure 5–23). For the latter process, in which the time delay replaces the first time constant of Figure 5–7:

$$N = K_L/(T_1s + 1)$$

$$1 + G = 1 + K_cK_p\, e^{-Ls}/(T_1s + 1)$$

Replacing the exponential term by the approximation,

$$1 + G = 1 + K_cK_p(1 - \tfrac{1}{2}Ls)/[(1 + \tfrac{1}{2}Ls)(T_1s + 1)]$$

whence

$$\frac{c}{u} = \frac{K_L}{T_1s + 1} \frac{(T_1s + 1)(1 + \tfrac{1}{2}Ls)}{(T_1s + 1)(1 + \tfrac{1}{2}Ls) + K(1 - \tfrac{1}{2}Ls)}$$

$$= K_L \frac{(1 + \tfrac{1}{2}Ls)}{\tfrac{1}{2}LT_1s^2 + (\tfrac{1}{2}L + T_1 - \tfrac{1}{2}KL)s + 1 + K}$$

The response to a step change of magnitude A, re-arranged into the usual form, is then given by

$$c(t) = \left(\frac{AK_L}{1 + K}\right) \mathscr{L}^{-1} \frac{(1 + \frac{1}{2}Ls)}{s(T^2s^2 + 2\zeta Ts + 1)} \qquad (5\text{-}28)$$

where $T = [\frac{1}{2}LT_1/(1 + K)]^{\frac{1}{2}}$

and $\quad \zeta = \frac{1}{2}(\frac{1}{2}L + T_1 - \frac{1}{2}KL)/[\frac{1}{2}LT_1(1 + K)]^{\frac{1}{2}}$

Equation (5–28) is of the same form as Equation (5–12) for proportional control of a two-stage process with an intermediate load change, the term $(1 + \frac{1}{2}Ls)$ in the numerator being equivalent to the $(T_1s + 1)$ term in Equation (5–12). The solution for the response to a step change will thus be of the same form as Equation (5–13), but the responses for the two cases will differ owing to the different definitions of T and ζ. In general the effect of the time delay in place of the time constant is to give a lower value of ζ when the delay and the time constant are of the same order of magnitude. Thus the system with the time delay is less stable, and to increase the stability a lower gain must be used with a consequent increase in offset and a longer stabilization time.

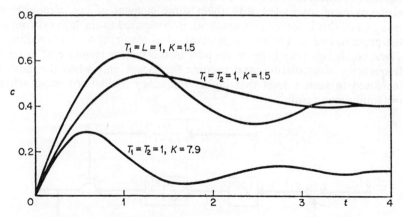

FIGURE 5–24. Response to load change of process with time delay (Example 5–6)

Example 5–6

Compare the damping ratios for proportional control of

(a) a time delay of 1 min followed by a first-order lag of 1 min, and
(b) a two-stage process with both time constants equal to 1 min, both systems having an overall gain of $K = 1.5$,
(c) determine the gain for the two-stage process which would produce the same damping ratio as for the system with the time delay.

(a) $T_1 = L = 1$ min, $K = 1.5$. From Equation (5–28), $\zeta = 0.34$.
(b) $T_1 = T_2 = 1$ min, $K = 1.5$. From Equation (5–7), $\zeta = 0.63$.
(c) Re-arranging Equation (5–7), for $\zeta = 0.34$, $T_1 = T_2 = 1$ min, $K = 7.9$.

The actual response equations for the three cases are

(a) $\quad c/K_L = 0.40[1 + 1.30\, e^{-0.75t} \sin(121°t - 50°)]$
(b) $\quad c/K_L = 0.40[1 + 1.58\, e^{-t} \sin(78°t - 39°)]$
(c) $\quad c/K_L = 0.11[1 + 2.98\, e^{-t} \sin(161°t - 19.5°)]$

These responses are plotted in Figure 5–24, from which it can be seen by comparison of (a) and (c) the serious effect of the time delay on the maximum deviation, offset and recovery time. It will also be noticed that the response does not show any actual time delay since the load change is not held back by the delay; in this instance only the control action is delayed.

Effect of Measuring Lag on System Response

In the systems whose transient responses have been considered up to this point, unity feedback ($H = 1$) has been assumed, i.e. no lag or delay has been considered to exist in the measurement feedback. Most measuring elements have associated with them some element of lag, although in many cases this is fortunately small when compared to the lags existing in the process, and no great error is involved in assuming a unity feedback. These small lags play little or no part in the system response at normal frequencies of oscillation, and become significant only when the gain is increased to such a level that higher frequency oscillations occur when instability may result.

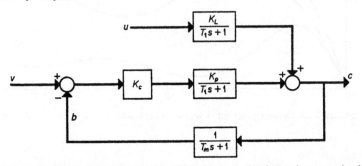

FIGURE 5–25. Proportional control of first-order process with first-order measuring lag

In some cases the measurement lag may be of the same order of magnitude as the process lags, and it is then necessary to consider the effect of a time constant lag in the feedback link, as in Figure 5–25. The initial effect mathematically is to raise the order of the characteristic equation of the system, since the denominator of the closed-loop function now becomes

$(1 + GH)$ where H is of the general form, $K_m/(T_m s + 1)$. For simplicity the measurement gain K_m is usually taken to be unity, which is equivalent to the usual practice of interpreting the reading of the measuring instrument in units of the controlled variable and not as the linear or angular displacement of a pointer. The actual dimensional gain (i.e. the pointer displacement per unit of the controlled variable) is then included in the controller gain K_c, which then has the dimensions of controller output per unit of the controlled variable.

Considering the system of Figure 5–25, where $H = 1/(T_m s + 1)$, the load gain is $N = K_L/(T_1 s + 1)$, and the denominator of the closed loop transfer function is given by

$$1 + GH = 1 + K_c K_p/[(T_1 s + 1)(T_m s + 1)]$$

Hence

$$\frac{c}{u} = \frac{N}{1 + GH}$$

$$= \frac{K_L}{T_1 s + 1} \cdot \frac{(T_1 s + 1)(T_m s + 1)}{(T_1 s + 1)(T_m s + 1) + K}$$

$$= K_L \frac{(T_m s + 1)}{T_1 T_m s^2 + (T_1 + T_m)s + 1 + K} \qquad (5\text{–}29)$$

The response to a step change of magnitude A will thus be given by

$$c(t) = \left(\frac{AK_L}{1 + K}\right) \mathscr{L}^{-1} \frac{T_m s + 1}{s(T^2 s^2 + 2\zeta T s + 1)} \qquad (5\text{–}30)$$

where $T = [T_1 T_m/(1 + K)]^{\frac{1}{2}}$
and $\zeta = \frac{1}{2}(T_1 + T_m)/[T_1 T_m(1 + K)]^{\frac{1}{2}}$.

The response of the first-order process with a first-order measuring lag is thus equivalent to that of a second-order process with an intermediate load change (Equation 5–12) and the response will be given by Equation (5–13), with T_m replacing T_1 in that equation. The effect of the measuring lag depends primarily on the ratio of the measurement time constant to that of the process, and the overshoot of the peak deviation beyond the final value may be several times the latter if T_m is large compared to T_1, as will be illustrated in the next example.

An important point, however, is that measuring lag is particularly insidious in that the effect tends to be obscured by its own existence. In process control the controlled variable is very rarely such as to be observable directly, and the variable which is observed in the display of the controller monitoring element is actually the measured feedback variable. The function of the measuring element is to transduce the controlled variable into a suitable signal to operate the controller, and it is this signal which is monitored by the indicator or recorder usually included in a process controller; the actual controlled variable is not displayed anywhere in the system. The effect of a large measuring lag is inevitably to cause oscillations of large amplitude in the controlled variable, but at the same time the measuring lag will attenuate these oscillations so that the displayed variable shows oscillations of much smaller amplitude. This effect can readily be demonstrated since the relationship between the measured and controlled variables is defined by the feedback transfer function, i.e. $b = Hc$. Thus, from Equation (5–29):

$$\frac{b}{u} = H\left(\frac{c}{u}\right)$$

$$= \frac{NH}{1 + GH}$$

$$= \frac{1}{T_m s + 1}\left[\frac{N}{1 + GH}\right]$$

The response to the step change of magnitude A thus becomes, from Equation (5–30):

$$b(t) = \left(\frac{AK_L}{1 + K}\right)\mathscr{L}^{-1}\frac{1}{s(T^2 s^2 + 2\zeta Ts + 1)} \tag{5–31}$$

where T and ζ have the same values as in Equation (5–29).

Equation (5–31) is now of the same form as Equation (5–8) for a second-order system with a load change before the first capacitance, and the response will thus be of the same form as Equation (5–11). This will differ from the response of the controlled variable (Equation 5–30) in that the peak overshoot is now less than twice the final value. The frequency of oscillation, subsidence ratio and offset are the same for the measured and controlled variables, but the overshoot of the controlled variable will be greater than that of the measured variable and the oscillations of the former will therefore take longer to come within the limits defining the recovery time.

Example 5–7

Compare the responses of the measured and controlled variables to a unit change in load for proportional control of a first-order process with a time constant of 20 s and a first-order measuring lag of (a) 2 s and (b) 20 s, the system being adjusted to give a damping ratio of the response of 0.25.

(a) $T_1 = 20$ s, $T_m = 2$ s, $\zeta = 0.25$. From Equation (5–30), $K = 48$, $T = 0.904$ s/rad.

The responses of both variables are of the general form

$$b, c = [K_L/(1 + K)][1 + A e^{-\alpha t} \sin(\beta t - \phi)]$$

For the measured variable b, Equation (5–11) applies, where

$$A = 1/(1 - \zeta^2)^{\frac{1}{2}} = 1.03$$

$$\alpha = \zeta/T = 0.275$$

$$\beta = (1 - \zeta^2)^{\frac{1}{2}}/T = 1.072 \text{ rad/s} = 61.4°/\text{s}$$

$$\phi = \tan^{-1} [(1 - \zeta^2)^{\frac{1}{2}}/ - \zeta] = 104.4°$$

$$1/(1 + K) = 0.020$$

whence

$$b/K_L = 0.020[1 + 1.03 \text{ e}^{-0.275t} \sin (61.4°t - 104.4°)]$$

For the controlled variable c, Equation (5–13) applies, where

$$A = [1 - 2\zeta T_m/T + (T_m/T)^2]^{\frac{1}{2}}/(1 - \zeta^2)^{\frac{1}{2}}$$
$$= 2.26$$

$$-\phi = \tan^{-1} \left[\frac{(1 - \zeta^2)^{\frac{1}{2}}T_m/T}{1 - \zeta T_m/T} \right] - \tan^{-1} [(1 - \zeta^2)^{\frac{1}{2}}/-\zeta]$$
$$= -26.2°$$

and the other values are the same as above. Hence

$$c/K_L = 0.020[1 + 2.26 \text{ e}^{-0.275t} \sin (61.4°t - 26.2°)]$$

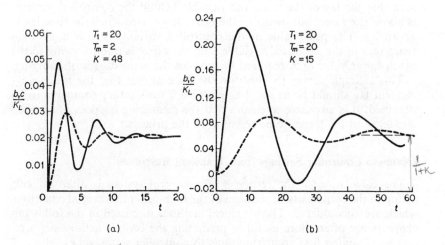

FIGURE 5–26. Response of first-order process with measuring lag

The two responses are plotted in Figure 5–26(a). As can be seen from the response equations, both responses have the same damping factor ($\alpha = 0.275$), frequency ($\beta = 61.4°/\text{s}$), and offset (0.020), but the measured variable lags behind the controlled variable by about 90° or approximately

1.5 s and the peak deviation of the controlled variable is about 70 per cent greater than that of the measured variable.

(b) $T_1 = T_m = 20$ s, $\zeta = 0.25$. From Equation (5–30), $K = 15$, $T = 5.0$ s/rad.

For the measured variable,

$$A = 1.03$$

$$\alpha = 0.05$$

$$\beta = 0.19 \text{ rad/s} = 11°/s$$

$$\phi = 104.4°$$

$$1/(1 + K) = 0.063$$

whence

$$b/K_L = 0.063[1 + 1.03 \, e^{-0.05t} \sin (11°t - 104.4°)]$$

For the controlled variable,

$$A = 4.0$$

$$\phi = 14.4°$$

whence

$$c/K_L = 0.063[1 + 4.0 \, e^{-0.05t} \sin (11°t - 14.4°)]$$

These responses are plotted in Figure 5.26(b), from which it will be seen that the lag of the measured variable behind the controlled variable is about the same, but owing to the much lower frequency the time lag is about 8 s. The peak value of the controlled variable is now more than twice that of the measured variable, and the offset is trebled owing to the much lower value of K required to produce the same degree of damping.

This example serves to illustrate the general rule that the measuring-element lag should be as small as possible if satisfactory control is to be obtained. It is advisable to ensure that the measuring lag does not exceed one-tenth of the largest time constant of the process.

Optimum Controller Settings from Transient Responses

The calculation of the response of a control system is often difficult owing to the complexity of solution of the high-order differential equations which are encountered. The graphical methods discussed in the following chapters are often more useful in predicting the control actions required and in obtaining first approximations to controller settings, i.e. values of proportional gain and integral and derivative times. The analogue computer is also a useful and powerful tool in situations where the nature of the process can be simulated. A difficulty often encountered in practice, however, is that the transfer function relating a pair of input and output variables is unknown and cannot be predicted with any accuracy by presently available theory, although in some cases a useful assessment can be made by experimental testing of the actual plant. A situation which is also not unusual in practice is the need to adjust a controller to suit the requirements of an existing plant where little is known of the dynamics of the process. Since there are three adjustable control parameters in K_c, T_i, and T_d, the setting of the controller to give an optimum performance by trial and error methods is likely to be, at the least, a time-consuming problem. There are, however, two empirical approaches to this problem which provide a systematic attack. Both are based on the transient response to a step change, in one case of the open loop process and in the other of the closed loop. Both methods aim at determining reasonably good initial settings for the controller parameters and provide little or no information about the process dynamics.

The Process Reaction Curve Method

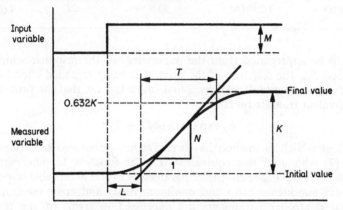

FIGURE 5–27. Process reaction curve

As a result of empirical tests on a wide variety of systems, Ziegler and Nichols [30] proposed a method of basing the controller settings required for reasonably good performance on the step response of the open loop system. In principle a step change in load is introduced into the system through the manipulated variable by manual operation of the control valve, the controller being disconnected from the valve to open the loop whilst the monitoring element of the controller is used to record the process reaction. This will almost invariably be the typical exponential response curve of a multi-capacitance process, and can be characterized by two parameters measured from the response curve. These are the maximum slope N, as a fraction of the total change in the variable per unit time, and an effective time delay L, given by the intercept of the maximum slope line with the initial value of the measured variable (Figure 5–27). The total change in the measured variable K and the maximum slope N are both proportional to the magnitude of the change in the input variable M, in units of the input variable.

The original recommendations which Ziegler and Nichols put forward for the controller settings are purely empirical relationships based on a large number of experimental tests on a variety of processes. These recommendations are listed in Table 5.1.

TABLE 5–1

RECOMMENDED CONTROLLER SETTINGS (ZIEGLER-NICHOLS PROCESS REACTION METHOD)

Control actions	K_c	Proportional band	T_i	T_d
P	$M/(NL)$	$100NL/M$	—	—
PI	$0.9M/(NL)$	$110NL/M$	L	—
PID	$1.2M/(NL)$	$83NL/M$	$2L$	$L/2$

It will be appreciated from the parameters of the response which are measured that the response of the process is being regarded effectively as that of a time delay followed by a first-order lag, i.e. that the process has an equivalent transfer function of

$$K_p \exp(-Ls)/(Ts + 1)$$

The Ziegler-Nichols method has been further elaborated by Cohen and Coon [7] who used this equivalent transfer function to determine the theoretical values of the controller parameters to give acceptable responses, i.e. a 4:1 subsidence ratio and minimum offset and error integral. The required controller parameters are expressed in terms of the transfer function constants K_p, L, and T, all of which can be measured from the process response curve. L is the time delay as already defined, i.e. the intercept of the maximum slope line with the initial value of the variable; K_p is the steady-state gain between the output and the input changes, i.e. K/M; and the time constant T is proportional to the change in the output variable and inversely proportional to the maximum rate of change, i.e. $T = K/N$.

A simple check of the accuracy of approximation to a first-order process and time delay is to measure the time interval for the variable to reach 63.2 per cent of the final value from the end of the time delay. This should be approximately equal to the value of the time constant (T) and agreement within 15 per cent is adequate. If agreement is not within this limit, this may be due to an incorrect slope line or to a non-linearity in the system. The step response should be applied in the opposite direction and the apparent time delay and time constant re-measured. If agreement within 10 per cent is obtained, the average values may be used to determine the controller settings; but if this measure of agreement cannot be obtained, there is a serious non-linearity in the system and the assumption of the equivalent transfer function is not valid.

The Cohen and Coon recommendations for controller settings are given in Table 5–2, where the additional symbol R is used for the 'lag

TABLE 5–2

RECOMMENDED CONTROLLER SETTINGS (COHEN AND COON)

Control actions	K_c	T_t	T_d
P	$\dfrac{M}{NL}\left(1 + \dfrac{R}{3}\right)$	—	—
PI	$\dfrac{M}{NL}\left(\dfrac{10}{9} + \dfrac{R}{12}\right)$	$L\left(\dfrac{30 + 3R}{9 + 20R}\right)$	—
PD	$\dfrac{M}{NL}\left(\dfrac{5}{4} + \dfrac{R}{6}\right)$	—	$L\left(\dfrac{6 - 2R}{22 + 3R}\right)$
PID	$\dfrac{M}{NL}\left(\dfrac{4}{3} + \dfrac{R}{4}\right)$	$L\left(\dfrac{32 + 6R}{13 + 8R}\right)$	$L\left(\dfrac{4}{11 + 2R}\right)$

ratio', i.e. L/T or NL/K. Since the method is based on a theoretical analysis of the assumed equivalent transfer function, it is also possible to relate the properties of the theoretical response to the measured parameters. Values of period of oscillation, offset and maximum deviation are also given by Cohen and Coon in terms of the measured parameters, and these are summarized by Eckman [12], who has also extended the method for application to integral and to two-position control [13].

Loop Tuning Method

The second method of basing controller settings on the transient response of the existing process was also proposed by Ziegler and Nichols [30]. The essential feature of this method is to determine the critical condition of stability of the closed loop under proportional control. The controller is set on proportional action only, integral and derivative actions (if present) are rendered inoperative. Starting with a small value of the gain (wide proportional band), the gain is progressively increased in stages until a small disturbance (usually introduced by a small shift in the desired value) produces a continuous oscillation of the controlled variable with a fixed amplitude. Usually only a few tests of this nature are required, since the subsidence ratio of the first response will indicate whether the gain is near to or far from the maximum value.

The proportional gain at which continuous oscillations are first set up corresponds to the maximum value for limiting stability ($K_{c\,max}$), or, if the controller is calibrated in terms of proportional band, the corresponding value will be the minimum band width, B_{min}. The period of the continuous oscillations is also observed as the 'ultimate' period, P_u. The controller settings suggested by Ziegler and Nichols, based on these experimentally determined critical values, are given in Table 5–3. These values were found

TABLE 5–3

RECOMMENDED CONTROLLER SETTINGS (ZIEGLER-NICHOLS
LOOP TUNING METHOD)

Control actions	K_c	Proportional band	T_i	T_d
P	$0.5K_{c\,max}$	$2B_{min}$	—	—
PI	$0.45K_{c\,max}$	$2.2B_{min}$	$P_u/1.2$	—
PID	$0.6K_{c\,max}$	$1.7B_{min}$	$P_u/2$	$P_u/8$

to give subsidence ratios of about 4:1 and reasonable amounts of overshoot or peak deviation for a variety of processes.

It will be seen that the gain is reduced by 10 per cent when integral action is added, since this makes the system less stable. The suggested gain of $0.45K_{c\,max}$ is about 50–70 per cent of the actual maximum gain for the suggested amount of integral action since the value of $K_{c\,max}$ has been determined by tests with proportional action only. When derivative is added, the phase lead of the action helps to stabilize the system, and a higher gain and lower integral time are recommended. The values given are rather conservative and do not reflect the large improvement that derivative action can make in some cases. The values of $0.6K_{c\,max}$ and $P_u/8$ would in fact be extremely conservative if used without the integral action, i.e. in a PD controller. The resulting system would be much too stable and a still higher gain should be possible.

The loop tuning method has been somewhat modified by Aikman [1] as a result of frequency response studies. In this technique the initial investigation is taken only as far as determining the proportional gain or bandwidth (without integral or derivative actions) to produce a damped oscillation at a suitable subsidence ratio, say 3–4:1. The appropriate value of the gain, K_c, or bandwidth, B, are noted and also the period of the damped oscillation P. The value of K_c or B is obviously that required for proportional control. For the addition of integral or derivative action, which would be indicated by too slow response or too large offset, Aikman suggests that derivative action should always be tried first, on the grounds that this additional action can only improve the quality of the control, whereas integral action has a destabilizing effect. He further suggests that the addition of derivative action should be carried out in two stages, firstly by setting the derivative time T_d equal to one tenth of the operating period (which will be about 25 per cent shorter than the period with proportional control due to the increased speed of response when derivative action is added). This will produce a 30° phase advance on the proportional action, which will be acceptable by almost all processes, and it will generally be possible to increase the gain by amounts up to 30 per cent. A further addition of derivative action to give a phase advance of 60° may then be attempted if the process response still requires improvement. This may be done by setting T_d to one-quarter of the operating period (which will now be about 60 per cent of the period with proportional control). At this stage, however, it is possible that the process will not accept this amount of derivative action and may become unstable without some reduction in gain, but this is a matter of trial and error.

With regard to integral action, Aikman suggests that this should only be added to proportional control when it is essential to eliminate rather than reduce an offset in view of the destabilizing effect of the addition. The integral time should be set equal to the operating period (which will now be some 10–30 per cent longer than the period with proportional control owing to the slower response of the PI combination) and the gain will have to be reduced by up to 40 per cent to restore stability. Aikman's proposals are summarized in Table 5–4. It will be noted that Aikman does not include

TABLE 5–4

RECOMMENDED CONTROLLER SETTINGS (AIKMAN)

Control actions	K_c	Proportional band	T_i	T_d
P	K_c	B	—	—
PD 30°	$1-1.5K_c$	$0.6-1B$	—	$0.07-0.08P$
PD 60°	$1-1.5K_c$	$0.6-1B$	—	$0.15P$
PI	$0.6-1K_c$	$1-1.6B$	$1.1-1.3P$	—

any recommendations for three-term control; this is on account of the differences in control design which call for special consideration when all three control actions are used together. It must be assumed that the Ziegler-Nichols suggestions refer to one particular design of instrument, and that a different design may require some modification of the values suggested.

There are, of course, several variations of the two basic methods discussed above, e.g. most controller manufacturers provide instructions for adjustment of their instruments which are similar in approach, and usually ease of application, to those discussed. It must be emphasized that no general conclusions about the accuracy or relative merits of these empirical methods can be drawn. Such methods are intended to be of general application and some factor of safety must therefore be included; the results are thus always on the conservative side. The only inference which it is reasonable to draw is that these techniques will give reasonable first estimates of the control parameters, and that the values obtained may require further 'trimming' in view of the special characteristics or the operating requirements of the process, or the design of the controller, until the optimum performance is found.

PROBLEMS

5-1 In most controllers there is an internal feedback loop as shown in Figure P5-1. What is the effective transfer function for the controller (p/e) and what is the closed-loop function (c/v)?

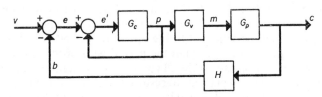

FIGURE P5-1

5-2 Determine the closed-loop transfer function for the multiple loop system of Figure P5-2.

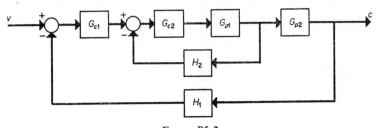

FIGURE P5-2

5-3 For a system consisting of a proportional-integral controller, a first-order linear valve, a first-order process and a first-order measuring lag, draw the block diagram for the system with the appropriate transfer functions, and determine the characteristic equation for the system.

5-4 Show that a level system without self-regulation will oscillate continuously if an integral controller is used. Would the system be more stable if a proportional-integral controller were to be used?

5-5 The major elements of a process control system have the following transfer functions, the time constants being in minutes:

Valve

$$G_v = 3/(0.1s + 1)$$

Process,
$$G_p = 2/[(2s + 1)(10s + 1)]$$
Load changes,
$$N = 3/[(2s + 1)(10s + 1)]$$
Controller,
$$G_c = K_c$$
Measuring element,
$$H = 1$$

Calculate the maximum controller gain and the transient response of a unit step change in load if the gain is set to half of the maximum value.

5–6 The control system of Figure P5–3 contains a three-term controller.

(a) Determine the values of the characteristic time, natural frequency and damping ratio in terms of the control parameters.

(b) If $T_i = T_d = 1$ and $T_1 = 2$, calculate ζ when $K_c = 0.5$ and 2 and sketch the transient response for a unit change in load for these values.

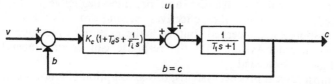

FIGURE P5–3

5–7 A proportional control system has a two-stage process with time constants of 1 and 0.5 min and a measuring element with a time constant of 0.25 min. Determine if the system is stable for values of $K_c = 8$, 11.25, 15.

5–8 A controller, plant, and measuring element have three isolated time constants of 30 s each. If proportional control only is applied and the proportional band is progressively reduced until continuous oscillations are set up by a disturbance, calculate the frequency at which the controlled variable oscillates.

5–9 A process with two major time constants of 2 and 5 min and process gains of 0.5 and 5 is controlled by a proportional controller. Determine the controller gain to produce a damping ratio of 0.3, give numerical values for the frequency, peak overshoot, offset and subsidence ratio, and sketch the response to a unit change in desired value.

5–10 A proportional-derivative controller is used in a system with a first-order process and a first-order measuring lag. Find expressions for T and ζ. If the process time constant is 1 min and the measuring time constant 10 s, find the gain to give $\zeta = 0.5$ for the two cases of $T_d = 0$ and 3 s. Compare the offset and period of oscillation for the two cases.

5–11 Water flows through three tanks in series with hold-up volumes of 0.1, 0.12, and 0.15 m³ (4, 5, 6 ft³) at a rate of 0.175 kg/s (1500 lb/h). The temperature is measured in the third tank and controlled by a proportional-derivative controller supplying 100 W to the first tank per kN/m² (500 Btu/min per lbf/in²) change in output pressure. There is no dynamic lag in measuring element, controller or valve.

(a) Draw a block diagram of the system with appropriate numerical transfer functions in each block.

(b) Determine the overall transfer function relating temperature of the outflow to a change in desired value at the controller.

(c) Find the offset for a unit step change in the inflow temperature when the controller gain is set at 20 kN/m² per °C (3 lbf/in² per °F) and the derivative time is 0.5 min.

Chapter 6: Root Locus Methods

The response required from a process control system after a finite change in load is a damped oscillation similar to the response of a second-order system. It is often convenient to assume that a control is dominated by a single pair of complex roots of the characteristic equation so that the behaviour is then basically second order. For example, as will be illustrated shortly, with a third-order system two of the roots may become complex, thus accounting for an oscillatory under-damped response, whilst the third root remains real and negative and becomes negligible as the gain is increased. The system is then effectively second-order as far as the oscillatory response is concerned, but with the essential difference from the true second-order system that the response can become unstable if a critical gain is exceeded.

For most typical process systems, this assumption of a dominant pair of complex roots giving a dominant second-order type of response is valid, but there are also many other systems which cannot be adequately described by a characteristic equation of less than third order. For such systems it is desirable to be able to identify the influence of adjustment of the system parameters on all the roots of the equation. The root locus method is a graphical procedure for identifying the roots of the characteristic equation, $1 + GH(s) = 0$, as one of the parameters of G, usually K_c, is varied continuously.

Concept of the Root Locus

Consider a three-stage process with proportional control, as in Figure 5–12, for which the open-loop transfer function is given by

$$GH = K/[(T_1s + 1)(T_2s + 1)(T_3s + 1)] \qquad (6\text{–}1)$$

The denominator of the closed-loop transfer function is then

$$1 + GH = 1 + K/[(T_1s + 1)(T_2s + 1)(T_3s + 1)]$$

$$= \frac{(T_1s + 1)(T_2s + 1)(T_3s + 1) + K}{(T_1s + 1)(T_2s + 1)(T_3s + 1)}$$

The transient response to any type of forcing is determined by the roots of the characteristic equation formed by equating the denominator of the closed-loop transfer function to zero, i.e. $1 + GH = 0$.

The characteristic equation in this case reduces to

$$(T_1s + 1)(T_2s + 1)(T_3s + 1) + K = 0$$

or

$$T_1T_2T_3s^3 + (T_1T_2 + T_2T_3 + T_3T_1)s^2$$
$$+ (T_1 + T_2 + T_3) + 1 + K = 0 \quad (6\text{--}2)$$

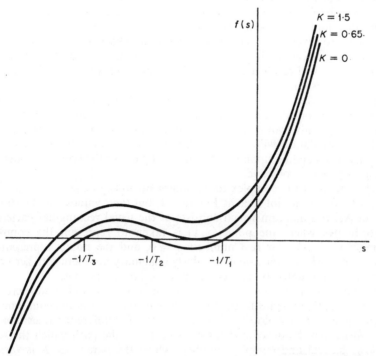

FIGURE 6–1. Plot of $(T_1s + 1)(T_2s + 1)(T_3s + 1) + K$ versus s for increasing values of K

The roots of this equation may be investigated graphically (Figure 6–1) by plotting the cubic equation in s against s for different values of K. When $K = 0$ the roots are $-1/T_1$, $-1/T_2$, and $-1/T_3$, which are of course the intercepts with the zero axis for values of the function. If K is gradually increased, the two roots nearest zero approach each other whilst the third becomes more negative. At a critical value of K the 'dominant' roots become equal at the condition of critical damping of the response. On further increasing K, these two roots cease to be real and are now a pair of conjugate complex numbers, the curve for the cubic function now shows only one intercept with the zero axis as there is only one real root. This latter remains a real negative number but increases numerically as K is increased and thus has a diminishing effect on the response (since $\exp(-\alpha t)$ decreases more and more rapidly with time as α increases). When $T_2 \gg T_3$, the effect of the third root soon becomes negligible and the response is determined essentially by the dominant complex roots; the system is then effectively second order in its response.

Actually in any practical system there are always a number of much smaller lags present, and the characteristic equation will be of a higher order and the plot of the characteristic function will strictly loop up and down to cut the s-axis a number of times. The corresponding additional roots will be far to the left of zero, since they are due to very small time constants, and consequently they have very little effect and can usually be ignored. However, if K is made sufficiently large a pair of complex roots will develop from each pair of real roots existing when $K = 0$. These further pairs will produce additional but very much weaker oscillations in the overall response.

When the first pair of roots becomes complex the overall response becomes oscillatory, all the other roots being real and negative and contributing only transient exponential terms to the overall response. The resulting oscillation will be under-damped or of constant or increasing amplitude depending upon whether the real parts of the complex roots are negative, zero, or positive.

The values of the complex roots cannot obviously be shown on a characteristic function plot such as Figure 6–1, but their values may be shown on an Argand diagram which consists of the usual rectangular cartesian co-ordinates where the point (x, y) is taken to represent the complex number $(x + jy)$. For real numbers $y = 0$, and the points representing real numbers lie on the x-axis; similarly imaginary numbers with zero real parts are represented by points on the y- or imaginary axis.

In the present case the characteristic equation is a cubic and there are three roots; there will thus be three values of s satisfying the equation for each value of K. By plotting these values of s for different values of K, a *root locus* may be developed for each root, i.e. the path which the root follows as real and complex numbers whilst the parameter K is varied. With the cubic function the locus of each root on the Argand diagram is evident, at least so long as they remain real. Starting from the values when $K = 0$ (i.e. $-1/T_1$, $-1/T_2$, $-1/T_3$, where $T_1 > T_2 > T_3$), the two dominant roots ($-1/T_1$ and $-1/T_2$) approach each other and eventually become coincident, i.e. equal, at a critical value of K, say K_1. The third root becomes increasingly negative and so moves away from the origin, since all three roots are real up to the value of $K = K_1$ the loci are on the real axis. There is no simple formula for deriving the roots of a cubic equation, so it is not possible to see immediately where the first two roots go when they become complex. However, as the roots must be conjugate their paths will be reflections of each other in the x-axis, since the imaginary parts must be equal but of opposite sign ($\pm jy$). The response of the system, being third-order, will ultimately exhibit zero damping and then instability as K is increased further. At the condition of zero damping the dominant pair of roots becomes conjugate imaginary, i.e. the real part of the roots becomes zero. Thus as K is increased above K_1, the real parts of the complex roots move towards zero whilst the conjugate imaginary parts increase; typical paths are shown in Figure 6–2. The loci of the two complex roots will ultimately cross the imaginary axis at a value of K which corresponds to zero damping (K_2), and any further increase in K will extend the loci beyond the imaginary axis. The complex roots then take on positive real parts and the system response becomes unstable.

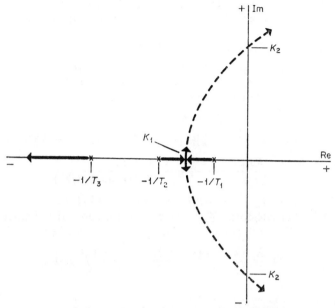

FIGURE 6–2. Partial root locus diagram for $(T_1s + 1)(T_2s + 1)(T_3s + 1) + K = 0$

The root locus diagram has distinct advantages in showing the nature of the system response as the controller gain is changed. As indicated, Figure 6–2 shows two critical values of K; K_1, where two of the roots become equal (critical damping), and K_2, where two of the roots become imaginary (limiting stability).

Plotting the Root Locus Diagram

The time and effort needed to plot the root locus diagram is reduced considerably by use of the rules introduced by Evans [15] which will now be reviewed.

To develop the root locus technique it is convenient to write the open loop transfer function in an alternate form:

$$GH = \frac{K}{C} \frac{(s - z_1)(s - z_2) \ldots (s - z_m)}{(s - p_1)(s - p_2) \ldots (s - p_n)} \tag{6–3}$$

where the numerator and denominator are polynomial functions of s and the constant C arises from the fact that in the usual definition of GH the factors appear as $(Ts + 1)$ and are now being written as $T(s + 1/T)$.

Each value such as z_i in Equation (6–3) is a *zero* of the open loop function, i.e. a value of s for which $GH(s) = 0$. A value such as p_i is a *pole* of the open-loop function and is any value of s for which $GH(s)$ approaches infinity. The factored terms arise naturally from the process time constants and control functions. Thus, for the example already cited (Equation 6–1), written in the alternate form, the open-loop function becomes

$$GH = (K/T_1T_2T_3)\{1/[(s + 1/T_1)(s + 1/T_2)(s + 1/T_3)]\}$$

where $C = T_1T_2T_3$, $p_1 = -1/T_1$, etc. and there are no zeros.

To illustrate a function where the numerator is not unity, the same system may be considered with proportional-derivative control, i.e.

$$G_c = K_c(1 + T_ds)$$

$$GH = [K(1 + T_ds)]/[(T_1s + 1)(T_2s + 1)(T_3s + 1)]$$

$$= \frac{KT_d}{T_1T_2T_3} \cdot \frac{(s + 1/T_d)}{(s + 1/T_1)(s + 1/T_2)(s + 1/T_3)}$$

where $C = T_1T_2T_3/T_d$, $p_1 = -1/T_1$, etc. and $z_1 = -1/T_d$.

Using the alternate form of the open-loop function, the characteristic equation, $1 + GH = 0$, is then written

$$1 + \frac{K}{C} \cdot \frac{(s - z_1)(s - z_2) \ldots (s - z_m)}{(s - p_1)(s - p_2) \ldots (s - p_n)} = 0$$

whence

$$\frac{(s - p_1)(s - p_2) \ldots (s - p_n)}{(s - z_1)(s - z_2) \ldots (s - z_m)} = -\frac{K}{C} \qquad (6\text{–}4)$$

In general the values of s satisfying Equation (6–4) which are of particular interest will be complex numbers. Any complex number, such as $w = x + jy$, can be represented by a vector in the complex plane of the Argand diagram. If P is the point (x, y), then the number $w = x + jy$ is represented by the vector $\overrightarrow{OP}$, of magnitude (or modulus, $|w|$) equal to $(x^2 + y^2)^{\frac{1}{2}}$, and with the direction (or argument, $\angle w$) measured from the positive real axis given by $\tan^{-1}(y/x)$. Further, if two numbers w_1 and w_2 are represented by two such vectors $\overrightarrow{OP_1}$ and $\overrightarrow{OP_2}$, then $\overrightarrow{P_1P_2}$ represents $(w_1 - w_2)$. In Figure 6–3 vectors representing the numbers $(s_a - z_1)$, $(s_a - p_1)$ and $(s_a - p_2)$ are shown.

The rules for combining vectors by multiplication or division are sufficiently well-known to be summarized. If

$$w = \frac{u_1u_2 \ldots}{v_1v_2 \ldots}$$

then

$$|w| = \frac{|u_1||u_2| \ldots}{|v_1||v_2| \ldots}$$

and
$$\angle w = (\angle u_1 + \angle u_2 + \ldots) - (\angle v_1 + \angle v_2 + \ldots)$$

Applying these rules for combination of vectors, Equation (6–4) may be written in vectorial form:

$$\frac{|s - z_1||s - z_2| \ldots |s - z_m|}{|s - p_1||s - p_2| \ldots |s - p_n|} = \frac{K}{C} \tag{6-5}$$

and

$$\sum_1^n \angle p_i - \sum_1^m \angle z_i = (2r + 1)\pi \tag{6-6}$$

where n is the number of poles, m is the number of zeros, and $r = 0$ or any integer. Note that the magnitude of the right hand side is K/C and the argument is $(2r + 1)\pi$, or an angle of 180° to $x+$, i.e. the argument of a vector on the negative real axis.

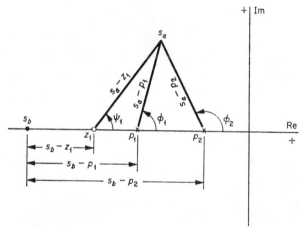

FIGURE 6–3. Use of the magnitude and angle criteria

The general criterion for a point to lie on a root locus is given by the angle criterion of Equation (6–6). The values of the corresponding gain can then be calculated from the magnitude criterion of Equation (6–5).

Rules for Sketching Root Loci

The following rules for sketching root loci may be derived in the main from the preceding discussion but are presented here without proof.

(1) All open loop poles and zeros are first plotted in the complex s-plane. It is conventional to indicate poles by X and zeros by 0.

(2) The loci are symmetrical about the real axis since all complex roots in a real system must appear as conjugate pairs.

(3) The number of loci or branches is equal to the number of poles n.

(4) Each locus begins (for $K = 0$) at a pole, and each zero is at the end of a locus (for $K = \infty$). In the case of a qth order pole (i.e. a term such as $(s - p_a)^q$ in the open loop function), q loci will emerge; for a qth order zero, q loci will terminate.

(5) A locus will lie on the real axis when the sum of the number of poles (including that from which the locus starts) and zeros to the right on the real axis is odd. Only real poles and zeros need be considered since complex poles and zeros must occur in pairs; qth-order poles and zeros must be counted q times.

(6) The other loci or branches (equal in number to $n - m$) which do not end at a zero will go to infinity either along the real axis (i.e. to $-\infty$), or will be asymptotic to straight lines which intersect the real axis at a 'centre of gravity' of the n poles and m zeros. The position of this point at which the asymptotes intersect the real axis is given by

$$s = [\sum_1^n p_i - \sum_1^m z_i]/(n - m) \qquad (6\text{--}7)$$

The asymptotes make angles of $[(2k + 1)/(n - m)]\pi$ with the real axis, where $k = 0, 1, 2 \ldots (n - m - 1)$, and are equally spaced to each other at angles of $2\pi/(n - m)$.

(7) The points at which two loci leave or join the real axis is given by

$$\sum_1^m [1/(s - z_i)] = \sum_1^n [1/(s - p_i)] \qquad (6\text{--}8)$$

The loci will leave or enter the real axis at angles of $\pm\pi/2$.

(8) The angles at which q loci emerge from a qth-order pole are given by

$$\phi = (1/q)[(2k + 1)\pi + \sum_1^m \angle(p_a - z_i) - \sum_1^n \angle(p_a - p_i)] \qquad (6\text{--}9)$$

where $k = 0, 1, 2, \ldots, (q - 1)$, and p_a is the particular pole of order q. Each of the loci which do not approach the asymptotes will terminate at one of the m zeros, approaching the particular zero at an angle given by

$$\psi = (1/v)[(2k + 1)\pi + \sum_1^n \angle(z_b - p_i) - \sum_1^m \angle(z_b - z_i)] \qquad (6\text{--}10)$$

where $k = 0, 1, 2, \ldots, (v - 1)$, and z_b is the particular zero of order v. For simple poles or zeros on the real axis, the angle of departure or entry will be 0 or π.

An analogy from potential theory which is useful in plotting root loci is that the loci correspond to the paths taken by particles with a positive charge moving in an electrostatic field established by positive charges at the poles and negative charges at the zeros. A locus is thus 'repelled' by the poles but 'attracted' by the zeros.

Most open-loop transfer functions in single-loop process-control systems will exhibit only real poles; complex poles can only arise when the loop contains a true second-order element, i.e. a process element of the mass-spring-damping type with a natural frequency and damping ratio, such as is found in many measuring instruments. Even in such cases the complex poles are located so far from the dominant poles that their presence can often be ignored.

Use of the Angle and Magnitude Criterion

The rules given in the previous section are a rapid guide to the general location of the root loci but it is usually necessary to confirm the position of at least a few points on one or more loci by use of the angle criterion of Equation (6–6). A trial and error procedure must be used by selecting a trial point and drawing the vectors to this point from each of the poles and zeros. If the trial point lies on the root locus, the angles of these vectors measured anti-clockwise from the real axis when tested by Equation (6–6) should yield an odd multiple of π radians, i.e. 180°, 540°, etc. For example, if the point s_a of Figure 6–3 is on the root locus, then

$$\phi_1 + \phi_2 - \psi_1 = 180°, \; 540° \text{ etc.}$$

where ϕ is the angle of a vector from a pole and ψ the angle from a zero. If the criterion is not satisfied, the trial point is in error and must be moved to a new location; moving it horizontally to the left increases the sum of the angles. The procedure is repeated until a sufficiently close approximation to an odd multiple of π is attained.

The gain corresponding to any point on the root loci may be found by applying the magnitude criterion of Equation (6–5), by measuring the magnitudes (lengths) of the vectors and substituting in the equation which is re-arranged to give the value of K, e.g. for the point s_a in Figure 6–3,

$$K = [C|s_a - p_1||s_a - p_2|]/(|s_a - z_1|)$$

The points at which the loci cross the imaginary axis can be obtained by application of the Routh test (pages 152–6) to the characteristic equation to find the value of K to make the elements of the nth row zero.

To find a point on the root locus corresponding to a given value of K also requires a trial and error procedure, i.e. a point must be selected and the magnitude of the vectors measured to determine the value of K by Equation (6–5) for checking against the given value. Alternatively, if the point lies on a locus on the real axis, such as s_b in Figure 6–3, the point can be found precisely for a given value of K by use of the characteristic equation and the algebraic theorem for the sum of the zeros of a polynomial.

It should be emphasized again that the root locus diagram is symmetrical about the real axis (due to complex roots occurring only in conjugate pairs). Thus it is necessary to locate points on the loci in one half-plane only, the loci in the other half-plane can then be drawn from symmetry.

The plotting technique for root locus diagrams will now be illustrated by two examples.

Example 6–1

Plot the root locus diagram for proportional control of a three-stage system with time constants of 1, 0.5, and 0.25 min and unit stage-gains, and with unity feedback.

The open-loop transfer function is given by

$$GH = K_c/[(s + 1)(0.5s + 1)(0.25s + 1)]$$
$$= K_c/[(0.5)(0.25)(s + 1)(s + 2)(s + 4)]$$
$$= 8K_c/[(s + 1)(s + 2)(s + 4)]$$

whence $C = \frac{1}{8}$, $p_1 = -1$, $p_2 = -2$, $p_3 = -4$.

The open-loop poles are first plotted by an X at -1, -2, and -4 on the real axis. Since there are three poles, there will be three loci, one starting from each pole, but since there are no open-loop zeros, all three loci will end at infinity. A portion of the locus is on the real axis to the left of -4, and also between -1 and -2, since in each case there is an odd number of poles to the right (3, and 1, respectively).

Since $n - m = 3$, there will be three asymptotes and the centre of gravity (Equation 6–7) is at

$$s = (-1 - 2 - 4)/3 = -2.33$$

The angles which these asymptotes make with the real axis will be $\pi/3$, $3\pi/3$, and $5\pi/3$, i.e. 60°, 180°, and 300°. For confirmation, the asymptotes make angles of

$$2\pi/(n - m) = 360°/3 = 120°$$

to each other. One asymptote is obviously the real axis to the left. At this stage the general shape of the diagram is emerging (Figure 6–4a); the real axis to the left of -4 is an asymptote, and one branch of the locus emerges from the pole at -4 and lies on this axis; this branch will then continue to infinity on the real axis. Two loci emerge from the poles at -1 and -2, and the real axis between the poles is part of the loci; the two loci must then approach along the real axis and meet at some point. Since two other asymptotes have been identified, the two loci meeting on the real axis must break away and eventually follow the asymptotes to infinity. If the breakaway point and the crossings of the imaginary axis are now determined, a reasonably accurate sketch of the root loci can be made.

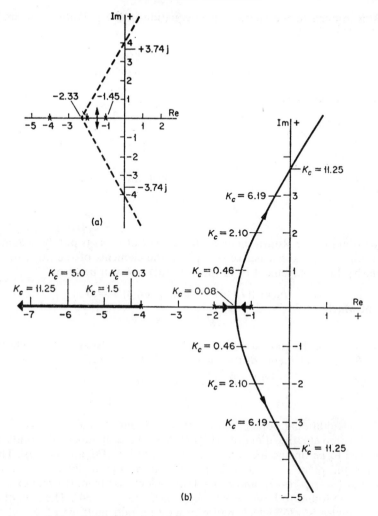

FIGURE 6–4. Construction of root locus diagram for Example 6–1

The breakaway point is determined by Equation (6–8), which, since there are no zeros, is written for the three poles:

$$1/(s + 1) + 1/(s + 2) + 1/(s + 4) = 0$$

which gives $s = -3.21$ or -1.45. Since the point must lie between -1 and -2, $s = -1.45$.

To find the points at which the loci cross the imaginary axis, the Routh test is applied to the characteristic equation in the polynomial form, i.e. $1 + GH = 0$, which gives

$$(s + 1)(s + 2)(s + 4) + 8K_c = 0$$

or $$s^3 + 7s^2 + 14s + 8 + 8K_c = 0$$

The Routh array is

Row		
1	1	14
2	7	$8 + 8K_c$
3	b_1	

According to the Routh theorem, if one pair of roots is purely imaginary and all others have negative real parts, the elements of the nth row will vanish; thus the element b_1 in the nth (3rd) row will be zero:

$$b_1 = [(7)(14) - (1)(8 + 8K_c)]/7 = 0$$

from which

$$K_c = 11.25$$

The values of the roots are given by the quadratic $As^2 + B = 0$, where A and B are the elements of the $(n - 1)$th (2nd) row, i.e.

$$7s^2 + 8 + 8(11.25) = 0$$

$$s = \pm 3.74j$$

These additional points are indicated on Figure 6-4(a).

To complete the diagram (Figure 6-4(b)), a number of points are checked to be on the locus by the angle criterion (Equation (6-6)). Thus, taking the point $s = -1 + 2j$ and measuring the angles of the vectors from p_1 (-1), p_2 (-2), and p_3 (-4) anti-clockwise from the real axis, the angles are found to be $\phi_1 = 90°$, $\phi_2 = 63°$, and $\phi_3 = 34°$. The sum of the three angles is $186°$ which is not exactly an odd multiple of π and this point is not then on the root locus. Moving the point horizontally to the right will reduce the sum of the angles and $s = -0.9 + 2j$ is taken for a second trial. The angles are now $89°$, $61°$, and $33°$ and the sum is $181°$ which is sufficiently close to π for most purposes. The point $-0.9 + 2j$ is thus taken to be on the locus, the conjugate point $-0.9 - 2j$ will obviously be on the locus in the lower half-plane. A sufficient number of additional points can be determined in a like manner and a smooth curve then drawn; that in the lower half-plane is simply the mirror image in the real axis.

The gain at various points along the loci is established by use of the magnitude criterion (Equation (6-5)). For example, taking the point $s = -0.9 + 2j$, the vector lengths from the three poles are measured (it is important to measure these lengths in units which are consistent with those used on the axes of the diagram) with the results

$$|s - p_1| = 2.00; \qquad |s - p_2| = 2.28; \qquad |s - p_3| = 3.69$$

Using the re-arranged Equation (6–5),

$$K_c = (\tfrac{1}{8})(2.00)(2.28)(3.69) = 2.10$$

Additional values at other points are determined as required to produce the full diagram of Figure 6–4(b).

Example 6–2

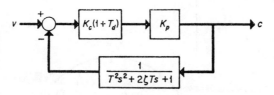

FIGURE 6–5. System for Example 6–2

To illustrate the effect of a zero and also that of complex poles the system of Figure 6–5 will be considered. This is a process with a negligible lag and an under-damped second-order measuring element with a proportional-derivative controller. This could be, for example, a flow control system with negligible lag and a measuring device such as a mercury manometer across an orifice plate. Assuming that $K_p = 1$ and that the transfer function of the measuring element is $1/(0.5s^2 + s + 1)$, plot the root locus diagram for various values of K_c when $T_d = 0.5$.

The open-loop transfer function is

$$GH = [K_c(1 + 0.5s)(1)]/(0.5s^2 + s + 1)$$
$$= [K_c(s + 2)]/(s^2 + 2s + 2)$$

The factors of the denominator (by the quadratic formula) are $(s + 1 \pm \text{j})$. Hence the open-loop function becomes

$$GH = [K_c(s + 2)]/[(s + 1 + \text{j})(s + 1 - \text{j})]$$

There are thus two poles which are complex at $-1 + \text{j}$ and $-1 - \text{j}$, and a zero at -2. A branch of the locus will emerge from each pole; part of the locus will be on the real axis to the left of the zero since there is only the one zero on the axis to the right; $(n - m) = 1$, and hence there will be only one asymptote at an angle of $180°$ to $x+$, i.e. on the real axis. The two branches from the poles must then join on the real axis to the left of -2, one branch must then go to the zero at -2 and the other to $-\infty$ on the real axis.

The point at which the two branches join the real axis is given by Equation (6–8):

$$1/(s + 2) = 1/(s + 1 + j) + 1/(s + 1 - j)$$

from which $s = -3.414$ or -0.586. Since the locus is to the left of the zero at -2, the point of entry will be -3.414.

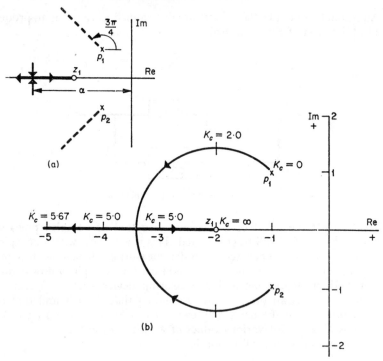

(a)

(b)

FIGURE 6–6. Root locus diagram for Example 6–2

The angles of departure from the poles of the two branches is given by Equation (6–6) in which $q = 1$ and $k = 0$:

$$(1)(\pi + \pi/4 - \pi/2) = 3\pi/4 = 135°$$

The position at this stage is shown in Figure 6–6(a); this locus is rather more difficult to plot than the previous example, since the asymptote on the real axis does not provide any guide to the plotting of the locus between the pole and the entry point to the real axis. It is therefore necessary to find a number of points by the trial and error procedure using the angle criterion, although the astute reader will notice that the loci is part of a circle centred on the zero at -2. The gains along the loci are found from the magnitude criterion as in the previous example, and the complete locus is shown in Figure 6–6(b).

It can be seen from the diagram that the control system never becomes unstable and, in fact, the responses become more damped as K_c is increased as is shown by the increasing negative real parts of the roots.

Exothermic Reactions

The root locus diagram is very revealing in the case of systems with a positive pole such as is found in temperature control of an exothermic reaction, a situation which is not uncommon in process control. Such exothermic systems have a heat generating source, usually due to chemical reaction in which the heat liberated is a function of temperature. The energy balance for such a system would be

$$q_i + rV(-\Delta H) - q_o = Mc_p(d\theta/dt)$$

where q_i and q_o are respectively heat supplied or removed by external agency, i.e. heating or cooling of the reaction vessel, r is the rate of reaction in moles converted per unit time per unit volume, V and M are the volume and mass of the reacting material, $-\Delta H$ is the heat of reaction per mole converted, and c_p is the specific heat capacity of the reaction mixture. Over moderate ranges of temperature, variations may be neglected in the heat capacity of the mixture, the heat of reaction, and the mass and volume of the mixture, which leaves the temperature as a function of time, the net heat transfer $(q_i - q_o)$ and the rate of reaction. The latter is generally a function of temperature and the concentration of the reactants, products and catalyst. The temperature relationship is of the form $\exp(-E/R\theta)$ which would be very difficult to 'linearize'. In many cases of continuous flow reactors, however, where the concentrations are kept constant by continuous addition of reactants and removal of products, the rate of reaction is limited by the rate of mass transfer from a fluid to a solid reactant or catalyst, and the rate may then be regarded as roughly proportional to temperature, i.e. $r \approx K_e\theta$. K_e is positive, thus showing an increase in the reaction rate with temperature and, although not a true constant, it will not vary greatly over moderate temperature ranges.

Substituting in the energy balance equation and transforming with the usual initial conditions:

$$(q_i - q_o) + K_eV(-\Delta H)\theta = Mc_ps\theta$$

from which a transfer function can be written:

$$G_p = \theta/(q_i - q_o)$$

$$= \frac{1/[K_eV(-\Delta H)]}{Mc_ps/[K_eV(-\Delta H)] - 1}$$

$$= K_p/(T_es - 1)$$

where $K_p = 1/[K_eV(-\Delta H)]$, and $T_e = Mc_p/[K_eV(-\Delta H)]$.

If ΔH is negative (which is the case for an exothermic reaction) both K_p and T_e are positive. For an endothermic reaction, ΔH is positive and G_p then becomes $K_p/(Ts + 1)$, which shows the self-regulation or inherent stability of the normal first-order system.

If the manipulated variable is a heating medium then q_i is the input variable for the process and $G_p = \theta/q_i(s)$; if, on the other hand, a coolant is manipulated, the input variable is q_o and K_p is then negative, i.e.

$$G_p = \theta/q_o$$
$$= -K_p/(T_e s - 1)$$
$$= K_p/(1 - T_e s)$$

In the latter case a reverse acting controller (with a negative value of K_c) would be used to make the overall gain positive.

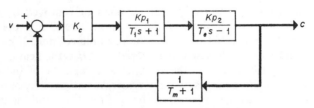

FIGURE 6–7. Block diagram for system with exothermal reactor

To plot the root locus diagram for a system of this type in a practical form, it is necessary to add a heat transfer lag in the heating or cooling coils and a temperature measuring lag, both of which would normally be present in a practical application. The system is as shown in Figure 6–7 and the open-loop transfer function is then

$$GH = K/[(T_e s - 1)(T_1 s + 1)(T_m s + 1)]$$

and in the alternative form this becomes

$$GH = [K/(T_e T_1 T_m)]\{1/[(s - 1/T_e)(s + 1/T_1)(s + 1/T_m)]\}$$

where $p_1 = +1/T_e$, $p_2 = -1/T_1$, and $p_3 = -1/T_m$. The system thus includes a positive pole at $+1/T_e$.

The root locus diagram for this system is plotted in Figure 6–8 for different values of the time constants. Considering Figure 6–8(a), for which $T_e = 5$, $T_1 = 2$, and $T_m = 1$, the general shape of the diagram is that of the three-stage process plotted in Figure 6–4, but the presence of the positive pole moves the loci of the complex roots to the right. A positive real root will start at the positive pole (for $K_c = 0$) and this will get smaller as the gain is increased until it becomes zero and then negative. The system is thus unstable at zero gain and this is, of course, the common property of an exothermal reaction in that in the absence of control the reaction will either become faster and faster as the temperature rises or will die away if the temperature falls. Increasing the gain will, however, stabilize the system as the real root from the positive pole becomes zero and negative. A further increase in gain follows the usual pattern; the roots from the positive and the nearest negative pole approach and meet so bringing the system to critical damping, the two roots then diverge along the two branches and become complex, so producing under-damping. Finally, the two complex roots cross the imaginary axis and take on positive real parts, and the system is again unstable. There is thus a limited range of gain, *below* and *above* which the system is unstable; the system of Figure 6–8(a) thus exhibits *conditional stability*.

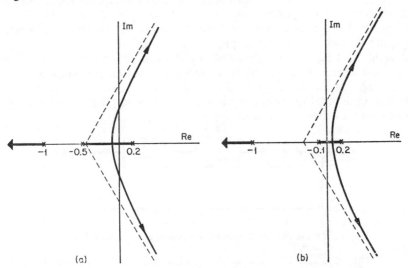

FIGURE 6–8. Root locus diagram for exothermal system: (a) conditionally stable, (b) unstable

If either of the poles nearest the origin are moved to the right, either by increasing the heat transfer time constant (T_1), or by decreasing the reaction time constant (T_e), this will effectively displace the breakaway point of the two complex root loci also to the right. This will reduce the range of gains in which the system is stable and, if the breakaway point is displaced to the right of the origin, the loci then lie completely on the positive side of the imaginary axis. This is shown in Figure 6–8(b), where the heat transfer time-constant has been increased to 10. The root from the positive pole now never reaches the origin but breaks away along one of the complex loci as the gain is increased; thus either one root is positive or the complex roots have positive real parts. The system is hence unstable at any value of controller gain, and in such a case proportional control alone will be useless. As will be shown in a later example, the addition of derivative action with a sufficiently large value of T_d will restore stability and so permit the system to be controlled.

Root Locus for a Time Delay

An advantage of the root locus technique is the ease with which the dynamic behaviour of a closed-loop system can be inferred from the open-loop poles and zeros, without the labour of solving polynomial equations of higher than second-order. The inference must necessarily depend upon the open-loop transfer function being readily expressed in a factored form, i.e. with the poles and zeros readily identifiable. A process system containing an appreciable time delay can be described precisely only by including in the transfer function the transcendental function, $\exp(-Ls)$, which effectively introduces an infinite number of roots into the characteristic equation for each value of the gain. A true root locus diagram would thus contain an infinite number of branches and would require considerable effort to prepare. A graphical procedure has been devised by Chu [6] and described by Truxall [25] based on a rather different technique, that of phase angle loci. In many practical cases, however, the true root-locus plot will contain only one branch relatively close to the origin, and the remaining branches will be so far removed that the roots on these branches can be ignored.

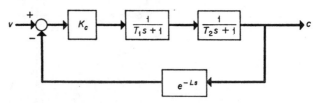

FIGURE 6–9. Block diagram for system of Example 6–3

It is thus possible to construct a root locus diagram by expanding the exponential function and retaining only a finite number of terms, or by use of the Padé approximation used in the previous discussion of the transient response when a time delay is present. Using the first-order Padé approximation it is first necessary to convert the function into a form suitable for use in the root locus method, i.e. so that the poles and zeros can be identified, as follows:

$$\exp(-Ls) \approx (1 - 0.5Ls)/(1 + 0.5Ls)$$
$$= (2/L - s)/(2/L + s)$$
$$= -(s - 2/L)(s + 2/L)$$

Adding this function to a two-stage system (as shown in Figure 6–9), the open-loop function is given by

$$GH = -K(s - 2/L)/[(s + 2/L)(T_1 s + 1)(T_2 s + 1)]$$

or, in the alternative form

$$GH = -(K/C)(s - z_1)/[(s - p_1)(s - p_2)(s - p_3)]$$

where $z_1 = +2/L$, $p_1 = -2/L$, $p_2 = -1/T_1$, $p_3 = -1/T_2$, and $C = T_1 T_2$.

Owing to the minus sign in the open-loop transfer function (which arises from the re-arrangement of the approximation of the time delay) a significant difference is now introduced. Equation (6–4) is fundamentally altered since the characteristic equation, $1 + GH = 0$, now becomes

$$1 - (K/C)(s - z_1)/[(s - p_1)(s - p_2)(s - p_3)] = 0$$

from which

$$(s - p_1)(s - p_2)(s - p_3)/(s - z_1) = +K/C \qquad (6\text{–}11)$$

The system can, in fact, be visualized as one with a *positive* feedback (since the negative feedback is itself now negative). If the feedback function is made positive (by a simple change of sign) and the closed-loop function for positive feedback, i.e. $G/(1 - GH)$ is used, the same characteristic equation as above is obtained.

Since K/C in Equation (6–11) is now positive instead of negative, the vector representation of this quantity lies on the real axis but now in the direction of $x+$, i.e. the vector angle is an *even* multiple of π, such as $2k\pi$ where $k = 0$ or any integer. This then requires that the angle criteria of Equations (6–6), (6–9), and (6–10), and the angles of the asymptotes of the loci must be modified to *even* multiples of π when plotting the root loci. The other rules for plotting are not affected, apart from that dealing with loci or branches on the real axis; this must be amended to an *even* number of poles and zeros, 'even' being interpreted as $0, 2, 4 \ldots$, so that the real axis to the right of all poles and zeros becomes part of the locus. The magnitude criterion is obviously unchanged.

Example 6–3

Construct the root locus diagram for proportional control of a system with time constants of 0.5 and 1 min and unit gain, with a time delay in the measurement feedback of 0.2 min, using the Padé approximation.

$$GH = \frac{K_c(1)}{(0.5s + 1)(s + 1)} \frac{(1 - 0.1s)}{(1 + 0.1s)}$$

$$= \frac{-2K_c(s - 10)}{(s + 10)(s + 2)(s + 1)}$$

Hence $C = \frac{1}{2}$, $z_1 = +10$, $p_1 = -10$, $p_2 = -2$, $p_3 = -1$.

Construction of the root locus diagram proceeds as follows: the three poles and single zero are plotted at the above values, and since there are three poles the locus will have three branches, one must end at the zero and the other two will end at infinity.

The number of poles and zeros is even to the right of z_1 (being nil), between p_2 and p_1 (being two, namely p_1 and z_1), and to the left of p_3 (all four poles and zero being to the right). Thus these sections of the real axis form part of the loci, and the two branches which do not end at z_1 must therefore go to $+\infty$ and $-\infty$ on the real axis. This is confirmed by $(n - m)$ being equal to 2, giving two asymptotes with angles to the real axis of $2k\pi/(n - m)$, i.e. 0 and π for $k = 0$ and 1.

The roots emerging from the poles at -1 and -2 must approach until they meet, since the real axis between the poles is part of the locus. They must then break away, at angles of $\frac{1}{2}\pi$, into the upper and lower half-planes, and will eventually cross the imaginary axis to rejoin the locus on the real axis to the right of the zero at $+10$. One locus will then go to the zero and the other to $+\infty$ on the real axis. The abscissae of the breakaway and re-entry points are given by Equation (6–8), i.e.

$$1/(s - 10) = 1/(s + 10) + 1/(s + 2) + 1/(s + 1)$$

Since this equation is a cubic the two required solutions are found by trial and error of appropriate values of s. The breakaway point is between -1 and -2 and by trial and error a value of $s = -1.475$ is found to give reasonable agreement between the two sides of the equation. Similarly the re-entry point is to the right of $+10$, and a value of $+17$ can be found to satisfy the equation.

To find the crossing points on the imaginary axis the Routh theorem is used. The characteristic equation gives

$$(s + 10)(s + 2)(s + 1) - 2K_c(s - 10) = 0$$

or $\qquad s^3 + 13s^2 + (32 - 2K_c)s + 20 + 20K_c = 0$

The Routh array is thus

$$1 \qquad (32 - 2K_c)$$

$$13 \qquad (20 + 20K_c)$$

$$b_1$$

$$b_1 = [13(32 - 2K_c) - 1(20 + 20K_c)]/13 = 0$$

whence

$$K_c = 8.61$$

The values of the roots are then given by

$$13s^2 + 20 + 20(8.61) = 0$$

$$s = \pm 3.85j$$

The asymptotes are of little value in assisting further with the plotting, and it is necessary to use the angle criterion on a trial and error basis to locate the curve between the breakaway and re-entry points, remembering that the sum of the angles is now an even multiple of π. Appropriate values of K_c at points on the curve are found by the magnitude criterion and the final diagram is shown in Figure 6–10. The diagram shows clearly that the system becomes unstable for values of K_c greater than 8.61.

The root locus diagram for a system involving a time delay is rather more difficult to derive, and is open to some inaccuracy in the approximation of exp $(-Ls)$ by the Padé function. The accuracy of the approximation depends mainly on the relative magnitude of the time delay and the time constants in the system. In the above example L was a fifth of the largest time constant, and the difference between the true and approximate transient responses is within 1–2 per cent of the final steady-state value, an amount which is within the accuracy or range of error of most commercial instruments. A greater error would be expected if L was increased relative to the largest time constant. As will be shown in the next chapter, the time-delay transfer function can be more easily handled by use of frequency response analysis.

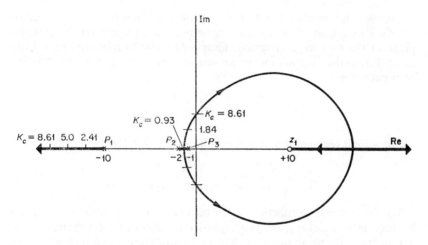

FIGURE 6–10. Root locus diagram for Example 6–3

Transient Response from the Root Locus Diagram

The root locus diagram can be used to determine the transient response of the control system to any simple input function, such as a step, ramp, or sinusoid. The method is to take from the diagram the roots of the characteristic equation for the given control parameters, and to use these in the usual partial fraction expansion of the response transform to provide the factors for the denominator of the transform, which is the step which provides the most difficulty in an algebraic solution. The numerators of the partial fraction expansion, which are the coefficients of the terms in the solution for the transient response, can also be determined graphically by vector combinations from the diagram.

Thus, considering a three-stage system such as that of Example 6–1, the response to a load change introduced between the first and second time-constants is given by

$$\frac{c}{u} = \frac{N}{1 + GH}$$

$$= \frac{K_L/[(T_2s + 1)(T_3s + 1)]}{1 + K/[(T_1s + 1)(T_2s + 1)(T_3s + 1)]}$$

$$= \frac{K_L(T_1s + 1)}{(T_1s + 1)(T_2s + 1)(T_3s + 1) + K}$$

$$= \frac{(K_L/C)T_1(s + 1/T_1)}{(s + 1/T_1)(s + 1/T_2)(s + 1/T_3) + K/C}$$

The denominator of this function when set equal to zero is the characteristic equation of the system whose roots are plotted on the complex plane in the root locus diagram. Thus values of the roots (r_1, r_2, r_3) can be read from the diagram for the particular value of K (or K_c). The equation then becomes

$$\frac{c}{u} = \frac{K_L T_1}{C} \cdot \frac{(s + 1/T_1)}{(s - r_1)(s - r_2)(s - r_3)}$$

For a load change of unit magnitude, $u(s) = 1/s$. Hence

$$c(s) = \frac{K_L T_1}{C} \cdot \frac{(s + 1/T_1)}{s(s - r_1)(s - r_2)(s - r_3)} \tag{6–12}$$

To obtain the time function of the response, $c(t)$, Equation (6–12) must be inversely transformed. The right-hand side can be expanded into partial fractions using the factors of the denominator, and each term, can then be inversely transformed into an exponential time function, i.e.

$$c(s) = \frac{K_L T_1}{C} \cdot \frac{(s + 1/T_1)}{s(s - r_1)(s - r_2)(s - r_3)}$$

$$= \frac{A_0}{s} + \frac{A_1}{s - r_1} + \frac{A_2}{s - r_2} + \frac{A_3}{s - r_3} \tag{6–13}$$

whence

$$c(t) = A_0 + A_1 \exp(r_1 t) + A_2 \exp(r_2 t) + A_3 \exp(r_3 t) \tag{6–14}$$

Whilst the coefficients $A_0, \ldots, A_3$ may be evaluated algebraically, they can also be obtained graphically from the root locus diagram. To evaluate any one coefficient, the usual partial fraction method is employed, i.e. for the coefficient A_1, from Equation (6–13), cross-multiply by the appropriate factor $(s - r_1)$ and put $(s - r_1) = 0$:

$$A_1 = (s - r_1)c(s)\big|_{s-r_1=0}$$

$$= \frac{K_L T_1}{C} \cdot \frac{(s + 1/T_1)}{s(s - r_2)(s - r_3)}\bigg|_{s=r_1}$$

$$= \frac{K_L T_1}{C} \cdot \frac{(r_1 + 1/T_1)}{(r_1 - 0)(r_1 - r_2)(r_1 - r_3)}$$

The factors in the brackets are now to be interpreted as vectors in the complex plane between the two points included in the brackets; the vector

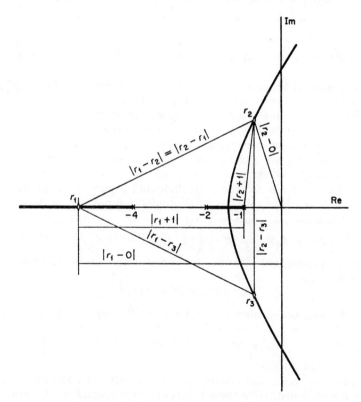

FIGURE 6–11. Vectors for determining A_1 and A_2

of the numerator ends at the pole $-1/T_1$ and that for the s term in the denominator ends at the origin. The magnitudes and angles of the vectors can be measured directly from the root locus diagram, as shown in Figure 6–11 for the coefficients, A_1 and A_2.

If the coefficient is assumed initially to be complex, as will be the case for two of the coefficients if the response is under-damped, the vector will have a magnitude m and an angle ϕ; thus for A_1, from above:

$$m = |r_1 + 1/T_1|/[|r_1 - 0||r_1 - r_2||r_1 - r_3|]$$

$$\phi = \angle(r_1 + 1/T_1) - [\angle(r_1 - 0) + \angle(r_1 - r_2) + \angle(r_1 - r_3)]$$

whence

$$A_1 = m \exp(j\phi) = m(\cos\phi + j\sin\phi) \qquad (6\text{–}15)$$

The other coefficients are evaluated in a similar way; in the case of A_0 the 'root' is zero.

For a three-stage system as discussed, the three roots for an under-damped response will be of the form

$$r_1 = -\alpha_1; \qquad r_2 = -\alpha_2 + j\beta; \qquad r_3 = -\alpha_2 - j\beta$$

where α_1, α_2, and β are real positive numbers, i.e. one root is real and negative and the other two are a conjugate complex pair with negative real parts.

In the event of a root being real, such as r_1, the corresponding coefficient (A_1) will also be real and the vector angle $(\angle A_1)$ will be an integral multiple of π, i.e. the resulting vector will lie on the real axis. This can easily be demonstrated by measurement from the diagram.

If two roots form a conjugate complex pair, e.g. r_2 and r_3, the co-efficients (A_2, A_3) will also be conjugate complex of the form $(a \mp jb)$ as seen previously, where $a = m \cos\phi$ and $b = m \sin\phi$ (from Equation 6–15) and m and ϕ are the magnitude and argument of the vector of either coefficient. The solution for the under-damped part of the response given by the complex roots will thus be

$$A_2\,e^{r_2 t} + A_3\,e^{r_3 t} = (a - jb)\,e^{(-\alpha_2 + j\beta)t} + (a + jb)\,e^{(-\alpha_2 - j\beta)t}$$

Using the exponential identities for the sine and cosine, the right hand side reduces to

$$2\,e^{-\alpha_2 t}(a \cos\beta t + b \sin\beta t)$$

and employing the identity for $\cos(A - B)$, further reduction is possible to

$$2\,e^{-\alpha_2 t}[(a^2 + b^2)^{\frac{1}{2}} \cos(\beta t - \psi)]$$

where $\psi = \tan^{-1}(b/a)$.

Since $a = m \cos\phi$ and $b = m \sin\phi$, $(a^2 + b^2)^{\frac{1}{2}} = m$, and $b/a = \tan\phi$, hence $\psi = \phi$. Substituting these values and converting from the cosine to the more usual sine gives

$$A_2 \exp(r_2 t) + A_3 \exp(r_3 t) = 2 \exp(-\alpha_2 t)[m \sin(\beta t - \phi + 90°)]$$

The amplitude term of the oscillation is thus twice the magnitude and the phase lag is the angle of the vector of either of the coefficients A_2 or A_3, both of which quantities are obtainable from the root locus diagram.

The full equation of the transient response is thus

$$c(t) = |A_0| + |A_1| \, e^{-\alpha_1 t}$$
$$+ 2|A_2| \, e^{-\alpha_2 t} \sin (\beta t - \angle A_2 + 90°) \qquad (6\text{-}16)$$

Figure 6–11 is the root locus diagram of the three-stage process of Example 6–1 with proportional control, and shows the vectors for determination of the coefficients A_1 and A_2.

Example 6–4

Determine the transient response to a unit load change with unit load gain $(K_L = 1)$ introduced after the first time constant for the process of Example 6–1 at a gain of $K_c = 3.0$.

The roots corresponding to $K_c = 3.0$ are found from the root locus diagram (Figure 6–4) by trial and error to be

$$r_1 = -5.5 \qquad r_2, r_3 = -0.75 \pm 2.3j$$

These values are plotted on the diagram (Figure 6–11).

The three time constants are 1, 0.5, and 0.25 min; hence $C = \frac{1}{8}$, $T_1 = 1$ min, $K_L = 1$, and $K_L T_1 / C = 8$.

The response will be given by Equation (6–16); the coefficient A_0 is the steady state error (final value) and is given by Equation (6–13) as

$$A_0 = sc(s)|_{s=0}$$
$$= \left. \frac{8(s + 1)}{(s - r_1)(s - r_2)(s - r_3)} \right|_{s=0}$$

The magnitude of A_0 is therefore given by

$$|A_0| = \frac{8|0 + 1|}{|0 - r_1||0 - r_2||0 - r_3|}$$

From the root locus diagram, the measured values of the vectors are

$$|0 + 1| = 1 \qquad |0 - r_2| = 2.42$$
$$|0 - r_1| = 5.5 \qquad |0 - r_3| = 2.42$$

Hence

$$|A_0| = 8(1)/[(5.5)(2.42)(2.42)]$$
$$= 0.25$$

Since this coefficient is real, the vector angle will be zero or an integral multiple of π. This can be checked thus:

$$\angle A_0 = \angle (0 + 1) - [\angle (0 - r_1) + \angle (0 - r_2) + \angle (0 - r_3)]$$

From the diagram,

$$\angle (0 + 1) = 180° \qquad \angle (0 - r_2) = 108°$$
$$\angle (0 - r_1) = 180° \qquad \angle (0 - r_3) = -108°$$

Hence

$$\angle A_0 = 180 - (180 + 108 - 108) = 0°$$

For the coefficient A_1, from Equation (6–13):

$$A_1 = (s - r_1)c(s)|_{s=r_1}$$

$$= \frac{8(s + 1)}{s(s - r_2)(s - r_3)}\Big|_{s=r_1}$$

$$= \frac{8(r_1 + 1)}{(r_1 - 0)(r_1 - r_2)(r_1 - r_3)}$$

From the diagram, the magnitudes of the vectors are

$$|r_1 + 1| = 4.5 \qquad |r_1 - r_2| = 5.28$$
$$|r_1 - 0| = 5.5 \qquad |r_1 - r_3| = 5.28$$

whence

$$|A_1| = 8(4.5)/[(5.5)(5.28)(5.28)] = 0.23$$

Since A_1 is real, $\angle A_1$ is also zero.
For the coefficient A_2, from Equation (6–13):

$$A_2 = (s - r_2)c(s)|_{s=r_2}$$

$$= \frac{8(s + 1)}{s(s - r_1)(s - r_3)}\Big|_{s=r_2}$$

$$= \frac{8(r_2 + 1)}{(r_2 - 0)(r_2 - r_1)(r_2 - r_3)}$$

From the diagram, the magnitudes of the vectors are

$$|r_2 + 1| = 2.31 \qquad |r_2 - r_1| = 5.28$$
$$|r_2 - 0| = 2.42 \qquad |r_2 - r_3| = 4.6$$

whence

$$|A_2| = 8(2.31)/[(2.42)(5.28)(4.6)] = 0.31$$

Since A_2 is complex, $\angle A_2$ will not be zero or a multiple of π.

$$\angle A_2 = \angle(r_2 + 1) - [\angle(r_2 - 0) + \angle(r_2 - r_1) + \angle(r_2 - r_3)]$$

From the diagram, the vector angles are

$$\angle(r_2 + 1) = 84° \qquad \angle(r_2 - r_1) = 26°$$
$$\angle(r_2 - 0) = 108° \qquad \angle(r_2 - r_3) = 90°$$

whence

$$\angle A_2 = 84 - (108 + 26 + 90) = -140°$$

By Equation (6–15)

$$A_2 = |A_2| \exp (j \angle A_2)$$
$$= 0.31 \exp (-j140°)$$
$$= |A_2|(\cos \angle A_2 - j \sin \angle A_2)$$
$$= 0.31(\cos 140° - j \sin 140°)$$
$$= -0.24 - 0.20j$$

The third coefficient A_3 is the conjugate of A_2. Hence

$$A_3 = -0.24 + 0.20j$$

The response is given by Equation (6–16):

$$c(t) = |A_0| + |A_1| e^{-\alpha_1 t} + 2|A_2| e^{-\alpha_2 t} \sin (\beta t - \angle A_2 + 90°)$$

From the values of the roots r_1 and r_2, $\alpha_1 = 5.5$, $\alpha_2 = 0.75$, and $\beta = 2.3$. Hence

$$c(t) = 0.25 + 0.23 e^{-5.5t} + 2(0.31) e^{-0.75t} \sin (2.3t - 140° + 90°)$$
$$= 0.25 + 0.23 e^{-5.5t} + 0.62 e^{-0.75t} \sin (132°t - 50°)$$

The response is plotted in Figure 6–12.

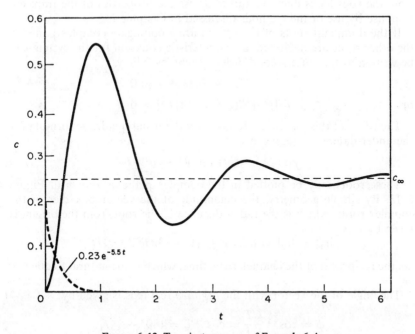

FIGURE 6–12. Transient response of Example 6–4

The Dominant Second-order Response

As can be seen from the plot of the response function in Figure 6–12, the over-damped response of the real root in this particular case (shown by the dotted line) becomes negligible after a very short period compared to the much longer time required for the under-damped oscillation to decay. Effectively the system in this example is dominated by the pair of complex roots producing the under-damped response, and this is generally true for most systems of third or higher order. As can be seen from the root locus diagram, the real root is situated some distance away from the complex pair and the value of the damping factor for this root is very much larger than that of the complex roots. Apart then from the short initial period whilst the over-damped response is decaying to zero, the overall response is basically second-order. This will still apply when there is more than one real root, and also when a second pair of complex roots is present, so long as the additional roots are reasonably far-removed from the dominant pair. These latter obviously are the roots on the locus or branch nearest to the origin of the root locus diagram.

The advantages of assuming that the system response is effectively second-order is that the terms in the transient response which decay relatively quickly may be neglected, and the work of evaluating the coefficients for the full response is much reduced. By regarding the system as second-order, it is possible to find the value of the damping ratio (ζ) from the root locus diagram and to assess the properties of the transient response by use of the methods discussed in Chapter 2.

If the dominant roots of the system are a conjugate complex pair and the other roots are neglected, a characteristic equation for the system can be written as that of a second-order system, as follows:

$$T^2 s^2 + 2\zeta T s + 1 = 0$$

or $$T^2[s^2 + 2(\zeta/T)s + 1/T^2] = 0$$

The roots of this equation are given by the usual quadratic formula for the under-damped case, when $\zeta < 1$:

$$r_1, r_2 = -(\zeta/T) \pm j(1 - \zeta^2)^{\frac{1}{2}}/T$$

These roots may be plotted in the complex plane as shown in Figure 6–13. By simple geometry, the magnitude of the vector of either of the complex roots, which is the radial distance of the root from the origin, is given by

$$|r_1| = |r_2| = [(\zeta/T)^2 + (1 - \zeta^2)/T^2]^{\frac{1}{2}} = 1/T$$

i.e. the reciprocal of the characteristic time, which is the natural frequency, ω_n.

The angle of the vector with the *negative* real axis is given by:

$$\phi = \cos^{-1}\frac{\zeta/T}{1/T} = \cos^{-1}\zeta = 180° - \angle r_1$$

Thus, by measurement of the magnitude and angle of the vector of one of the complex roots, the damping ratio and natural frequency of the oscillation can be obtained, i.e.

$$\omega_n = |r_1| \qquad \zeta = \cos{(180° - \angle r_1)}$$

Alternatively, a root corresponding to a given value of ζ can be located by a line through the origin at $\cos^{-1}\zeta$ to the negative real axis, and from this the value of ω_n can be found. Knowing these two parameters, most of the properties of the transient response can be predicted. It will also be seen that the imaginary part of the complex root is the radian frequency of the damped response, $\omega = (1 - \zeta^2)^{\frac{1}{2}}/T$.

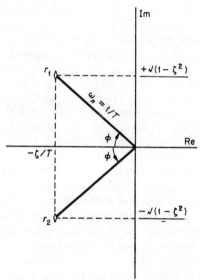

FIGURE 6–13. Roots of second-order system

Example 6–5

Determine the approximate response to a unit step change in load for the system of Example 6–4 if it is assumed that the system is dominated by the complex roots.

For $K_c = 3.0$ there is one real root (r_1) at -5.5 and the complex pair $(r_2, r_3 = -0.75 \pm 2.3\text{j})$. The root r_1 is sufficiently far removed from r_2 and r_3 for the response of r_1 to be highly damped compared to the response of r_2 and r_3 ($\exp{(-5.5t)}$ *versus* $\exp{(-0.75t)}$), as shown in Figure 6–12. The system may then be reasonably assumed to be dominated by the complex roots.

For the root $r_2 = -0.75 + 2.3\text{j}$, the angle of the vector $\angle r_2$ is $108°$, the angle with the negative real axis is $72°$ and

$$\zeta = \cos 72° = 0.31$$

The magnitude of the vector, $|r_1| = 2.42$, hence the natural frequency ω_n is 2.42 rad/min. The actual frequency of oscillation, ω, is the imaginary part of the complex root, i.e. 2.3 rads/min.

The steady-state value of the response may be determined by the final value theorem of the Laplace transform, or more simply by use of the expression $AK_L/(1 + K)$ of page 140 *et seq.* For $A = 1$ (unit step) and $K_L = 1$ (unit load gain), with $K = 3.0$ (since the stage gains are also unity), the final value is $\frac{1}{4}$ or 0.25.

Knowing ζ, ω and the final value, the response of the system is completely described. The step response curve of Figure 2–25 for the appropriate value of ζ defines the nature of the response; the ordinate is multiplied by the final value to reduce the amplitude to the required value, and the time scale is interpreted in terms of the value found for T. Overshoot, decay ratio, etc. can be found from Figure 2–27.

There are, however, some discrepancies between the approximate response and the true response as calculated in Example 6–4, e.g. the overshoot of the variable beyond the final value for the value of $\zeta = 0.31$ determined from Figure 2–27 is 37 per cent, whilst that of the solution to Example 6–4 is about 120 per cent. This is due to the fact that the approximate method does not take into account the point of incidence of the load change. In many practical problems these discrepancies can usually be disregarded, since the usual purpose of such calculations is to compare the effects of changes in the parameters of the system on the transient response and as a rule only the values of ζ and ω need be considered for this purpose. The graphical technique provides a reasonable method of making these assessments without involved calculations but care must be taken that the terms which are neglected in assuming that a pair of roots is dominant have only a small effect of the transient response.

The Effect of Control Actions as Shown by Root Loci

The use of the root locus diagram in studying the effect of the additional control actions, integral and derivative, will now be considered. The three-stage process of Example 6–1, whose root locus diagram was plotted as Figure 6–4, for proportional control, will be taken as the basic case.

Proportional-derivative Control

With a PD controller the open-loop transfer function for the three-stage time-constant process with unit gain and unity feedback becomes

$$GH = \frac{K_c(1 + T_d s)}{(T_1 s + 1)(T_2 s + 1)(T_3 s + 1)}$$

or, in the alternative form,

$$GH = \frac{K_c T_d (s + 1/T_d)}{T_1 T_2 T_3 (s + 1/T_1)(s + 1/T_2)(s + 1/T_3)}$$

The root locus diagram will thus have three poles at $-1/T_1$, $-1/T_2$, and $-1/T_3$, with a zero at $-1/T_d$. The constant C will now include T_d:

$$C = T_1 T_2 T_3 / T_d$$

The addition of a single zero reduces the difference between the numbers of poles and zeros, $(n - m)$, from 3 to 2; there will thus now be two asymptotes instead of three and these will be at angles of $\pm\frac{1}{2}\pi$ to the real axis, i.e. parallel to the imaginary axis. One of the three branches of the locus, starting from each of the three poles, must now end at the zero, the other two will go to infinity on the asymptotes as before. The intersection of the asymptotes with the real axis is the centre of gravity of the poles and the zero; the position of the latter depends on the value of T_d. As T_d is increased, the zero will be moved towards the origin; the centre of gravity and hence the asymptotes will be moved to the left. Depending upon the value of T_d and the position of the zero with respect to the poles, three situations can be distinguished.

With a relatively small derivative time ($T_d <$ smallest time constant), the zero is to the left of the poles and the branch from the smallest time constant pole (that furthest to the left) will end at the zero. In the case illustrated in Figure 6–14(b), the asymptotes are displaced to the right of the origin, the centre of gravity given by Equation (6–7) for a value of $T_d = 0.1$ min being at $+1.5$. The two loci from the poles at -1 and -2 will meet and break away from the real axis as complex roots, and will cross the imaginary axis to reach the asymptotes. The system can thus become unstable since the complex roots can take on positive real parts, yet it will be considerably more stable than without the derivative action. For the process of Example 6–1, with time constants of 1, 0.5, and 0.25 min and a derivative time of 0.1 min, the intersection of the loci with the imaginary axis occurs at a proportional gain of $K_c = 37.5$, as compared to a value of 11.25 without the derivative (Example 6–1; see also Figure 6–4 which is reproduced as Figure 6–14(a) for comparison). The value of the imaginary root at the intersection is 6.63j compared with 3.74j; thus the limiting proportional gain is trebled and the frequency of oscillation of the dominant roots is almost doubled. The dominant roots will not take on positive real parts until a much higher gain is applied, with a correspondingly smaller offset.

For $K_c = 3.0$ (as used in Example 6–4 for the same process), the roots without derivative action were -5.5 and -0.75 ± 2.30j; with the added derivative action at the same value of K_c the roots are -5.0 and -1.05 ± 2.35j. Considering only the second-order part of the response as given by the complex roots, the damping factor has been increased from -0.75 to -1.05, the values of ζ, determined by the method of the preceding section, being 0.31 and 0.41 respectively. The frequency of the damped oscillation is only slightly increased at the relatively low gain for the PD control. The responses of the two cases are compared in Figure 6–18(a) and (b), for a unit change in desired value, calculated by the method used on pages 205–209.

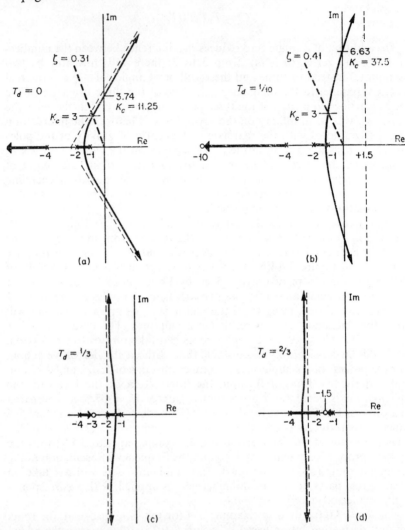

FIGURE 6–14. Root locus diagram for proportional-derivative control

As T_d is increased, so bringing the zero nearer to the left-hand pole, the asymptotes will approach the imaginary axis and ultimately will be coincident for a unique value of T_d at which the centre of gravity of the poles and the zero is at the origin. In the present case this will occur when the zero is at -7, for $T_d = 1/7$. At this value, and for any further increase in T_d, the system is unconditionally stable since the loci do not intersect the imaginary axis and the complex roots cannot take on positive real parts. The zero now effectively cancels one of the poles and makes the system second-order.

Further increases in T_d do not affect the stability of the system but cause significant changes in the root loci. As the zero moves towards the origin it will ultimately cross the first pole; the branch from that pole will still go to the zero but now in the opposite direction (Figure 6–14c). After passing the second pole, however, the complex roots will start from the left-hand pair of poles and the branch on the real axis is from the right hand pole to the zero (Figure 6.14d). The real root is now less damped than the complex roots and the response will be predominantly over-damped as the under-damped response will decay more quickly.

These findings confirm the stabilizing effect of the addition of derivative action to the control system.

Proportional-integral Control

Adding integral action to the system of Example 6–1 produces an open-loop function of

$$GH = \frac{K_c(1 + 1/T_i s)}{(T_1 s + 1)(T_2 s + 1)(T_3 s + 1)}$$

$$= \frac{K_c(s + 1/T_i)}{T_1 T_2 T_3 s(s + 1/T_1)(s + 1/T_2)(s + 1/T_3)}$$

The root locus diagram for the three-stage time-constant system with PI control thus has an additional pole at the origin ($s = 0$), and a zero at $-1/T_i$. As $(n - m)$ is unchanged by the addition of both a pole and a zero, the asymptotes remain three in number as with proportional control. The loci will now have four branches; starting from the four poles, one branch must end at the zero and the other three at infinity on the asymptotes, but one of the latter lies on the negative real axis as with proportional control.

A large value of T_i, corresponding to a small integral action and slow recovery of offset, places the zero close to the pole at the origin; in fact the two will coincide and cancel each other when T_i is infinite. For this condition the branch locus from the pole at the origin will end at the zero; the rest of the diagram is apparently unaltered from that of proportional control, but the presence of the zero shifts the centre of gravity of the poles and zeros slightly to the right and the asymptotes therefore intersect the imaginary axis at smaller values. It follows that the loci of the complex roots going to infinity on the asymptotes will also cut the imaginary axis at smaller values than with proportional control, i.e. the gain for limiting stability is reduced and the frequency will also be less. Comparing Figure 6–15(a) and (b), for $T_i = 1.4$, the limiting proportional gain is reduced from 11.25 to 7.4 and the imaginary root (frequency) from 3.74j to 3.10j.

The value of T_i is approximately that indicated by application of the Ziegler-Nichols rules, i.e. $P_u/1.2$, where P_u is the period of oscillation at the limiting gain with proportional control. For $K_c = 3.0$, the value of ζ is reduced to about 0.21, indicating less damping and a more oscillatory response (see Figure 6–18(c)).

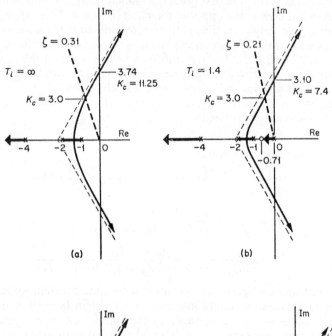

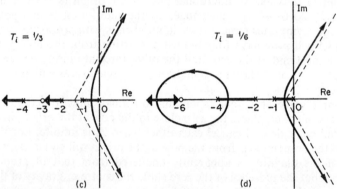

FIGURE 6–15. Root locus diagram for proportional-integral control

As has already been seen in previous discussions, the sole advantage of addition of integral action to proportional control is the elimination of offset and only the minimum amount should be used to overcome the offset in reasonable time. The Ziegler-Nichols recommended value is a suitable compromise between eliminating the offset in a reasonable time and undue decrease of stability. Any further increase in the integral action, by reducing T_i, moves the zero to the left and the centre of gravity to the right, thus reducing still further the limiting proportional gain and the frequency of oscillation. As seen in Figure 6–15(c), when the zero passes the pole nearest the origin, that provided by the largest time constant, the complex roots are then provided by this pole and that at the origin, the branch ending at the zero coming from the pole of the second largest time-constant in the direction determined by the position of the zero with respect to that pole.

An interesting situation arises when T_i is sufficiently small to bring the zero to the left of all the poles, as in Figure 6–15(d). A branch of the loci must now lie on the real axis between the two left-most poles, since the number of poles and zeros to the right is odd. The two branches from these poles must then meet and break away to form a second pair of complex roots, but the two branches must re-enter the real axis for one branch to go to the zero and the other to infinity on the real axis. This secondary pair are thus oscillatory over a limited range of gain, but the damping factor will be relatively high compared to that of the dominant complex roots and the latter will take on positive real parts at a relatively low gain and so produce instability before the second pair of complex roots has much effect.

Proportional-integral-derivative Control

To obtain the benefit of integral action in elimination of offset with the greater stability and faster response of a reasonably high gain, it is necessary to add derivative action to the PI controller. For the three stage process with unit gain and PID control, the open-loop function becomes

$$G = \frac{K_c(1 + T_d s + 1/T_i s)}{(T_1 s + 1)(T_2 s + 1)(T_3 s + 1)}$$

$$= \frac{K_c T_d(s^2 + s/T_d + 1/T_i T_d)}{T_1 T_2 T_3 s(s + 1/T_1)(s + 1/T_2)(s + 1/T_3)}$$

There are thus four poles on the root locus diagram from the denominator but there are also two zeros provided by the factors of the quadratic term in the numerator. As such the zeros may be real (equal or unequal) or complex, but always with negative real parts, depending on the relative values of T_i and T_d. For example, when $T_i = T_d$ the roots are conjugate complex; when $T_i = 4T_d$ the roots are real and equal and the two zeros are coincident (i.e. a second-order zero). The number of poles in excess of the number of zeros, $(n - m)$, is reduced to two, there are thus only two asymptotes at angles of $\pm\frac{1}{2}\pi$ to the real axis. The locus will have four branches since there are four poles; two of these branches will end at the zeros and the other two at infinity on the asymptotes.

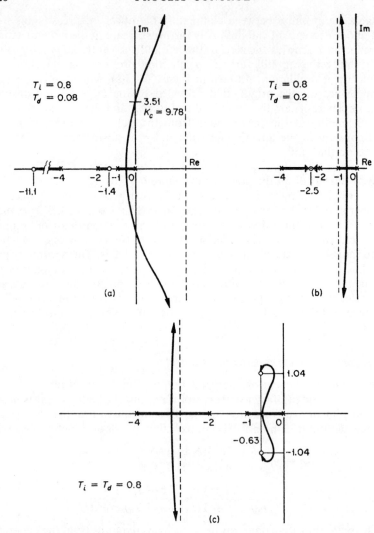

FIGURE 6–16. Root locus diagram for proportional-integral-derivative control with constant T_i

Nothing more can be said in general terms about the root locus diagram since the latter can now be altered considerably by changing the values of both T_i and T_d and the problem of studying the effect of controller settings requires a little more effort. In Figure 6–16, the integral time is set approximately to the Ziegler-Nichols recommendation for three term control, $\frac{1}{2}P_u$ and the effect of different ratios of $T_i:T_d$ is examined. In Figure 6–16(a), for a relatively small amount of derivative action ($T_d = 0.1T_i$), the roots of the quadratic term are real and unequal and the zeros are thus widely spaced and 'bracket' the left-hand pair of poles. The centre of gravity of the poles and zeros is such that the asymptotes lie to the right of the imaginary axis. The system can thus be unstable since the complex roots from the right-hand pair of poles intersect the imaginary axis and the roots can take on positive real parts. The stabilizing effect of the derivative action is evident in that the intersection of the complex root loci with the imaginary axis comes at a higher value than with PI control, even though more integral action is being used than in Figure 6–15(b), and both the gain and frequency are increased by the addition of the derivative action.

Increasing T_d brings the zeros closer together and ultimately will bring the asymptotes to the left of the imaginary axis so that the system becomes unconditionally stable for all values of K_c. At the Ziegler-Nichols recommendation of $T_i = 4T_d$, the zeros coincide; for the particular value of T_i in Figure 6–16(b) this occurs between the left-hand pair of poles; the two zeros effectively cancel the two poles and the system is basically second-order in response. If T_d is made equal to T_i, the zeros become complex, as in Figure 6–16(c). There are now two sets of complex roots, with those from the right-hand poles ending at the complex zeros, but the system is still completely stable since none of the loci can cut the imaginary axis.

Figure 6–17 shows the alternative approach of maintaining a fixed ratio of $T_i : T_d$, the value chosen being the Ziegler-Nichols recommendation of 4:1 so that the zeros are coincident throughout, and changing the values of both T_i and T_d simultaneously. This means that in the extreme cases of Figures 6–17(a) and (c), a large integral action is combined with a small derivative and *vice versa*. Only in the case with a large integral action and small derivative can instability now occur.

Figure 6–18 compares the response to a unit change in desired value for the four cases whose root locus diagrams are given in Figures 6–4, 6–14(b), 6–15(b), and 6–16(b), for a value of $K_c = 3.0$ in each case, the responses being determined by the graphical method. Taking proportional control as the basic case, proportional-derivative shows a smaller overshoot, greater damping and faster recovery; proportional-integral shows greater overshoot, less damping and a much slower recovery, but the offset is eliminated. Three-term control also shows the elimination of offset and the response is roughly intermediate between that of proportional-integral and proportional-derivative. Both proportional and proportional-derivative show the same offset since the same value of gain is used in each case; these are not, of course, the 'optimum' responses since the values of K_c would in practice be altered to suit the combination; the value would be decreased for PI control to give a little more damping at the expense of increased overshoot and would be increased for PD and PID controls to give a much smaller offset without unduly affecting the recovery time in the first case and to reduce the overshoot in the second case.

The root locus method has been used in this discussion to consider variations in controller gain and discrete changes in T_i and T_d. The method can be extended to cover continuous changes in parameters other than controller gain, e.g. in T_d or T_i, or in one of the time constants, and also to multi-loop systems, although the procedures are more tedious. These further applications are presented in several texts on servo-mechanisms [28].

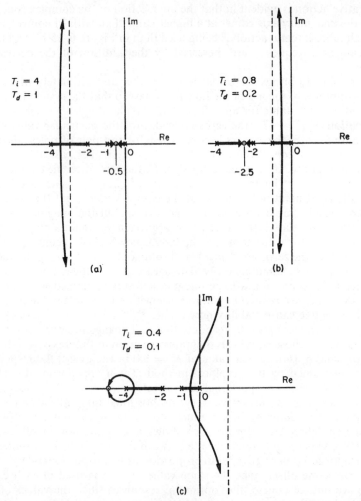

FIGURE 6–17. Root locus diagram for proportional-integral-derivative control with constant ratio of T_i to T_d

The root locus technique, along with frequency response analysis which will be discussed in the next chapter, is a basic tool for solution of control problems of which the main advantage is the ability to distinguish the stability characteristics of the system by inspection.

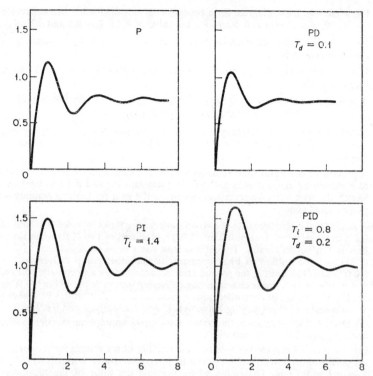

FIGURE 6–18. Response of proportional, proportional-derivative, proportional-integral, and proportional-integral-derivative controls to unit change in desired value ($K_c = 3$ in each case)

PROBLEMS

6-1 Draw the root locus diagrams for the following equations, showing the poles, zeros, and asymptotes, and determine the value of K for limiting stability by a graphical method:

(a) $(s + 1)(3s + 1) + K = 0$
(b) $s(s + 1)(2s + 1) + K = 0$
(c) $(s + 1)^2(4s + 1) + K = 0$
(d) $(s + 1)(s + 2) + K(2s + 1) = 0$
(e) $s(s + 1)(2s + 1) + K(s + 2)^2 = 0$

6-2 Draw the root locus diagram for a system with the two time constants of 1 and 2 min and a controller with a transfer function of

$$G_c = K_c(1 + \tfrac{1}{2}s + 1/s)$$

6-3 Draw the root locus diagram for integral control of a process with time constants of 2 and 4 min.

6-4 Draw the root locus diagram for a process with a transfer function of

$$20/[(3s + 1)(5s + 1)]$$

and a measuring device with a transfer function of $1/(0.5s + 1)$ for proportional control. Determine the values of K_c for critical damping and limiting stability.

6-5 Draw the root locus diagram for a process with four first-order time-constants of 60, 30, 20, and 15 min and determine the value of K for $\zeta = 0.5$ and 0.2.

6-6 Draw the root locus diagram for the system of Problem 5-5 and determine the maximum controller gain.

6-7 Determine the maximum controller gain for the system of Problem 5-11 by root locus methods.

6-8 A temperature controlled polymerization process has a transfer function of

$$k/(s - 40)(s + 80)(s + 100)$$

Sketch the root locus diagram and show by the Routh test that the system is conditionally stable. Determine the upper and lower values of K_c for proportional control.

6-9 A process of two first-order elements with time constants of 10 min is connected to a measuring element with a 4 min time constant lag and a 1 min time delay. Sketch the root locus diagram and determine the gain for limiting stability with proportional control.

6-10 A liquid level system consists of two non-interacting tanks in series. The first tank has a cross-sectional area of 0.2 m² (2 ft²) and a level of 1 m (5 ft) when the outflow is 2 m³/s (10 ft³/min). The second tank has a cross-sectional area of 0.1 m² (1 ft²) and the outflow is by a constant displacement pump delivering 2 m³/s (10 ft³/min). The level in the second tank is controlled by a pneumatic controller and a valve with a linear characteristic delivering 0.01 m³/s per kN/m² (5 ft³/min per lbf/in²) change in controller output pressure to the first tank, which also has an independent inflow varying between 0.75 and 1.25 m³/s (4 and 6 ft³/min).

(a) Draw a block diagram of the system showing the appropriate transfer functions and numerical values in each block.

(b) Using root locus methods discuss the stability of the system for proportional, proportional-derivative, and proportional-integral-derivative controls with T_d and T_i equal to 0.2 min. Determine where possible the value of the proportional gain at which the system becomes unstable and the corresponding frequency of oscillation.

Chapter 7: Frequency Response Analysis

The response of a control system, or of the individual elements of a system, to a sinusoidal input of fixed amplitude over a range of differing frequencies of oscillation is the basis of frequency response analysis, which is one of the most valuable tools of control systems analysis and design. The principal advantage of the technique is that the response of a complete system can be derived from the response of the individual elements of the system by a simple graphical construction. In contrast, the calculation of the transient response is at the best tedious, and is increasingly difficult as the order of the system equation increases. The frequency response method can also be applied experimentally to existing systems by the use of a suitable sine-wave generator to inject a sinusoidal signal into either the open or closed loop, and in this way the frequency response characteristics of the complete system or of the elements of a system can be obtained when the transfer functions are not known.

Frequency Response of the First-order System

When a sinusoidal input is applied to a linear system, the output has the usual transient and steady-state terms. The transient expires after relatively few cycles of oscillation and the steady-state response is then a sine wave of the same frequency as the input but of a different amplitude, and is subject to a lag in time which is evident as an angular difference in phase between the input and output oscillations. Both the change in amplitude and the phase angle are functions of the frequency of oscillation.

These points can be conveniently illustrated by means of the first-order system response previously considered (pages 55–7). The steady-state response of such a system is shown in Figure 7–1 for two different frequencies of oscillation. At a low frequency there is a small reduction (attenuation) of the amplitude and a small phase lag; at higher frequencies the response amplitude is reduced still further and the phase angle approaches a maximum lag of 90°.

The steady-state response of the first-order system with time constant T, and with a unit non-dimensional gain (i.e. a transfer function of $1/(Ts + 1)$, to a sinusoidal input of $\theta_1 = A \sin \omega t$ is given by Equation (2–16):

$$[\theta_2]_{ss} = \frac{A}{\sqrt{(1 + \omega^2 T^2)}} \sin (\omega t - \psi)$$

where $\psi = \tan^{-1} (\omega T)$.

The response for any given frequency of oscillation is characterized by the ratio of the amplitudes of the output and input sine waves (B/A in Figure 7–1) and the phase angle, the latter always being a *lag* for a first-order system.

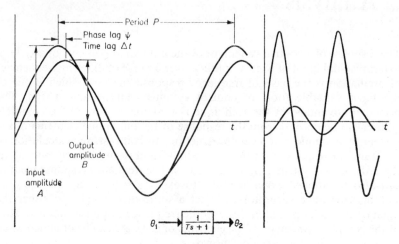

FIGURE 7–1. Frequency response analysis

In the present case of a system with a unit gain, the output amplitude is $B = A/\sqrt{(1 + \omega^2 T^2)}$. Hence the ratio of the amplitudes is given by

$$B/A = 1/\sqrt{(1 + \omega^2 T^2)} \tag{7–1}$$

The ratio obviously decreases with increasing frequency, from a value of one at $\omega = 0$ to zero as $\omega \to \infty$.

The phase lag is given directly by the angle, ψ:

$$\text{Phase lag } \psi = \tan^{-1}(\omega T) \tag{7–2}$$

or, alternatively, the phase angle is $-\tan^{-1}(\omega T)$. The phase lag is zero at zero frequency and approaches a maximum of $90°$ ($\tan^{-1} \infty$) as the frequency is increased.

Equation (2–16) was derived by the usual technique of substituting the Laplace transform of the sinusoidal input into the transfer function equation for the first-order stage with unit gain, followed by the subsequent inversion to give the full time function of the response (Equation (2–14)). This treatment gives the full solution, i.e. the transient and steady-state terms of the response; Equation (2–16) is a shortened version in which the transient term has been omitted (since this term becomes zero in a relatively short time) to leave only the steady-state portion of the response on which the frequency response analysis is based. The required frequency-response characteristics of any process element can, however, be obtained directly and more simply from the transfer function by the expedient of substituting $j\omega$ for s.

Substitution of $j\omega$ for s in any transfer function such as $G(s)$ gives a complex number, $G(j\omega)$, which may be considered as a vector in the complex plane with a magnitude of $|G|$ and an argument of $\angle G$. It can be shown that the magnitude is equal to the ratio of the amplitudes, and that the argument is equal to the phase angle of the frequency response to a sinusoidal oscillation of frequency ω. The proof of this method of deriving frequency response characteristics from the transfer function is given in many texts on servo-mechanisms, and no proof will be offered here beyond a simple demonstration that the correct result will be obtained for the first-order system under consideration.

In more general terms, the transfer function of a first-order element will be given by

$$G(s) = K_p/(Ts + 1)$$

where K_p is the gain of the particular element. K_p is not in practice usually equal to one, and is not necessarily dimensionless as was considered to be the case in the derivation of Equation (2–16). Substituting $j\omega$ for s in the transfer function gives the complex number $G(j\omega)$, and this can be written in the usual form of $(x + jy)$ by clearing the imaginary term from the denominator by multiplying both numerator and denominator by the conjugate of the latter:

$$
\begin{aligned}
G(j\omega) &= K_p/(j\omega T + 1) \\
&= K_p(1 - j\omega T)/(1 + \omega^2 T^2) \\
&= K_p/(1 + \omega^2 T^2) - j[K_p\omega T/(1 + \omega^2 T^2)]
\end{aligned}
$$

For a complex number $x + jy$, the magnitude is given by $\sqrt{(x^2 + y^2)}$ and the argument by $\tan^{-1}(y/x)$. Hence

$$
\begin{aligned}
|G(j\omega)| &= \{[K_p/(1 + \omega^2 T^2)]^2 + [-K_p\omega T/(1 + \omega^2 T^2)]^2\}^{\frac{1}{2}} \\
&= K_p/\sqrt{(1 + \omega^2 T^2)}
\end{aligned}
\tag{7–3}
$$

and

$$
\begin{aligned}
\angle G(j\omega) &= \tan^{-1}\left[\frac{-K_p\omega T/(1 + \omega^2 T^2)}{K_p/(1 + \omega^2 T^2)}\right] \\
&= \tan^{-1}(-\omega T)
\end{aligned}
\tag{7–4}
$$

These values of the magnitude and argument of the complex number $G(j\omega)$ may now be compared with the ratio of amplitudes and the phase lag derived from the steady-state response equation (Equation 2–16) for a sinusoidal input to a first-order system with unit gain, as given in Equations (7–1) and (7–2). It will be seen that the argument, $\angle G(j\omega)$ is strictly the *phase angle* (since $\tan^{-1}(-\omega T) = -\tan^{-1}(\omega T)$), the minus sign in Equation (7–4) confirming that the angle is actually a lag.

The expression for the ratio of output to input amplitude of Equation (7–1) differs from the magnitude of Equation (7–3) by the factor K_p, and this raises a rather more fundamental issue. The factor K_p is the steady-state gain but the system used to derive Equation (2–16) and that illustrated in Figure 7–1 has a unit (dimensionless) gain, and the output is equal to the input at zero frequency at the steady state. If the process gain is not equal to one, as is often the case, the output is not equal to the input at zero frequency, and a more comprehensive definition of 'amplitude ratio' is needed. This is provided by 'normalizing' the ratio of output to input amplitudes, i.e. by defining *amplitude ratio* (which it will be convenient to refer to as A.R.) as the ratio of the output to input amplitudes at a given frequency divided by the ratio at zero frequency. The latter quantity is the steady-state gain K_p; hence dividing the magnitude, $|G(j\omega)|$, by K_p makes the amplitude ratio identical with the value given in Equation (7–1). This procedure has the additional advantage of making the A.R. dimensionless, and the expressions of Equations (7–1) and (7–2) are thus of general application to any first-order system.

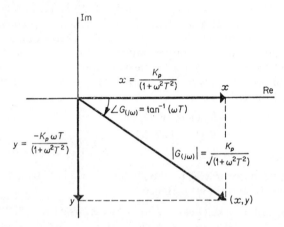

FIGURE 7–2. Representation of $G(j\omega)$ as a vector in the complex plane for a first-order system

The quantity, $|G(j\omega)|$, is also a ratio of output and input amplitudes but is not necessarily dimensionless, since the gain term K_p relates the dimensions of the input and output which may well be different physical quantities, and it will also not be unity at zero frequency unless $K_p = 1$. To avoid confusion, ratios of output to input amplitudes containing the gain term K_p, when the latter is not equal to one, are referred to as *magnitude ratios* (M.R.), and amplitude ratio (A.R.) is reserved for the normalized dimensionless value.

When there are several elements in series in a system, the overall transfer function is the product of the individual transfer functions. Similarly the overall frequency response characteristics can be obtained by multiplying the individual complex numbers for each element obtained by substituting $j\omega$ for s in each transfer function. This requires the magnitudes of the complex numbers to be multiplied and the phase angles to be added, following the usual rules for vector quantities, i.e. for a multiple system,

$$G(s) = G_1 G_2 G_3 \ldots$$

$$|G(j\omega)| = |G_1||G_2||G_3| \ldots$$

and
$$\angle G(j\omega) = \angle G_1 + \angle G_2 + \angle G_3 + \ldots$$

This combination of the responses of individual elements can be justified by direct reasoning, as was done on pages 60–61. With a sinusoidal input to the first of a series of elements, the output of any one element is a sine wave of the same frequency forming the input to the next element in the series; the attenuation of the amplitude and the phase lag are therefore cumulative.

The overall A.R. is obtained by dividing the overall magnitude ratio by the product of the gain terms of each element, and is thus the product of the individual amplitude ratios, i.e.

$$\begin{aligned}
\text{A.R.}_{\text{overall}} &= \frac{|G_1||G_2||G_3|}{K_{p1}K_{p2}K_{p3}} \ldots \\
&= \frac{|G|}{K_{p1}K_{p2}K_{p3}} \ldots \\
&= (\text{A.R.}_1)(\text{A.R.}_2)(\text{A.R.}_3) \ldots
\end{aligned}$$

Frequency Response Diagrams

The frequency response of an individual system element or of a complete system is characterized by three related properties, the amplitude (or magnitude) ratio, the phase angle, and the frequency. As already seen, the two former are dependent upon the value of the latter, and the relationship can be shown graphically by means of a *frequency response diagram* in a number of different ways. The first two quantities (A.R. and phase angle) may be plotted each against the third (frequency), or they may be plotted against each other with the third as a parameter (since it has already been noted that both A.R. and phase angle are uniquely determined by the frequency). In view of the range of A.R. and frequency it is convenient to use logarithmic scales, and this is additionally helpful in the multiplication of A.R.s when combining separate elements into a system. Since one of the quantities is an angle, it is also convenient in some cases to use a polar co-ordinate plot, in which case the latter is effectively a vector locus plot of the transfer function in the complex plane. Finally, for some purposes it is advantageous to plot the reciprocal of the amplitude or magnitude ratio on these diagrams.

Figure 7–3 illustrates examples of the more commonly used frequency response diagrams which are briefly as follows:

(a) The arithmetic rectangular co-ordinate plot of A.R. (or $|G|$) and $\angle G$ *versus* frequency ω.

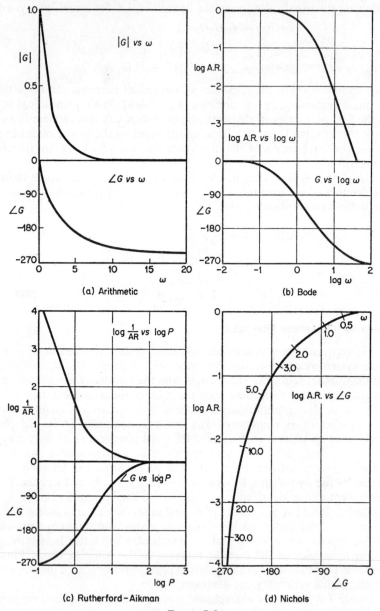

(a) Arithmetic

(b) Bode

(c) Rutherford–Aikman

(d) Nichols

FIGURE 7–3

(b) The *Bode* diagram, which is a rectangular plot of log (A.R.) *versus* log ω, and phase lag *versus* log ω.

(c) The *Rutherford-Aikman* diagram, which is an inversion of the Bode diagram, log (1/A.R.) being plotted against the log of the period of oscillation, and also phase lag *versus* log (period).

(d) The *Nichols* diagram, which is a rectangular plot of log (A.R.) *versus* phase lag (or alternatively, *versus* phase margin, which is $180° - \angle G$).

(e) The *Nyquist* diagram, a polar co-ordinate plot of $|G|$ *versus* $\angle G$; this is effectively a plot of the function $G(s)$ in the complex plane.

(f) The *inverse Nyquist diagram* which is the polar plot of $1/|G|$ *versus* $\angle G$.

In the last three diagrams, the values of the frequency can be marked as a parameter on the curve if required, although this is not necessary for many of the purposes for which these diagrams are used.

In general these different methods of representing frequency response characteristics can be used alternatively, but one may often be more convenient for certain applications than the others. The Bode diagram is of greatest value for determination of the overall response of a system of several individual elements and for a graphical procedure of determining stability characteristics, and will be greatly used later in this chapter. The Rutherford-Aikman diagram is a variant of the Bode diagram, although somewhat less flexible, and is used principally to correlate controller action times with the period of oscillation. The Nichols diagram is designed to permit the phase margin (a quantity used to define system performance) to be read off directly for a given A.R. The Nyquist diagram is the basis of the Nyquist stability criterion by which the stability of servo-mechanisms is commonly assessed, although possibly of less general utility in process control. Finally, the inverse Nyquist diagram can be used for a graphical

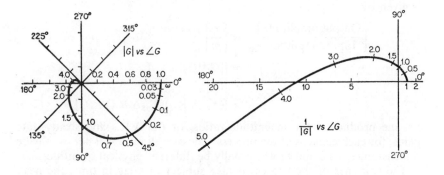

(e) Nyquist   (f) Inverse Nyquist

FIGURE 7–3. Frequency response diagrams: (a) Arithmetic, (b) Bode, (c) Rutherford-Aikman, (d) Nichols, (e) Nyquist, (f) inverse Nyquist

approach to control stability and for consideration of the influence of integral and derivative actions on the controlled system.

A Graphical Approach to Control-loop Stability

In view of the basic importance of control loop stability, this topic will now be discussed by use of the inverse Nyquist diagram.

In Chapter 5 a mathematical explanation of the stability or otherwise of the control loop was discussed when it was seen that the limiting condition of stability, beyond which the system becomes unstable, is the zero-damped oscillation which occurs when two of the roots of the characteristic equation of the system are conjugate imaginary. A physical explanation of this condition will now be developed.

First it must be appreciated that consideration is being given to the oscillating response which follows some finite disturbance to the system such as a step change in load or desired value. After a short transient period the response becomes a continuous steady-state oscillation of the output variable. This oscillation is maintained entirely by the action of the controller in the control loop and is not due to any influence external to the loop. It is true that such oscillations are usually set up initially by some external disturbance such as a change in load, but the oscillations continue indefinitely after the load has become constant or even returned to its original value.

If the controlled variable oscillates at a constant amplitude and frequency, then each signal between the individual elements of the loop must also oscillate at the same frequency and at a fixed amplitude. In particular, an oscillating input is being applied to the process elements by the manipulated variable m; the output oscillation of the controlled variable c will be of the same frequency but of reduced amplitude, as seen in Figure 7–4(a). The ratio of the output to input amplitudes at the particular frequency ω is given by

$$\left[\frac{\text{Output amplitude}}{\text{Input amplitude}}\right]_{\omega} = \left[\frac{c}{m}\right]_{A\omega}$$
$$= [|G_1||G_2| \ldots]_{\omega}$$
$$= |G_p|_{\omega}$$
$$= [(K_{p1}\text{A.R.}_{.1})(K_{p2}\text{A.R.}_{.2}) \ldots]_{\omega} \quad (7\text{--}5)$$

i.e. the product of the magnitude ratios, or the A.R.s and steady-state gains, for each element of the process. The overall ratio will not usually be dimensionless as c and m will usually be different physical quantities.

The response of the process is also subject to a lag in time, the peak output amplitude arrives some time after the peak input is applied, and this is shown by a lag in phase between the output and input sine waves. Thus an input of peak amplitude m_1 entering the process will produce a peak output amplitude c_1 at a time Δt later, corresponding to a lag in phase of ψ, which is equal to the sum of the individual phase lags of the separate process elements at the particular frequency, i.e.

$$\psi = \angle G_1 + \angle G_2 + \cdots$$

Turning now to the controller, as seen in Figure 7–4(b), the input is now the oscillation of the controlled variable c, and the output is an amplified oscillation of the manipulated variable m, of the same frequency but subjected to a change in phase of γ. The amplification is due to the gain of the controller which is the ratio of the output to input amplitudes at the steady state, the latter now being the continuous oscillation. Thus

$$\left[\frac{\text{Output amplitude}}{\text{Input amplitude}}\right]_\omega = \left[\frac{m}{c}\right]_{A\omega} = [K_g]_\omega \qquad (7\text{–}6)$$

As will be seen in due course, the controller gain K_g is not the same as the proportional gain K_c when derivative or integral actions are also present.

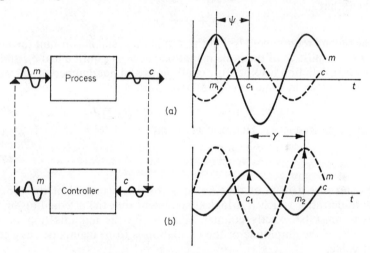

FIGURE 7–4. Continuous oscillation (stability limit)

The change in phase over the controller arises from the corrective nature of the control action. The manipulated variable m is a continuous function of the error e, where $e = v - c$ (assuming a unit measurement feedback). In the present case of the continuous oscillation at limiting stability, there are no external disturbances affecting the system; the desired value must then be constant and $v = 0$. The variable m is thus a function of $-c$, i.e. $m = f(-c)$, so that a negative deviation of the controlled variable produces a positive correction to the process, and *vice versa*. The controller introduces an inherent phase change between input and output waves of $-180°$, the output wave conventionally lagging behind the input, although this may be increased or decreased by the effects of integral or derivative actions, as will be seen later. A peak amplitude input to the controller, such as c_1, will produce a corresponding peak amplitude of the output m_2, subject to a lag in phase of γ between the two waves.

If now the controller is to maintain the continuous oscillation, the peak output of the controller (m_2) must become the new peak input to the process. It must therefore be applied in the correct amplitude and phase relationship to reproduce the oscillation of the controlled variable; in other words, the response diagrams of Figures 7–4(a) and (b) must fit together. A first criterion for stability of the continuous oscillation can now be seen. Considering only peak amplitudes, the peak input m_1 to the process produces the peak output c_1, with a phase lag of ψ; the controlled variable peak c_1 produces, through the controller, the next peak input to the process m_2, with a phase lag of γ. If m_2 is to be in the correct phase relationship to prolong the oscillation, it must occur exactly one cycle of the oscillation, or 2π radians, later than m_1. The first essential condition is, then, that

$$\psi + \gamma = 2\pi \tag{7-7}$$

The second requirement is that the amplitudes remain constant throughout, i.e. m_2 must be of the same magnitude as m_1, c_2 must equal c_1, and so on, if the oscillation is to be truly continuous and not damped or unstable. As already seen in Equations (7–5) and (7–6),

$$c_1/m_1 = |G_p| \quad \text{and} \quad m_2/c_1 = K_g$$

at any given frequency. Hence, if $m_2 = m_1$,

$$K_g = \frac{1}{|G_p|} = \frac{1}{K_p(\text{A.R.}_{\text{overall}})} \tag{7-8}$$

The controller gain must then be equal to the reciprocal of the process magnitude ratio, which is a logical statement that the attenuation of the process must be exactly counter-balanced by the amplification of the controller if the amplitude of the oscillation is to be maintained at a constant value.

From Equation (7–8), it follows that

$$K_g \times K_p = 1/\text{A.R.}_{\text{overall}}$$

where $K_g \times K_p$ will be recognized as the product of the steady state gains of every element in the loop and is therefore the overall gain of the loop, previously designated K, and in the present circumstances of limiting stability is the maximum value K_{max}. This alternative statement is then that the maximum overall gain of the loop is equal to the reciprocal of the overall amplitude ratio of the process elements at the condition of limiting stability.

Equation (7–8) can also be written in a further alternative form:

$$K_g|G_p| = K_{\text{max}}(\text{A.R.}_{\text{overall}}) = 1 \tag{7-9}$$

The product of the controller gain and the process magnitude ratio, or the overall gain and the process amplitude ratio, is sometimes referred to as the 'loop gain', and is the magnitude ratio of the open-loop transfer function $|G|$.

If the second condition defined by Equations (7–8) and (7–9) is not fulfilled, the amplitude of the oscillation will not be constant, e.g. if K_g is less than the reciprocal of $|G_p|$, then the reduction in amplitude by the process is greater than the amplification of the controller and $m_1 > m_2$; hence $c_1 > c_2$ and so on, so that the oscillation has a decreasing amplitude and is therefore damped and transient. If, on the other hand, $K_g > 1/|G_p|$, the amplification of the controller is greater than the attenuation of the process, i.e. $m_2 > m_1$ and $c_2 > c_1$, and the wave has an increasing amplitude and is unstable.

Equation (7–8) equates the controller gain to the reciprocal of the magnitude ratio, or the overall gain to the reciprocal of the amplitude ratio, of the process. The ratio of input to output amplitudes $(1/|G_p|)$ is the *attenuation ratio* which can be normalized like the magnitude ratio. An equivalent statement to Equation (7–8) is then that the controller gain equals the attenuation ratio, or the overall gain equals the normalized attenuation ratio, for the condition of limiting stability. This is the reason for plotting these reciprocal quantities, 1/A.R. in the Rutherford-Aikman diagram, and $1/|G|$ in the inverse Nyquist diagram.

The Inverse Nyquist Diagram and Stability

The inverse Nyquist diagram is a plot on polar co-ordinates of the attenuation ratio of the process $(1/|G|)$ *versus* the phase lag $(\angle G)$, where G is the open-loop transfer function of the process. Since each time-constant stage in the process can contribute up to 90° of phase lag at high frequencies, the curve is essentially a spiral about the origin for any system containing more than four such stages. The value of the attenuation ratio, being the reciprocal of the magnitude ratio, is unity at zero frequency and increases to infinity with increasing frequency. The curve thus starts at a value of one on the $\angle G = 0$ axis and spirals out with increasing frequency. The general shape of the curve is shown in Figure 7–5; in practice it is not usually possible to attain a phase lag approaching 270° as the response is then too attenuated to be measurable, i.e. the attenuation ratio reaches a very large magnitude before the lag attains 270°.

Any point on the inverse Nyquist curve, such as B in Figure 7–5, now defines a vector at a particular frequency of oscillation, ω_ψ; the length of this vector (OB) is the attenuation ratio at this frequency, and the $\angle BOA$ clockwise to the zero axis is the corresponding phase lag, ψ. Effectively OB is the amplitude of the input wave to the process which will produce a response wave of unit amplitude, the unit response vector (OA), at the particular frequency. It can immediately be seen that the $\angle AOB$ measured clockwise from the zero axis is $(2\pi - \psi)$, which is the value of the controller phase lag required by Equation (7-7) for the condition of limiting stability. The diagram thus illustrates one complete cycle of oscillation, from OB to OA with a phase lag of ψ through the process, and from OA to OB with a phase lag of γ over the controller.

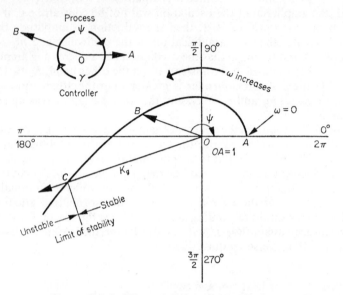

FIGURE 7-5. Stability and the inverse Nyquist diagram

In practice the frequency of the oscillation is determined by the controller adjustment of integral and/or derivative actions which define a controller output vector, such as OC, with a phase lag of γ and whose length is the controller gain K_g. The phase lag over the process can then only be the angle required to complete one cycle of the response wave, i.e. $(2\pi - \gamma)$. It should be noted that when the control loop is broken, oscillations of any frequency can be imposed on the system, but when the loop is closed the phase shift of the controller is effectively fixed at a certain value by the controller adjustment, and oscillation can now only occur at the single frequency which will permit a phase lag over the process given by

$$\psi = 2\pi - \gamma$$

This comment is rather over-simplified in that γ is also dependent on frequency when integral and/or derivative actions are present, and in practice the frequency is likely to vary over the first few cycles, but the main premise remains the same, there is only one frequency which can satisfy the $\psi + \gamma = 2\pi$ requirement when the control loop is closed.

The second condition for stability of the oscillation is that the controller gain should not exceed the attenuation ratio. The value of the latter is defined by the curve, and this condition is fulfilled if the controller vector OC does not cross the curve; the system will then be stable. If the vector tip just touches the curve then the gain is equal to the attenuation ratio and continuous oscillations will be set up by any finite disturbance.

It is useful to continue this discussion in considering the application of the various continuous control actions.

Proportional Control

Proportional control is defined by the equation $m = K_c e$, where $e = v - c$. In the absence of desired value changes, $v = 0$ and $m = -K_c e$. For a continuous oscillation of the controlled variable, $c = A \sin \omega t$, the controller output will be $m = -A K_c \sin \omega t$. Thus the amplification or gain of the controller given by the ratio of amplitudes is the proportional sensitivity K_c, and the phase angle between input and output waves is π radians, since the input and output are always of opposite sign.

The proportional control vector will thus lie on the π-axis of the diagram, the phase angle γ being independent of frequency, and the phase lag over the process can only be $2\pi - \gamma$, which is also π. The corresponding frequency is thus a unique or critical value and may be denoted by ω_c.

It is interesting to note the confirmation of the previous discussions regarding the effect of the proportional sensitivity on the stability of the system with proportional control. K_c is the length of the vector on the π-axis and there is a maximum value $(K_{c\,max})$ at which the vector just touches the curve at the limit of stability. For values less than the maximum, the gain is less than the attenuation and the system response will be stable and damped; for greater values the response will be unstable.

Integral Control

For integral control with a constant desired value, $m = -C \int e \, dt$, where the constant has the dimension of reciprocal time. For an input of $c = A \sin \omega t$, the output is $m = A(C/\omega) \cos \omega t$. The output/input amplitude ratio is thus (C/ω) and the phase difference between the waves is $\pi/2$ lead or $3\pi/2$ lag. The latter is more logical in the circumstances and the controller phase angle γ is thus 270° clockwise from the zero axis. The integral control vector then lies on the $\pi/2$-axis of the inverse Nyquist diagram. This in turn fixes the phase lag in the process for continuous oscillations at $\pi/2$ and the frequency of such oscillations at a unique value of $\omega_{\pi/2}$. The controller gain, the length of the control vector, is C/ω.

Comparing the two cases for proportional and integral control (Figure 7-6), the attenuation ratio at a phase lag in the process of $\pi/2$ is very much less than that at a phase lag of π; hence the control vector required for integral control is much smaller than that for proportional. Of much greater importance is that the frequency of continuous oscillation for integral control $(\omega_{\pi/2})$ will be very much less than the frequency for proportional control (ω_π or ω_c), since frequency increases anti-clockwise around the curve. Although the stability limit is a continuous oscillation, whereas in practical applications a damped oscillation is required, the frequencies of the damped oscillations with the same degree of damping will be of the same order of magnitude as the undamped oscillations. The response of a system with integral control will therefore be much slower than that of the same system with proportional control. For integral control to be preferable to proportional control requires that the system should have a very fast response so that a reasonably short recovery time can be obtained, and integral control used alone is, in fact, almost restricted to cases where there is little or no capacitance in the system and the response is therefore fast.

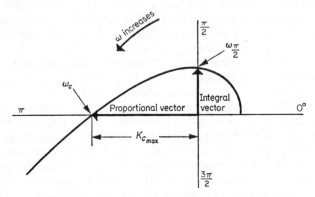

FIGURE 7-6. Proportional and integral control vectors

Proportional-integral Control

For the combination of proportional with integral control, the two control vectors on the π- and $\pi/2$-axes are combined to give a resultant which must lie in the second quadrant ($\psi = 90 \rightarrow 180°$) of the diagram. The transfer function of the combined action is

$$G_c = K_c[1 + 1/(T_i s)]$$

and substituting $j\omega$ for s gives the complex number

$$G_c(j\omega) = K_c[1 - j/(\omega T_i)]$$

from which

$$|G_c| = K_c[1 + 1/(\omega^2 T_i^2)]^{\frac{1}{2}}$$

and

$$\angle G_c = \tan^{-1}(-1/\omega T_i)$$

The resultant PI vector thus lags behind the proportional vector, the integral action increasing the controller phase angle γ from the π of proportional control to $\pi + \lambda$ for PI control, where λ is given by $\tan^{-1}(1/\omega T_i)$.

The two important features can readily be seen from the diagram of Figure 7–7. Firstly, the length of the proportional vector must be reduced to less than that with proportional control alone if the tip of the resultant vector is not to be brought outside the curve and so make the system unstable. Secondly, the frequency of the continuous oscillation is reduced below that with proportional control since the phase lag of the process is now $(\pi - \lambda)$. This again confirms the findings of previous discussions, that when integral action is added to proportional control the response of the combined action is slower for a given degree of damping than with proportional control alone, and also that, to maintain stability in the system, a reduction in the proportional gain (K_c) may be required. It can also be seen that instability can be caused by too much integral action, since increase in the integral vector can also bring the resultant outside the curve.

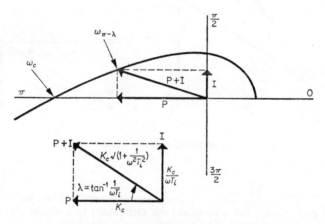

FIGURE 7–7. Proportional-integral control vectors

Since the only beneficial effect of integral action is the elimination of offset, it follows that the minimum amount should be used to limit the relative loss of stability and the slowing down of the response. In practice a phase lag of only some 10° is found to be necessary to eliminate offset in a reasonable time without requiring undue reduction of the proportional gain.

Proportional-derivative Control

In this case the controller transfer function is

$$G_c = K_c(1 + T_d s)$$

and substitution of $j\omega$ for s leads to the complex number

$$G_c(j\omega) = K_c(1 + j\omega T_d)$$

The gain and phase angle are thus

$$|G_c| = K_c(1 + \omega^2 T_d^2)^{\frac{1}{2}}$$

and

$$\angle G_c = \tan^{-1}(\omega T_d)$$

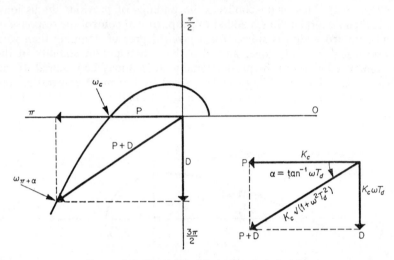

FIGURE 7–8. Proportional-derivative control vectors

Consideration of the response of the derivative action alone ($m = -K_c T_d(dc/dt)$ for a constant desired value) to a sinusoidal input of $c = A \sin \omega t$ shows that the output is $m = -AK_c\omega T_d \cos \omega t$. The derivative response thus lags $\pi/2$ on the input wave and the vector will accordingly lie on the $3\pi/2$-axis of the inverse Nyquist diagram, with a length of $K_c\omega T_d$. Combining this with the proportional control vector gives a resultant which lies in the third quadrant ($\psi = 180 \to 270°$) of the diagram. The result is basically the complete reversal of the PI case; the controller phase angle is reduced from π to $\pi - \alpha$, where α is a phase advance on the proportional action given by $\tan^{-1}(\omega T_d)$. The length of the proportional vector can in general be increased, without loss of stability, to a value much greater than that with proportional control alone, although there will be a maximum value since the curve must ultimately turn back to cross the $3\pi/2$ axis. In practice most process systems will accept up to 30° phase advance by derivative action and often more, up to about 60°. The larger the phase advance which can be introduced, the greater is the phase lag ($\pi + \alpha$) over the process, and thus the higher the frequency of the continuous oscillation. The response of the PD combination is thus faster and more stable than that of proportional control alone.

Proportional-integral-derivative Control

For PID control three control vectors have to be combined to give a resultant which theoretically could lie in the second or third quadrant, depending on the relative magnitudes of the integral and derivative vectors which are directly opposed. However, the derivative action is added to give faster response and improved stability, and to provide this the resultant must lie in the third quadrant as with PD control. The derivative vector must then be larger in magnitude than the integral so that the net result is a phase advance on the proportional; the phase advance of the derivative is, however, opposed by the phase lag of the integral action so that the increase in frequency and proportional gain are not quite so marked as with PD control, but offset will, of course, be eliminated by the integral action.

In practice the mechanism of the three-term controller often limits the amount of phase advance from the derivative component, and the non-ideal behaviour of the controller must be taken into account.

Integral-derivative Control

It is interesting at this stage to consider the possibility of a third dual-function action, the combination of integral with derivative action. This combination is theoretically possible since the integral action provides the correlation with the value of a constant error which is lacking in the derivative action. However, study of the combination on the inverse Nyquist diagram reveals that the combination is hardly a practical possibility. The two control vectors are directly opposed and, depending upon which is the larger, the resultant would show a controller phase angle of $\pi/2$ or $3\pi/2$ and the corresponding frequency of oscillation would be either extremely fast or extremely slow.

A more serious disadvantage is that both control vectors are dependent on frequency, the integral being effectively proportional to $1/\omega T_i$ and the derivative to ωT_d. There will then be a frequency (proportional to $1/(T_i T_d)^{\frac{1}{2}}$) at which the vectors are equal and therefore give a zero correction; at values near to this critical frequency only small corrections will be produced. Such frequencies may be impressed on the control system by external disturbances such as pressure pulsations, mechanical vibration, etc. and this is a much too serious risk to face in most applications.

Bode Diagrams for Open-loop Frequency Response

The frequency response of the open-loop system is best considered by means of the Bode diagram, since this provides a simple graphical means of combining the frequency response characteristics of individual elements into the overall response of the system.

First-order Systems

The frequency response characteristics of a first-order system with transfer function $K_p/(Ts + 1)$, have already been derived, the pertinent results being (Equation 7–3)

$$|G| = K_p/(1 + \omega^2 T^2)^{\frac{1}{2}}$$

and (Equation 7–4)

$$\angle G = \tan^{-1}(-\omega T)$$

Dividing the magnitude ratio by the steady-state gain gives the amplitude ratio (Equation 7–1) as

$$\text{A.R.} = 1/(1 + \omega^2 T^2)^{\frac{1}{2}}$$

The Bode diagram consists of two parts, a logarithmic plot of the A.R. *versus* frequency, and a semi-logarithmic plot of phase angle *versus* log frequency. Whilst it is not unusual for some users to superimpose the two plots with a common log-frequency axis, there is some virtue in plotting the A.R. curve above the phase-angle curve, retaining the same log frequency scale, particularly when the responses of several elements are to be combined.

It will be seen from Equation (7–1) that the amplitude ratio is asymptotic to one at low frequencies ($\omega T \rightarrow 0$) and to $1/\omega T$ at high frequencies ($\omega T \gg 1$). On the logarithmic co-ordinates these asymptotes are two straight lines, A.R. = 1 and A.R. = $1/\omega T$, the latter having a slope of -1, and which intersect at a value of $\omega T = 1$, i.e. at a 'corner' frequency of $\omega = 1/T$. Substituting this value of $\omega T = 1$ in the expression for the amplitude ratio (Equation 7–1) gives a value of $1/\sqrt{2}$ or 0.707 at the corner frequency. The A.R. curve thus passes through this value, which represents a maximum departure from the two asymptotes of 30 per cent. For many purposes the asymptotes may often be taken as representative of the curve with reasonable accuracy; if greater accuracy is desired, a smooth curve passing through the 0.707 value and tangential to the two asymptotes is generally adequate.

In the lower half of the diagram, the phase angle (which is, of course, a lag and is therefore plotted downwards from zero) is asymptotic to zero at low frequencies ($\tan^{-1}(\omega T) \rightarrow 0$) and to 90° at high frequencies, passing through a value of 45° at the corner frequency ($\tan^{-1} 1$). The curve is axially symmetrical about the 45° point. Additional points on the curve can be easily obtained by use of a table of natural tangents.

It will be appreciated that the A.R. and phase-angle curves are identical for all first-order elements, being displaced along the frequency scale according to the value of the time constant which determines the corner frequency. Figure 7–9 shows the general case of a first-order system Bode diagram with A.R. and phase lag plotted against the parameter ωT. Since all first-order curves are the same when plotted on the same scale, it is possible to use a template for drawing the curves; the construction and use of such a device is described by Caldwell, Coon, and Zoss [4]. It is not, of course, essential to use logarithmic or semi-logarithmic graph paper for drawing Bode diagrams. The logarithmic decades can be ruled on normal graph paper and the 2 and 5 points ruled off at 0.3 and 0.7 with sufficient accuracy for most purposes.

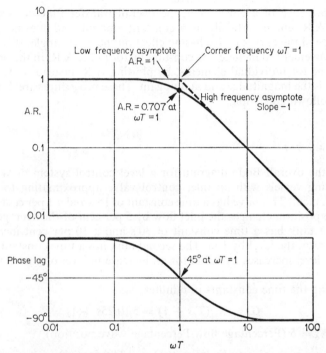

FIGURE 7–9. Open-loop Bode frequency response diagram for a first-order element

Combination of Elements

Before proceeding to develop further Bode diagrams for other process elements, it is desirable to consider the graphical technique for combining the frequency response characteristics of a number of elements in series into those of the overall system. It has already been noted that the overall A.R. of a series of elements is the product of the individual A.R.'s and that the A.R. scale of the Bode diagram is logarithmic. Hence

$$\log (A.R._1) + \log (A.R._2) + \ldots = \log (A.R._{overall})$$

The overall A.R. curve is then obtained by simply summing the individual curves, i.e. by adding the individual log (A.R.) values at the same frequency, values above unity being taken as positive and those less than unity as negative (since $\log (A.R.) = 0$ when A.R. $= 1$). Cases of the A.R. exceeding unity will be seen in due course.

The overall phase angle is the sum of the individual phase angles:

$$\angle G_1 + \angle G_2 + \ldots = \angle G_{\text{overall}}$$

Since the phase angle curve is arithmetic, the values at a particular frequency can be summed without difficulty, phase leads being regarded as positive and phase lags as negative.

The presence of a constant in the overall transfer function shifts the overall A.R. curve vertically by a constant amount, and does not change the shape of the curve; it also has no effect on the phase angle. It is usually more convenient to include any constant factor in the A.R. in the ordinate, as is done for individual elements, by plotting A.R. instead of M.R., i.e. the plot is the magnitude ratio at unit gain. These procedures are illustrated in the following example.

Example 7-1

Plot the overall Bode diagram for a level control system of two non-interacting vessels with an inlet control valve approximating to a first-order element. The valve has a time constant of 15 s and a 1 per cent change in valve position changes the inlet flow by 5 per cent of the average value. The first tank has a time constant of 30 s and a 10 per cent increase in inflow raises the level by 1 m. The second tank has a time constant of 60 s and the level increases by 0.8 m for an increase in level of 1 m in the first tank.

Writing the time constants in minutes,

$$G_v = K_v/(T_v s + 1) = 5/(0.25 s + 1)$$

where $K_v = 5$ (Percentage flow/Percentage valve position)

$$G_{p1} = K_{p1}/(T_1 s + 1) = 0.1/(0.5 s + 1)$$

where $K_{p1} = 0.1$ (m/Percentage flow)

$$G_{p2} = K_{p2}/(T_2 s + 1) = 0.8/(s + 1)$$

where $K_{p2} = 0.8$ (m/m).

On the Bode diagram (Figure 7–10), the corner frequencies for the three individual elements are given by $\omega = 1/T$, and are thus at 4, 2, and 1 rads/min. The individual A.R. and phase-lag curves are identical with those of Figure 7–9 moved horizontally to the appropriate corner frequency; the overall curves are the sums of the individual curves. The slope of the overall asymptote is seen to be a good indication of the shape of the overall A.R. curve. The slope changes by -1 at each corner frequency; in the present case the slope is 0 up to $\omega = 1$, -1 for $\omega = 1 \rightarrow 2$, -2 for $\omega = 2 \rightarrow 4$, and -3 above $\omega = 4$.

The advantage of plotting the normalized ratio (i.e. $|G_p|/K_p$) in place of the magnitude ratio is worthy of note. By plotting A.R., the low frequency value is unity in each case and the individual curves do not overlap or intersect. If the magnitude ratio is plotted, the low frequency value is the gain of the individual element and the individual curves start at different levels (in this case, 5, 0.1, and 0.8) and may intersect, as would happen in this case where the order of increasing magnitude of the gains differs from that of the time constants. This would make the plotting and the graphical addition rather more difficult.

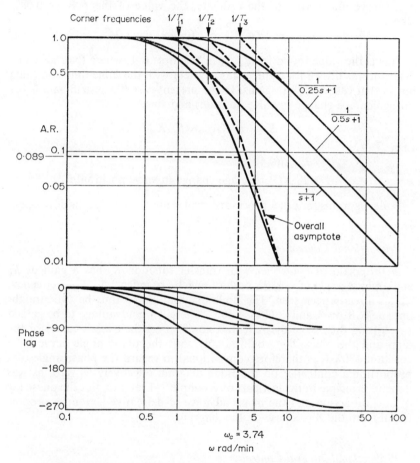

FIGURE 7–10. Combination of elements on the Bode diagram (Example 7–1)

The overall A.R. curve is now effectively the magnitude ratio curve at unit gain, i.e. the values of $|G|$ divided by the product of the individual stage-gains, $(5)(0.1)(0.8) = 0.4$.

An initial application of the Bode method of plotting may now be demonstrated by use of the discussion on pages 234–7. From Figure 7–10, the frequency at which a phase lag of 180° occurs in the three-stage system is 3.74 rads/min. This is the critical frequency ω_c at which continuous oscillations are set up at the maximum value of the proportional sensitivity when proportional control is applied to the system. The overall gain of the system with proportional control has already been seen (Equation 7–8) to be equal to the reciprocal of the A.R. at the critical frequency of continuous oscillation; by projecting up from the phase lag of 180° to the A.R. curve and across to the ordinate, the value of the A.R. is 0.089. Hence

$$K_{max} = 1/(A.R.) = 1/(0.089) = 11.24$$

This is the value found for $K_{c\,max}$ in Example 6–1, where the root locus diagram was drawn for a three-stage system with the same time constants but in that case with unit gains. In the present case the system gain is 0.4 (m/percentage change in valve position) and since

$$K_{max} = K_{c\,max}K_vK_{p1}K_{p2}$$

then

$$K_{c\,max} = K_{max}/(0.4)$$
$$= 28.1 \text{ (Percentage change in valve position/m)}$$

Bode diagrams for other process control elements will now be developed.

Proportional Controller

A proportional controller with transfer function K_c has a gain of K_c and a phase angle of $-180°$, both of which are independent of frequency, as has already been seen. The Bode diagram would thus be two straight lines at A.R. = 1 and $\angle G = -180°$. There is no advantage to be gained by plotting this, since the A.R. has no effect on the overall curve for the loop and the phase lag would simply shift the phase angle curve by a constant 180°. It is therefore conventional to regard the phase angle of a proportional controller on the Bode diagram effectively as zero, as was done in principle in the last part of Example 7–1, and to plot the phase lag of integral action and the phase advance of derivative action on proportional from the zero phase on the diagram.

Proportional-integral Controller

The ideal transfer function of the PI controller is $K_c(1 + 1/T_i s)$; substituting $j\omega$ for s, the frequency response characteristics are

$$A.R. = |G(j\omega)|/K_c$$
$$= [1 + 1/(\omega^2 T_i^2)]^{\frac{1}{2}}$$
$$\angle G(j\omega) = \tan^{-1}[-1/(\omega T_i)]$$

Using the parameter ωT_i as abscissa, the Bode diagram is shown in Figure 7-11(a). The A.R. is asymptotic to 1 at high frequencies and to $1/(\omega T_i)$ at low frequencies, the asymptotes intersecting at the 'reset' corner given by $\omega T_i = 1$. Since the element is a controller, the A.R. is an amplification and the maximum value is infinity. The phase angle is $-90°$ at low frequencies, $-45°$ at the corner frequency, and approaches zero at high frequencies. The curves are the mirror images of those of the first-order system of Figure 7-9.

Proportional-derivative Controller

The transfer function for the PD controller is $K_c(1 + T_d s)$, and the amplitude and phase behaviour are the complete inversion of the first-order system with transfer function, $1/(Ts + 1)$. The derivative action introduces a *linear lead*, $(1 + Ts)$, as compared to the linear lag, $(1 + Ts)^{-1}$, of the first-order system. Substitution of $j\omega$ for s gives an A.R. of $(1 + \omega^2 T_d^2)^{\frac{1}{2}}$ and a phase angle of $\tan^{-1}(+\omega T_d)$. The Bode diagram is shown in Figure 7-11(b); since the phase angle is now an advance, the addition of derivative action will reduce the phase lag of the process at all frequencies. In particular, the 180° phase lag will now occur at a higher frequency, and this introduces the stabilizing influence and permits a lower value of the overall gain (see Example 7-4).

Proportional-integral-derivative Control

For the transfer function, $K_c(1 + 1/T_i s + T_d s)$, the system exhibits infinite gain at both high and low frequencies. If T_d is much less than T_i, the net response is basically that of proportional-integral control at low frequencies and proportional-derivative control at high frequencies. The A.R. curve has a slope of -1 at low frequencies to the reset corner, zero at intermediate values, and a slope of $+1$ at high frequencies from the derivative corner. The two corner frequencies are given by $\omega T_i = 1$ and $\omega T_d = 1$, and the curve is symmetrical about an intermediate frequency given by $\omega = 1/(T_i T_d)^{\frac{1}{2}}$. The phase curve goes from $-90°$ to $+90°$ and has inflections at the two corner frequencies and at the point between where the angle is zero at $\omega = 1/(T_1 T_2)^{\frac{1}{2}}$. Figure 7-11(c) shows a typical case where $T_d = 0.25T_i$.

When T_d is near to or greater than T_i, the A.R. and phase angle must be calculated from the expressions obtained by substituting $j\omega$ for s in the transfer function, i.e.

$$\text{A.R.} = [1 + (\omega T_d - 1/\omega T_i)^2]^{\frac{1}{2}}$$

and

$$\angle G = \tan^{-1}(\omega T_d - 1/\omega T_i)$$

A large value of the ratio T_d/T_i puts the derivative corner to the left of the reset corner, and the two asymptotes of the A.R. curve cross each other.

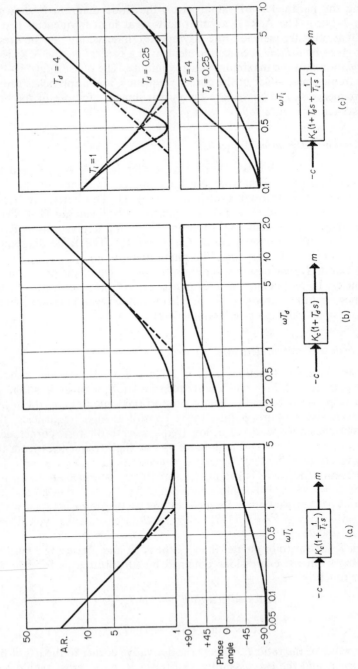

FIGURE 7-11. Bode diagrams for controllers: (a) proportional-integral, (b) proportional-derivative, (c) proportional-integral-derivative

For most frequencies the gain, as shown by the A.R. curve, is much greater than the nominal but drops sharply near the intersection of the asymptotes. This is shown in Figure 7–11(c) for $T_d = 4T_i$. Such large values of T_d/T_i are very rarely used in practice since the sharp drop in gain is undesirable.

The diagrams of Figure 7–11 are derived from the ideal transfer functions for the three cases. Practical instruments usually have a gain which is limited to a maximum value by the internal feedback mechanism of the controller, and the phase angles at high and low frequencies may also differ from the ideal values. These departures from ideality are discussed in Chapter 8.

Second-order System

The frequency response characteristics of the second-order system with a transfer function of

$$G = 1/(T^2 s^2 + 2\zeta T s + 1)$$

is obtained by substituting $j\omega$ for s, giving the complex number

$$G(j\omega) = \frac{(1 - \omega^2 T^2)}{(1 - \omega^2 T^2)^2 + 4(\zeta \omega T)^2} - j\frac{2(\zeta \omega T)}{(1 - \omega^2 T^2)^2 + 4(\zeta \omega T)^2}$$

from which

$$\text{A.R.} = 1/[(1 - \omega^2 T^2)^2 + 4(\zeta \omega T)^2]^{\frac{1}{2}} \qquad (7\text{–}10)$$

$$\angle G = \tan^{-1}[-2\zeta \omega T/(1 - \omega^2 T^2)]$$

If the parameter ωT is used as the abscissa on the Bode diagram it is evident that there will be a family of curves for different values of ζ. For small values of ωT the A.R. approaches unity, and for large values is asymptotic to $1/(\omega T)^2$, the asymptote having a slope of -2 and intersecting A.R. $= 1$ at $\omega T = 1$. The important feature of the second-order frequency response (Figure 7–12) is that for values of $\zeta < 1/\sqrt{2}$, the A.R. passes through a maximum value greater than unity, in the vicinity of $\omega T = 1$. The frequency at which this peak A.R. occurs is the *resonant frequency*, ω_r. This is a well-known property of under-damped second-order systems in which an input oscillation is amplified over a range of frequencies around the resonant frequency. The value of the maximum A.R. and the resonant frequency can be found by differentiating Equation (7–10) and equating to zero, which gives

$$\text{A.R.}_{\text{max}} = 1/[2\zeta(1 - \zeta^2)^{\frac{1}{2}}] \qquad (7\text{–}11)$$

for $\zeta < 0.707$ and

$$\omega_r = (1 - 2\zeta^2)^{\frac{1}{2}}/T = \omega_n(1 - 2\zeta^2)^{\frac{1}{2}}$$

where $\omega_n \, (= 1/T)$ is the natural frequency of the second-order system. It can be seen from Equation (7–11) that the value of the peak A.R. and the resonant frequency are dependent on the value of ζ.

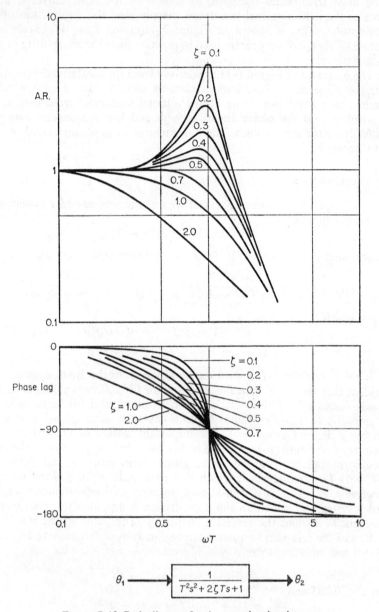

FIGURE 7–12. Bode diagram for the second-order element

The curves of Figure 7–12 are not easily constructed when the second-order system is under-damped ($\zeta < 1$), but fortunately most such systems for which Bode diagrams are required are composed of two first-order elements in series and are thus critically or over-damped. The curves of Figure 7–12 are of much greater value in analysing the frequency response of the closed loop which also shows a peak value of amplitude or magnitude ratio. By comparison with the resonant peak of the second-order system, remembering that the response of most process systems is predominantly second-order, it is possible to estimate values of ζ for the closed loop which in turn may be used to evaluate the transient response of the closed loop.

The phase-angle curves of the second-order system are not of such utility; it can be seen that they also differ for different values of ζ but have a number of points in common. For small values of ωT, the angle is approximately $\tan^{-1}(-2\zeta\omega T)$ and is thus zero at low frequencies. For $\omega T = 1$, the angle is 90° irrespective of ζ, and as ωT increases the angle approaches a maximum of $-180°$. Thus the phase curves all start at zero, pass through the value of 90° at $\omega T = 1$, and approach the maximum of 180°. Since a second-order system with $\zeta \geqslant 1$ is equivalent to two first-order elements in series, this is consistent with a maximum of 180° phase lag and an asymptote of -2 on the A.R. curve at high frequencies.

Time Delay

The frequency response of a time delay can be deduced quite simply. A true time delay has no effect on the magnitude of the signal but only delays the transmission. The amplitude of a sinusoidal signal is therefore unchanged by a time delay and the A.R. is unity at all frequencies. The time delay, however, is evident as a lag in time between output and input, there will thus be a lag in phase between the two sinusoids whose magnitude will be dependent on the frequency of the oscillation. For a time delay of length L, the phase lag will be $2\pi L/P$ radians, where P is the period of oscillation; since $P = 2\pi/\omega$, the phase angle will be ωL radians, or $57.3\omega L°$.

These results are confirmed by substituting $j\omega$ for s in the transfer function of the time delay, $\exp(-Ls)$, which gives the complex number

$$G(j\omega) = (1)\exp(-j\omega L)$$

i.e. a magnitude of 1 and an angle of $-\omega L$ radians, the minus sign indicating that the angle is a lag.

Since ω can take on any value from 0 to ∞, the phase lag for a time delay has no limit, and any system with a time delay must have a finite critical frequency since the phase angle must always pass through 180° at some value of the frequency. Figure 7–13 shows the phase angle for a time delay with ωL as the abscissa.

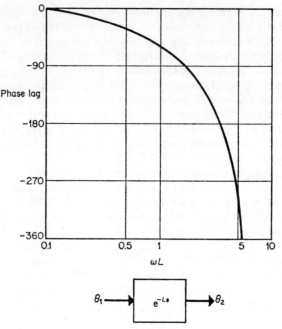

FIGURE 7–13. Bode diagram for a time delay

Capacitance Element

The transfer function of a capacitance element, i.e. the first-order element without self-regulation, is $1/Ts$, and substituting $j\omega$ for s gives a complex number with an imaginary part only, i.e.

$$G(j\omega) = -j/\omega T$$

The A.R. is thus $1/\omega T$, and the phase lag is 90° at all frequencies. The A.R. is a straight line of slope -1, passing through the value of 1 at $\omega T = 1$, as shown in Figure 7–14(a).

Exothermic Reactions

As seen on pages 199–200, an exothermic reaction has a transfer function of $K_p/(1 - Ts)$ when heat is extracted from the system by cooling, or $K_p/(Ts - 1)$ when heat is added. Substituting $j\omega$ for s in these functions gives complex numbers which differ only in the signs of the real and imaginary parts, those for the first case (cooling) both being positive $(+ x + jy)$ and those for the second (heating) being negative $(-x - jy)$. The amplitude ratio is thus the same in each case and is actually the same as that for a first-order system, i.e.

$$\text{A.R.} = 1/(1 + \omega^2 T^2)^{\frac{1}{2}}$$

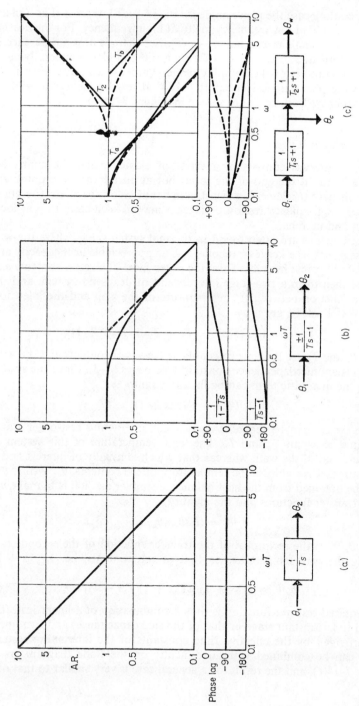

FIGURE 7–14. Bode diagram for: (a) capacitance (integrating) element, (b) exothermal systems, (c) side capacity

For the cooling case the phase angle is $\tan^{-1}(+\omega T)$, there is thus a phase advance from $0°$ at low frequency to $90°$ at high frequency. For the heating case the phase angle is also $\tan^{-1}(+\omega T)$, but there is now a difference in phase of $180°$ due to the vectors forming the complex numbers lying in opposite quadrants of the complex plane. Thus for a heated exothermic reaction the phase angle is a lag of $-180°$ at low frequency to $-90°$ at high frequency. Both cases are shown in Figure 7–14(b).

Side Capacity Effect

When a process consists of a series of elements and the controlled variable is not at the end of the series but at an intermediate point, the further element or elements downstream act as side capacities. The phase angles of the frequency response diagram may in such cases exhibit local maxima and minima.

Side capacities are often found in thermal systems, e.g. where the wall of a vessel such as a kettle or reboiler has an appreciable heat capacity and is insulated against heat loss from the outer face. The vessel contents and the wall then form a two-stage interdependent dead-end system, and the transfer function between the temperature of the wall and the inlet flow to the vessel will be given by

$$\theta_w/\theta_i = 1/[T_1 T_2 s^2 + (T_1 + T_2 + T_{12})s + 1]$$

where T_1 and T_2 are the time constants of the vessel contents and the wall, and T_{12} is an interdependence constant (see pages 61–3). Since the system is real, the quadratic term can be factorized into

$$1/[(T_a s + 1)(T_b s + 1)]$$

i.e. the system can be regarded as two non-interacting first-order stages with time constants T_a and T_b. The 'end' temperature of this system is, however, that of the wall, whereas that which is usually of interest and is the controlled variable, is the temperature of the contents, θ_c. Whilst this could be obtained from the heat balance, a simpler method is to make use of the transfer functions and the identity

$$\theta_c/\theta_i = (\theta_c/\theta_w)(\theta_w/\theta_i)$$

where θ_c/θ_w is the reciprocal of the transfer function of the second stage relating the wall temperature to that of the contents, i.e. $1/(T_2 s + 1)$. Thus

$$\theta_c/\theta_i = (T_2 s + 1)/[(T_a s + 1)(T_b s + 1)]$$

The required transfer function is thus a combination of a linear lead (the second time constant stage or that of the side capacitance) and two linear lags provided by the effective time constants of the interacting system. These can be combined without difficulty on the Bode diagram, as in Figure 7–14(c), and the result for a general case is very similar to that of a first-order system with a maximum 90° lag, as would be expected, but the A.R. curve shows an inflection due to the change in the slope of the asymptote from -1 to 0 and back to -1, since T_2 must always lie between T_a and T_b. The phase angle curve shows a considerable 'flattening' around the centre, and if T_a and T_b are very different in value, the phase lead contributed by T_2 may make the phase curve pass through a maximum value [11].

For any thermal system such as a kettle, reboiler, heat exchanger, etc., the outer wall will act as a side capacity in this way. The top plate of a distillation column shows similar unusual phase and amplitude curves, owing to the lower plates in the column acting as side capacities which interact with the top plate.

Distributed Parameters

As discussed in Chapter 2, a distributed system, such as a thick wall in a heat transfer path or a pressure transmission line, is equivalent to an infinite number of interacting elements in series. The simplest case to consider is that in which the distributed element does not interact with preceding or succeeding elements, which is the case if the input is constant and there is no temperature or pressure gradient at the output end. For a thermal system with an insulated outer face or a pneumatic transmission system where the volume at the end of the line is small compared to the total volume, the transfer function which can be derived is that of Equation (2–33), i.e.

$$G = 1/\cosh (Ts)^{\frac{1}{2}}$$

where $T = X^2/K$ for the thermal system (X being the wall thickness and K thermal diffusivity), and RC for the transmission line where R and C are the total resistance and capacitance of the system.

The frequency response characteristics are obtained by substituting $j\omega$ for s in the transfer function giving the complex number

$$G(j\omega) = 1/\cosh (j\omega T)^{\frac{1}{2}} \qquad (7\text{--}12)$$

Substituting the positive form of $\sqrt{j}$, i.e. $[(1 + j)/\sqrt{2}]$ in Equation (7–12) (since use of the negative form gives a result with A.R. > 1 and a phase lead, which is contrary to the response of a real system),

$$G(j\omega) = 1/\cosh [(1 + j)(\tfrac{1}{2}\omega T)^{\frac{1}{2}}]$$
$$= 1/\cosh [(\tfrac{1}{2}\omega T)^{\frac{1}{2}} + j(\tfrac{1}{2}\omega T)^{\frac{1}{2}}]$$

Using now the expansion for $\cosh (x + jy)$, this equation can be resolved into the usual form of a complex number:

$$G(j\omega) = \frac{\cosh \phi \cos \phi}{\cosh^2 \phi - \sin^2 \phi} - j \frac{\sinh \phi \sin \phi}{\cosh^2 \phi - \sin^2 \phi} \qquad (7\text{--}13)$$

where $\phi = (\tfrac{1}{2}\omega T)^{\frac{1}{2}}$.

From Equation (7–13), the magnitude and argument of the complex number can be obtained in the usual way, thus:

$$A.R. = 1/[\cosh^2 (\tfrac{1}{2}\omega T)^{\tfrac{1}{2}} - \sin^2 (\tfrac{1}{2}\omega T)^{\tfrac{1}{2}}]^{\tfrac{1}{2}}$$

and

$$\angle G = \tan^{-1} [\tanh (\tfrac{1}{2}\omega T)^{\tfrac{1}{2}} \tan (\tfrac{1}{2}\omega T)^{\tfrac{1}{2}}]$$

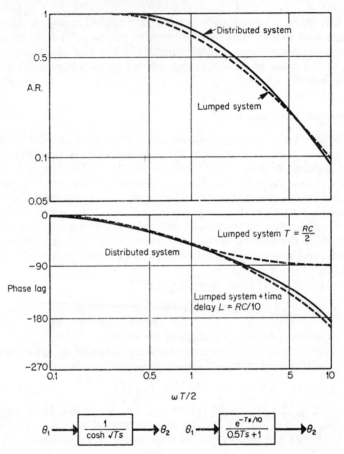

FIGURE 7–15. Bode diagram for distributed parameter system and lumped parameter model

These characteristics are plotted on the Bode diagram of Figure 7–15; the A.R. is one at low frequencies and decreases as the frequency is increased. The phase lag increases indefinitely with increasing frequency, thus showing the non-minimum phase behaviour typical of a distributed system.

It was shown on pages 87–8 that the step function response of a distributed system could be approximated by that of a lumped parameter first-order system with a time constant equal to half the equivalent time constant of the distributed system, i.e.

$$T = \tfrac{1}{2}X^2/K \quad \text{or} \quad \tfrac{1}{2}RC$$

The frequency response of such a model is also plotted in Figure 7–15, and the two responses are seen to be very similar. The A.R.s are reasonably close at all frequencies but the phase lag of the lumped parameter model, being first-order, is limited to 90° and thus only approximates to that of the distributed system at low frequencies up to a phase lag of about 50°. Thus the distributed element can be replaced by a single time-constant model with reasonable accuracy if the distributed element contributes less than 50° phase lag at the frequencies of interest. Adding a time delay to the model will provide additional phase lag at higher frequencies and improve the approximation if the phase lag is over 50°. Adding the time delay will not affect the A.R. curve in any way since the A.R. for the time delay is one at all frequencies. This is also shown in Figure 7–15, where a time delay equal to one tenth of the equivalent time constant is added ($L = X^2/10K$ or $RC/10$).

If heat or a fluid flows through a distributed element in series with an upstream and/or downstream element, interaction will occur and there is no simple or graphical solution. Fortunately, in most examples of heat transfer the plant design is such as to promote transfer, and accordingly tends to high conductivity (low resistance) and a short transfer path. The resistance of a heat transfer wall is then generally small compared to that of the fluid films on either side, and there is usually no need to consider the wall as a distributed system; the lumped system will give negligible error at all frequencies likely to be encountered in process control. However, for boundary zones such as lagging and thermal insulation, the design conditions are reversed, and the effective time constant is likely to be so large that the lumped parameter model is valid only over a limited range; it is then advizable to employ the distributed system analysis.

Comparison of System Controllability

The purposes of process control is to minimize the effect of disturbances on the output of the process. A logical measure of the system performance in attaining this end is the time integral of the absolute error, i.e. $\int |e| \, dt$, for a given disturbance. It is reasonable to assume that any economic losses caused by deviation of the controlled variable from the desired value will be proportional to the error and the time during which the error persists; the absolute error is used since it is also reasonable to assume that positive and negative deviations are equally important.

The response of a control system to a disturbance should normally be an under-damped oscillation with a subsidence ratio of about 4:1; the frequency of the damped oscillation will be somewhat less (10–30 per cent) than the critical frequency, i.e. ω_c, the frequency of the continuous oscillation at limiting stability with proportional control alone. The error integral is the sum of the areas enclosed by the peaks of the oscillation (see Figure 5–10) with the final steady-state value taken as the base-line. For systems which are adjusted to the same subsidence ratio only the first peak need be considered, and the error integral will be roughly proportional to the height of the first peak and the reciprocal of the critical frequency (the latter as a measure of the period). The height of the first peak depends on a number of contributing factors, e.g. the size of the load change and the point of imposition, and also on the system parameters such as the load gain, K_L, and the overall gain of the loop, K. For comparison of the performances of two or more systems, the first two factors would obviously be maintained the same, and it is unlikely that there would be any changes in the load gain K_L. If the load change is imposed at the start of the process, with a subsidence ratio of 4:1, the height of the first peak following a load change is about 50 per cent more than the steady-state error or offset (if proportional control alone is used). The steady-state error with proportional control is $K_L/(1 + K)$ and the height of the first peak will then be roughly $1.5K_L/(1 + K)$. Adding integral action will eliminate the offset but will not appreciably affect the height of the first peak, since the integral action has little effect during the first half-cycle. The estimate of the peak error can then be based on the offset that would occur with the particular value of K if only proportional control were used.

In attempting to improve the performance of a control system, changes could be made in the number or magnitude of time constants, delays, etc. which will change the critical frequency and the recommended overall gain, but it is not likely that there will be any major changes in the load gain K_L. This will probably remain a common factor between the different systems, and will thus cancel out if the error integrals are compared as an error integral ratio; the peak error then is left inversely proportional to the factor $(1 + K)$. The error integral has already been suggested as roughly proportional to the peak error and the reciprocal of the critical frequency, it thus becomes inversely proportional to both $(1 + K)$ and ω_c. To compare the performance of two systems A and B in overcoming the effect of a given disturbance, the ratio of the error integrals is given approximately by the relationship

$$\frac{[\int|e|\ dt]_A}{[\int|e|\ dt]_B} \approx \frac{[(1 + K)\omega_c]_B}{[(1 + K)\omega_c]_A} \qquad (7\text{–}14)$$

This relationship can be further simplified by using the maximum gain K_{max} in place of $(1 + K)$. This may be somewhat less accurate, but for processes without time delays or distributed elements, K_{max} usually exceeds a value of 10 and the recommended value of K is about half of the maximum, so that the additional inaccuracy is not usually material. The relationship then becomes that the ratio of the error integrals is approximately that of the inverse ratio of the products of the critical frequency and maximum gain for limiting stability, i.e. $(\omega_c K_{max})$. Both of these parameters can readily be extracted from the Bode diagram and also, of course, from the root locus diagram. The former is however somewhat more flexible, particularly when changes in time constants, etc., are being investigated.

It will be appreciated that Equation (7–14), or its further simplification, is not intended for accurate comparisons between systems performance, for which it would be necessary to compare the actual transient responses to a given disturbance. Equation (7–14) is intended for quick order-of-magnitude comparisons only. The method is not applicable when measuring lag is significant, since the actual error would then exceed the measured error by different amounts, and also tends to over-estimate the effect of controller gain if major load disturbances are towards the end of a series of elements. In the latter case it is advizable to base comparisons on the transient responses obtained from the closed-loop frequency response. The following example illustrates the use of the method.

Example 7–2

Compare the effect on controllability of the three-stage system of Example 7–1 (time constants of 1, 0.5, and 0.25 min) of halving the first or second time constant.

The Bode plots for the original system and the two modifications are shown in Figure 7–16, these being drawn from the combination of the appropriate first-order curves for time constants of 1, 0.5, and 0.25 min, as in Example 7–1. The results are summarized in Table 7–1.

TABLE 7–1

System	Time constants			ω_c rads/min	A.R.	K_{max}	$\omega_c K_{max}$
	T_1	T_2	T_3				
A	1	0.5	0.25	3.74	0.089	11.24	42
B	0.5	0.5	0.25	4.50	0.110	9.13	41
C	1	0.25	0.25	4.90	0.080	12.51	61

Halving the first time constant increases the critical frequency by about 20 per cent but reduces the maximum gain by about the same figure, and the error integrals are then about the same. The system response would cycle rather more quickly, but with a larger peak error, and there would be little significant difference between the responses. Halving the second-largest time constant increases the critical frequency by about 30 per cent and also increases the maximum gain by about 10 per cent; the response would thus cycle faster and with a slightly smaller peak error, so reducing the error integral. The ratio of the products of $(\omega_c K_{max})$ is 61/42, showing that the modification reduces the error integral to two-thirds of that of the original system.

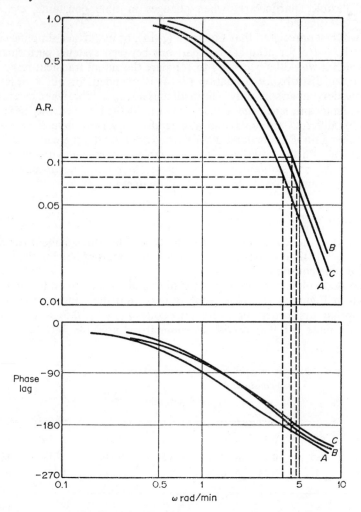

FIGURE 7–16. Bode diagram for Example 7–2

This example illustrates a general rule that reducing the second-largest time constant in a multiple system improves the performance more than a reduction in the largest time constant. The latter contributes the largest amount of phase lag at any frequency, and at the critical frequency this may well approach 90°. Reducing this time constant will have relatively little effect on the total phase angle and thus on the critical frequency, but may have a large effect on the A.R. Reducing the second-largest time constant has a relatively greater effect on ω_c since the phase angle curve for this time constant will be much more steep in the region of the critical frequency. If the largest time constant is much greater than the others, a reduction might in fact lead to poorer performance since the lower gain might more than offset the relatively small increase in frequency. This suggests that in some cases an *increase* in the largest time constant might improve the control but, unfortunately, this is usually the most expensive and least practical method of improvement. The largest time constant is associated with the major process element in the system, being determined by the size or volume of the vessel, tank, reactor, heat exchanger, etc. The second-largest time constant is usually associated with a valve, heating coil, steam jacket, measuring element, etc. and can often be reduced at much less expense, by use of a valve positioner, higher steam velocity or reduced hold-up, change of measuring element, etc.

With a three-stage time-constant system, reducing the smallest time constant has a large effect on both critical frequency and maximum gain; as the smallest time constant approaches zero the system becomes theoretically second-order, and both ω_c and K_{max} approach infinity since the other two constants provide 180° phase lag only at infinite frequency. A real system has additional small lags and delays which become evident under these conditions, and the advantage of reducing the smallest time constant is not very great.

Example 7–3

Compare the controllabilities of the process of Example 7–1 and that of a process with time constants of 1 and 0.5 min and a time delay of 0.25 min.

The Bode diagrams for the two systems are shown in Figure 7–17; replacing the smallest time constant by a time delay of equal magnitude reduces ω_c by about 10 per cent and K_{max} by about 50 per cent and the error integral is effectively doubled.

TABLE 7–2

System		ω_c	A.R.	K_{max}	$\omega_c K_{max}$
A	$T_3 = 0.25$ min	3.74	0.089	11.24	42
B	$L = 0.25$ min	3.30	0.150	6.67	22

It can also be demonstrated by this method that reducing the magnitude of the time delay improves the performance of the system by a relatively greater amount than a similar reduction in the magnitude of a time constant. Halving a time delay produces about a three-fold improvement as compared to the 50 per cent increase shown in Example 7–2.

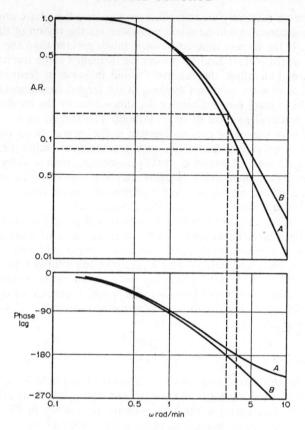

FIGURE 7–17. Bode diagram for Example 7–3

Gain and Phase Margins

As has been seen previously, a control system reaches the limit of stability when two of the roots of the characteristic equation of the system $1 + G(s) = 0$ are a conjugate imaginary pair, i.e. $\pm j\omega$. For a control loop verging on to instability there will thus be a critical frequency at which $1 + G(j\omega)$ becomes equal to zero, i.e. a frequency at which s in the characteristic equation becomes conjugate imaginary. Under these conditions, $G(j\omega)$, the open-loop transfer function with $j\omega$ substituted for s, is equal to -1; the magnitude ratio of the function is then, $|G(j\omega)| = 1$ and the phase angle, $\angle G(j\omega) = -180°$, and the frequency is the critical frequency, ω_c.

A criterion for stability for a closed-loop system can thus be based on the open-loop frequency response characteristics, that the system will be unstable if the magnitude ratio exceeds unity at the frequency (ω_c) for which the phase lag is 180°.

The ordinate employed on the Bode frequency-response diagram is the amplitude ratio (A.R.), and the above criterion applies to the magnitude ratio (M.R. or $|G|$), which differs from the A.R. when the gain is not unity. The situation is illustrated in Figure 7–18 for the three time constant process of Example 7–1 in which the A.R. (unit process gain) is plotted as curve 1 (see also Figure 7–10). The actual gain of the process of Example 7–1 was 0.4 m/percentage change in valve position, hence the A.R. curve must be multiplied by 0.4 to give the M.R. curve (curve 2). As the ordinate is logarithmic, multiplication does not change the shape of the curve but only moves it bodily from the low frequency gain of 1 for the A.R. curve to the new value of 0.4 for the M.R. curve.

In Example 7–1 the maximum gain (at limiting stability) of a proportional controller used with the process was found to be 28.1 per cent change in valve position/m. The M.R. curve for a proportional controller is a straight line at the value of K_c, in this instance $K_{c\,max} = 28.1$. The overall M.R. for the open-loop process and controller is then the product of the M.R.s of the process and the controller (curve 3); since the M.R. of the controller is not affected by the frequency, the shape of the curve is again unchanged.

According now to the criterion for stability outlined above, the value of the overall M.R. for the loop will be unity at the frequency corresponding to 180° phase lag. The proportional controller does not introduce any change in phase (or, more strictly, no change in phase is shown on the Bode diagram) and the phase-angle curve for the system is the same throughout. The critical frequency of $\omega_c = 3.74$ rads/min found in Example 7–1 is thus the critical frequency for 180° lag for the system with $K_c = 28.1$ As can be seen the corresponding value for the overall M.R. for process and controller (curve 3) is unity, showing that the closed loop with $K_c = 28.1$ is at the condition of limiting stability.

It will be appreciated that the above argument is a graphical statement of Equation (7–9), i.e.

$$K_g|G_p| = |G| = 1$$

This stability criterion based on the Bode diagram, and often referred to as the Bode criterion, is not completely general but applies readily to systems whose gain and phase decrease continuously with increasing frequency. If, however, the phase curve shows inflections about the 180° value, the Bode criterion is not applicable and the more general Nyquist criterion must be used. Other exceptions may also occur, but for most process systems the simple criterion finds wide application.

If the controller gain is now reduced below the maximum value, the system will be completely stable and the magnitude ratio at 180° lag will be less than unity. This is shown by curve 4 in Figure 7–18 for a value of

$$K_c = 0.5K_{c\,max} = 14$$

obtained by multiplying the process M.R. (curve 2) by the new value of K_c.

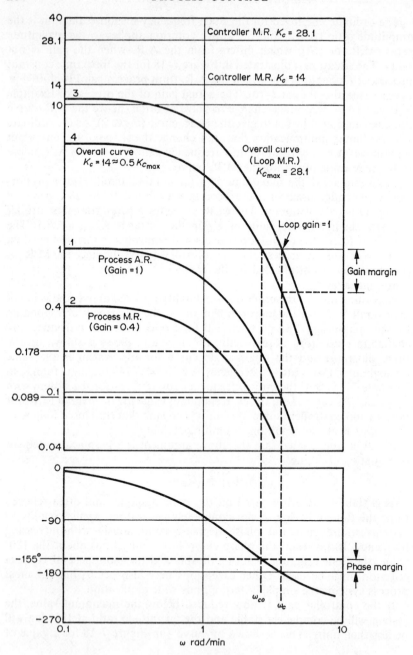

FIGURE 7–18. Bode stability criterion and gain and phase margins applied to system of Example 7–1

It is obviously advantageous to know how stable a system has become as a result of reducing the controller gain below the maximum value. Intuitively, it can be deduced that the system will be only slightly stable if the M.R. at 180° lag is only slightly less than unity, i.e. the response to a disturbance will be highly oscillatory with a very small damping. If, on the other hand, the M.R. at 180° lag is very much less than unity then the system will be very stable and may be even over-damped. It must be remembered that the stability of a system depends upon the values of the physical parameters of the system, such as the time constants, etc. These parameters have to be measured or estimated with some possible inaccuracy or even uncertainty, and furthermore the values may not remain constant but may change with time during the life of the system due to corrosion and similar effects in the actual plant. A design which gives a value of M.R. very close to unity at 180° lag may not then have an adequate safety factor against the possible inaccuracies or changes in the parameters, and the system may later become unstable.

To give a quantitative assessment to the relative stability, the concepts of *gain* and *phase margins* are introduced. These are a quantitative measure of the safety margin between the stable conditions as designed and the onset of instability; the design values of these quantities are basically safety factors and, as such, may vary considerably with the nature of the application and with the views of the individual designer.

The gain margin is usually defined as the reciprocal of the magnitude ratio of the open loop system (which will be less than one for a system which is stable) at the frequency corresponding to 180° phase lag; thus for curve 4 of Figure 7–18, the M.R. at $\omega_c = 3.74$ rad/min is 0.498 and the gain margin is 2.01.

By further consideration of the Bode diagram, the gain margin can be seen to be the factor by which curve 3 (for $K_{c\,max} = 28.1$) must be divided to obtain curve 4 (for $K_c = 14$). The gain margin is thus the ratio of $K_{c\,max}/K_c$, or the ratio of the overall gains K_{max}/K. A typical design specification for gain margin would be a minimum value of 1.7 [10], although other authors suggest values between 2.0 and 2.5 [19].

Phase margin is a measure of the additional phase lag required to destabilize the loop, and is the difference between 180° (the phase lag of the stable loop with M.R. < 1) and the phase lag when the loop is on the verge of instability at M R. = 1. The frequency corresponding to the latter condition (M.R. = 1 for the stable system) is the *gain cross-over frequency* (ω_{co}), i.e. the frequency at which the gain (M.R.) crosses the value of unity. It should be noted that some authors refer to the critical frequency (ω_c) as the *phase cross-over frequency* i.e. the frequency at which the phase angle curve crosses the value of −180°. One advantage of considering the phase margin and defining the cross-over frequency at M.R. = 1 for the stable loop is that the frequency of the damped oscillations of the stable system (which must obviously be less than the critical frequency) is approximately the same as the cross-over frequency.

Since multiplication of the gain curves of the Bode diagram in the case of proportional control does not affect the phase angle curve in the lower part of the diagram, the single phase-angle curve applies to each of the gain curves. For the stable system (curve 4 for $K_c = 14$), the intercept with the unit M.R. occurs at a frequency of 2.64 rad/min and this then is the cross-over frequency (ω_{co}). The corresponding phase lag from the lower diagram is 155° and the phase margin is thus $180 - 155 = 25°$.

A typical design specification for phase margin is a minimum value of 30°, although again other authors recommend higher values (up to 45°). The present example does not quite satisfy the 30° requirement and the value of K_c could be reduced to about 12 at the expense of slightly increased offset.

By considering a two stage system where the damping ratio ζ can be related to the time constants and proportional gain, it is possible to derive a simple relationship between the phase margin and the damping constant of the oscillations following a disturbance. As would be expected, the response becomes less oscillatory as the phase margin is increased. Although such a system (being second-order) cannot exhibit instability with proportional control, the result can be generalized to more complex systems, and the phase margin is a useful design tool in cases where the transient response of a system cannot be easily determined.

Similar remarks can also be applied to the gain margin; as the latter is increased so the system response becomes less oscillatory and more damped. Both margins do, of course, exist simultaneously and both should be considered and the minimum requirements met. In the example discussed and illustrated in Figure 7–18, whilst the gain margin exceeds the minimum of 1.7, the phase margin is less than the suggested 30°. This result is quite typical; the values are determined by the shapes of the gain and phase curves and thus by the system parameters. If the system contained a time delay instead of a small time constant, the gain curve would be less steep and the phase curve steeper over the range of frequencies of interest; the relation of the gain and phase margins to the suggested minimum might well be reversed. The additional phase angles introduced by integral and derivative controls may also have a similar effect.

There are obviously practical upper limits to the gain and phase margins, since large values will tend to make the system response less oscillatory and thus more sluggish. In general the damping ratio should be such as to produce an oscillation with a short rise time without excessive overshoot of the final value or a long recovery time. The system designer is invariably faced with a compromise between speed of response and degree of oscillation. It is usual, therefore, to select or adjust the control parameters so that either the gain or the phase margin is equal to the lowest acceptable value, with the other margin probably greater than the minimum value. In addition, if integral action is not to be used, the amount of offset must be considered.

Gain and phase margins present little difficulty in determining the value of K_c for proportional control, but the application becomes more difficult when the additional control parameters (T_i and/or T_d) are added and a certain amount of trial and error procedure becomes inevitable. Fortunately there are some simple rules for directly establishing values of the parameters which usually give reasonably satisfactory, although possibly conservative, values of gain and phase margins. These are the Ziegler-Nichols values originally suggested for use in the loop tuning method of determining control parameters for an existing process (see pages 181–2).

Before considering further examples, it should perhaps be pointed out that it is not necessary in practice to re-plot the Bode diagram in terms of magnitude ratios, as was done in Figure 7–18 for purely illustrative purposes. The alternative statement of Equation (7–9) relates the overall amplitude ratio to the overall gain, i.e.

$$K_{max}(\text{A.R.}_{\text{overall}}) = 1$$

and the data can be extracted from curve 1 of Figure 7–18. This is the overall A.R. curve for the process, and also for the open loop if only proportional control is used. Thus the overall A.R. at the critical frequency corresponding to 180° lag is 0.089, whence the overall gain for limiting stability is $K_{max} = 11.24$. Since the process gain (Example 7–1) is 0.4 m/percentage change in valve position, $K_{c\,max} = 28.1$ percentage change in valve position/m. For a gain margin of 2, the overall gain is reduced to 11.24/2 = 5.62, or an A.R. of 1/5.62 = 0.178. The latter corresponds to a frequency of 2.64 rad/min, which is the cross-over frequency, and to a phase lag of 155° with a phase margin of 25° as derived previously.

Example 7–4

Compare the application of P, PI, PD, and PID controls to the system of Example 6–3, i.e. a process with two first-order time constants of 1 and 0.5 min and a time delay of 0.2 min with unit gain.

Figure 7–19 shows the development of the Bode diagram for the process without the controller, or with proportional control alone. The critical frequency for 180° phase lag is 3.75 rad/min and the corresponding A.R. is 0.121. For proportional control the value of $K_{c\,max}$ (which is the same as K_{max} since the process has unit gain) is 1/0.121 = 8.25.

The value of $K_{c\,max}$ is also the value of K_u (ultimate proportional sensitivity) as would be determined by a Ziegler-Nichols loop-tuning experiment on the actual process. The value of K_c suggested by Ziegler and Nichols for proportional control alone is $0.5K_u$, giving a value of gain margin of 2.0 by definition. The value of K_c is then 4.125, an A.R. of 0.242, corresponding to a cross-over frequency of 2.43 rads/min and a phase angle of 146°. The phase margin is thus 34°.

For proportional-integral control the Ziegler-Nichols recommendation is $K_c = 0.45K_u$ and $T_i = P_u/1.2$, where P_u is the period of the continuous

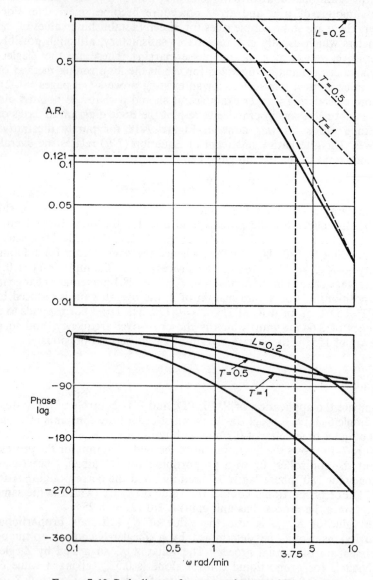

FIGURE 7–19. Bode diagram for process elements of Example 7–4

oscillations set up when $K_c = K_u$. From the critical frequency of Figure 7–19, $\omega_c = 3.75$ rad/min, and $P_u = 2\pi/\omega_c = 1.68$ min. Hence $T_i = 1.4$ min.

Figure 7–20 shows the effect of adding PI control with three different integral times, of $T_i = 5$, 1.4, and 0.5 min, the curves being those for the process + controller. As can be seen, the additional phase lag of the integral action gives an immediate reduction in the critical frequency and also the maximum gain. The values for the four cases (taking that for proportional control (curve 1) as the datum) are as follows:

TABLE 7–3

Curve	T_i min	ω_c rads/min	$A.R._{overall}$	K_{max}	$\omega_c K_{max}$
1	∞	3.75	0.121	8.25	31
2	5	3.60	0.130	7.68	28
3	1.4	3.20	0.162	6.18	20
4	0.5	2.17	0.386	2.59	6

The addition of a small amount of integral action ($T_i = 5$ min, producing 3° additional lag) has little effect on frequency or gain, but the system will take much too long for offset to be eliminated. With T_i set to the Ziegler-Nichols value of 1.4 min, the additional phase lag is about 13°, the gain is reduced by 25 per cent and the frequency by 15 per cent, producing a 50 per cent increase in the error integral. With a large integral action ($T_i = 0.5$, producing an excessive additional lag of 44°) the gain is reduced by about 80 per cent and the frequency by 40 per cent, and the increase in error integral of some six-fold is too great a penalty to pay for faster elimination of the offset.

Considering curve 3 for the Ziegler-Nichols recommendation, the suggested value of K_c is $0.45K_u$, i.e. $0.45 \times 8.25 = 3.71$. The gain margin is then given by

$$K_{max}/K = 6.18/3.71 = 1.7$$

The A.R. for $K = 3.71$ is 0.270, whence ω_{co} is 2.33 rad/min with a phase lag of 160° and a phase margin of 20°. Since this is lower than the suggested minimum of 30°, a smaller value of K_c is indicated. A phase lag of 150° occurs at a frequency of 1.95 rad/min, with a corresponding A.R. of 0.345, whence $K = 2.90$. The gain margin is therefore increased to $6.18/2.90 = 2.1$.

The specification of K_c and T_d for proportional-derivative control cannot be based on the ultimate values of K_u and P_u. The values of $0.6K_u$ and $P_u/8$, corresponding to the Ziegler-Nichols recommendations for PID control and omitting the integral action, are in general much too conservative and the system will be too stable. An alternative approach is to select a value for T_d which will permit a maximum value of K_c to provide

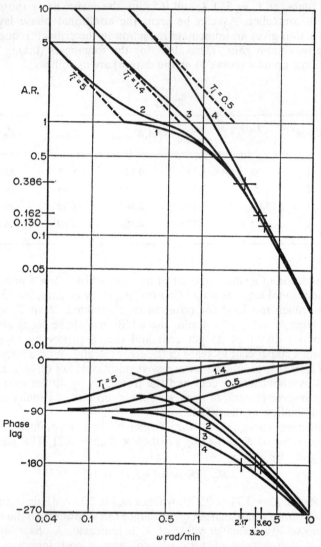

FIGURE 7–20. Bode diagram for PI control (Example 7–4)

a minimum offset with an adequate phase margin (30°). Figure 7–21 is the Bode diagram for three values of T_d of 0.1, 0.25, and 0.4 min for the process with PD control, the curve for proportional control being included for comparison. The phase advance of the derivative action gives an

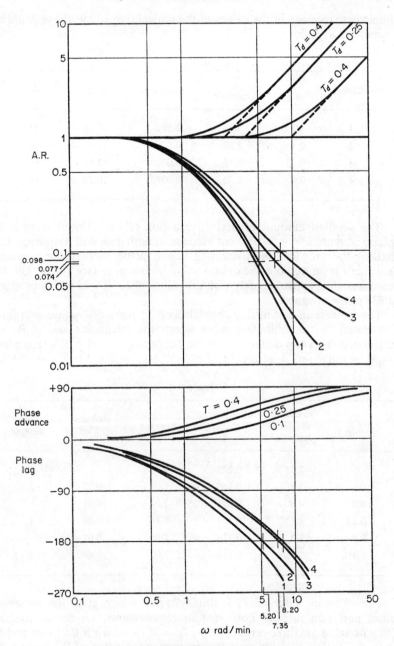

FIGURE 7-21. Bode diagram for PD control (Example 7-4)

immediate increase in the values of the critical frequency; the appropriate values are as follows:

TABLE 7–4

Curve	T_d min	ω_c rad/min	$A.R._{overall}$	K_{max}	$\omega_c K_{max}$
1	0	3.75	0.121	8.25	31
2	0.1	5.20	0.0765	13.07	68
3	0.25	7.35	0.0740	13.52	99
4	0.4	8.20	0.0982	10.18	84

The smallest amount of derivative action ($T_d = 0.1$ min) gives a $28°$ phase advance and a 50 per cent increase in both gain and frequency, thus halving the error integral. Increasing T_d will further increase the frequency but there is an obvious maximum value of the gain (see pages 239–40), corresponding approximately with the values of $T_d = 0.25$ which produces a $63°$ phase advance.

The process of optimizing the value of K_c for a $30°$ phase margin is illustrated in the following table where the frequency and A.R. are calculated from the definitions of the phase angles and A.R.s (e.g. phase angle $= \tan^{-1}(\omega T_1) + \tan^{-1}(\omega T_2) + 57.3\omega L - \tan^{-1}(\omega T_d) = 150°$).

TABLE 7–5

T_d min	ω_{co} at 150° rad/min	$A.R._{overall}$	$K(=K_c)$	Relative offset	Gain margin
0	2.55	0.225	4.44	1	1.9
0.1	3.2	0.166	6.03	0.77	2.2
0.2	4.3	0.126	7.94	0.61	
0.25	4.8	0.123	8.16	0.59	1.7
0.3	5.2	0.126	7.96	0.61	
0.4	5.7	0.143	7.02	0.68	1.5

The optimum value of T_d is thus 0.25 min which gives the minimum offset and also satisfies both margin requirements. Using the Ziegler-Nichols value for three-term control, $T_d = P_u/8$, which is 0.21 min and so near the optimum value, but the recommended value of $K_c = 0.6K_u$ is 4.95 which is considerably smaller than any value indicated above, and which yields a gain margin of 2.6.

Figure 7–22 is the Bode diagram for three-term control using the Ziegler-Nichols recommended values of $T_i = P_u/2$ and $T_d = P_u/8$, i.e. 0.84 and 0.21 min respectively. The critical frequency is then 6.60 rad/min with an A.R. of 0.068 and a maximum gain of 14.71. Thus the critical frequency is somewhat less than that of the optimum PD control but more than twice that for PI control. The maximum gain is slightly larger than with PD control, giving an error integral of about the same magnitude but, owing to the lower frequency and the integral lag, the phase advance is less than with PD control and the value of K_c will be lower. With a value of $K_c = 0.6K_u$, the gain margin is very nearly 3 but the phase margin is only 33°.

The results for the four control modes are summarized in Table 7–6 below and the open loop gain and phase curves are shown in the Bode diagram of Figure 7–23.

TABLE 7–6

Control	T_i min	T_d min	ω_c rad/min	K_{max}	K_c	Gain margin	Phase margin
P	—	—	3.75	8.25	4.13	2	34°
PI	1.4	—	3.20	6.18	2.90	2.1	30°
PD	—	0.25	7.35	13.52	8.16	1.7	30°
PID	0.84	0.21	6.60	14.71	4.95	3	33°

Closed-loop Frequency Response

The discussion of the previous sections has been based on the frequency response characteristics of the open control loop, i.e. with the loop broken, usually between the controller and the valve, and the gain and phase relationships determined between the input and output sinusoidal signals.

The frequency response of the closed loop is the response of the controlled variable to sinusoidal changes in the desired value or a load variable; in a similar way to the open-loop response, the gain and phase relationships of the input and output signals are dependent on the frequency of oscillation, and can be investigated by means of the Bode type of diagram. The additional information provided by closed-loop frequency response is often helpful in judging the performance of a system, since a reasonably accurate prediction of the transient response of the system can be obtained from the closed-loop response diagram. In addition the closed-loop characteristics are essential in plotting the frequency response of a multi-loop system such as cascade control.

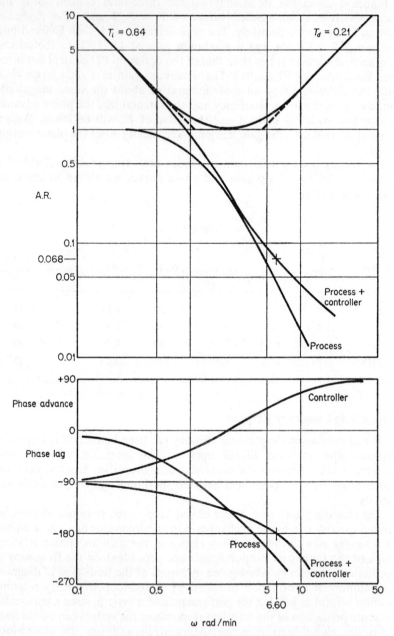

FIGURE 7–22. Bode diagram for PID control (Example 7–4)

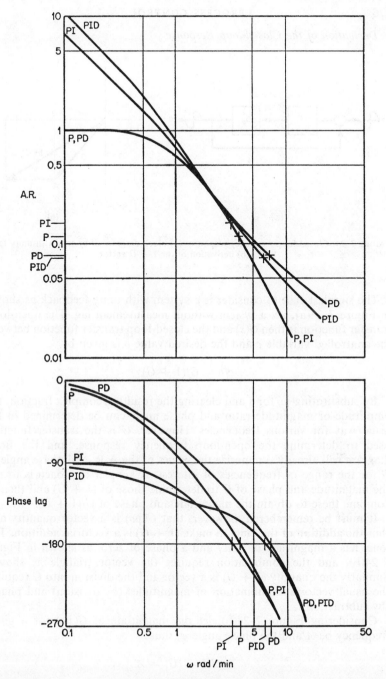

FIGURE 7-23. Bode diagrams for P, PI, PD, and PID controls based on optimum values of Example 7-4

Derivation of the Closed-loop Response

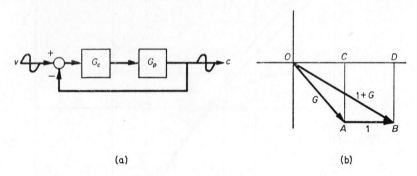

(a)   (b)

FIGURE 7–24. Closed-loop frequency response: (a) block diagram of system with unity feed-back, (b) derivation of the 1 + G vector

The simplest case to consider is a system with unity feedback as shown in Figure 7–24(a), i.e. a system without measurement lag. The open-loop transfer function is then $G(s)$ and the closed-loop transfer function between the controlled variable c and the desired value v is given by

$$c/v = G/(1 + G)$$

By substituting $j\omega$ for s and clearing the resulting complex fraction, the amplitude or magnitude ratio and phase angle can be determined in the usual way for various frequencies. However, G is the transfer function used to determine the open-loop frequency response, and the Bode diagram will already summarize the values of the A.R. and phase angle of G for the range of frequencies of interest. A simpler approach is to use the magnitude and phase of G to determine those of $(1 + G)$ and then to combine these to obtain the magnitude and phase of $G/(1 + G)$.

It must be remembered, however, that $G(j\omega)$ is a vector quantity and thus the addition of the 'one' to make $(1 + G)$ is a vectorial addition. The 'one' has a magnitude of unity and a phase of zero, as shown in Figure 7–24(b), and the combination requires the vector triangle as shown. Similarly the quantity $(1 + G)$ is a vector and the division into G requires the usual vectorial combination of magnitudes (by division) and phases (by subtraction).

Considering Figure 7–24(b), let the magnitude of $G(j\omega)$ for a given frequency be A and the phase angle ψ, thus:

$$OA = |G| = A$$

and
$$\angle AOC = \angle G = \psi$$

From the triangle BOD, the magnitude of $(1 + G)$ is given by

$$|1 + G| = OB$$
$$= [(OD)^2 + (DB)^2]^{\frac{1}{2}}$$
$$= [(OC + CD)^2 + (DB)^2]^{\frac{1}{2}}$$
$$= [(A \cos \psi + 1)^2 + (A \sin \psi)^2]^{\frac{1}{2}}$$

since $OC = A \cos \psi$, $DB = AC = A \sin \psi$, and $CD = AB = 1$. Further,

$$\angle(1 + G) = \angle BOD$$
$$= \tan^{-1}[(DB)/(OD)]$$
$$= \tan^{-1}[(A \sin \psi)/(1 + A \cos \psi)]$$

Let the magnitude of $G/(1 + G)$ for the same frequency be M and the phase be α; then

$$M = \left| \frac{G}{1 + G} \right|$$

$$= \frac{|G|}{|1 + G|}$$

$$= \frac{A}{[(A \cos \psi + 1)^2 + (A \sin \psi)^2]^{\frac{1}{2}}}$$

$$= 1 \bigg/ \left[\left(1 + \frac{\cos \psi}{A}\right)^2 + \left(\frac{\sin \psi}{A}\right)^2 \right]^{\frac{1}{2}} \tag{7-15}$$

$$\alpha = \angle G/(1 + G)$$
$$= \angle G - \angle(1 + G)$$
$$= \psi - \tan^{-1}[(A \sin \psi)/(1 + A \cos \psi)]$$
$$= \tan^{-1}(\tan \psi) - \tan^{-1}[(A \sin \psi)/(1 + A \cos \psi)]$$

which, by use of the identity

$$\tan^{-1} C - \tan^{-1} D = \tan^{-1}[(C - D)/(1 + CD)]$$

reduces to

$$\alpha = \tan^{-1}\left[\left(\frac{\sin \psi}{A}\right) \bigg/ \left(1 + \frac{\cos \psi}{A}\right) \right] \tag{7-16}$$

Since A and ψ are the quantities plotted as A.R. and phase angle on the Bode diagram for the open loop response, Equations (7–15) and (7–16) provide a means of calculating the magnitude and phase of the closed loop response. It should be noted that the Bode diagram is a plot of amplitude ratio and not magnitude ratio (i.e. the Bode diagram is plotted for unit gain), and the appropriate value of the overall gain K for the system must be used to obtain the value of A as a magnitude ratio for the given frequency, i.e. $A = K(\text{A.R.})$.

Calculations of M and α are somewhat tedious, and where great accuracy is not required, the Nichols (or Black-Nichols) chart is often used. This is a plot of contours, or constant values, of M and α on rectangular co-ordinates of A and ψ. For values of A and ψ at a selected frequency the corresponding values of M and α can be read off by interpolation between the contours. The Nichols chart is included in most texts on servo-systems and is not reproduced here.

The Closed-loop Response Diagram

Example 7–5

Determine the closed-loop frequency response characteristics for the system of Example 7–1 with three time constants of 1, 0.5, and 0.25 min at overall gains of $K = 6$ and $K = 3$.

The open-loop Bode diagram for this system was plotted in Figure 7–10. The closed-loop response diagram is plotted in Figure 7–25, with values of M.R. and phase calculated from Equations (7–15) and (7–16). The diagram also shows the open loop M.R. and phase for the gain of $K = 6$.

The general features of the closed-loop response are fairly evident from the diagram. At low frequencies the time constants provide little attenuation and phase lag, and the magnitude is effectively the value $K/(1 + K)$. This is the value of the final steady-state following a unit change in desired value as seen in Chapter 5, and the difference between this value and unity is the steady-state offset. At a frequency somewhat less than the critical ($\omega_c = 3.75$ rad/min), the gain curve shows a resonant peak similar to that shown by the open loop response of an under-damped second-order system (Figure 7–12). At high frequencies the gain curve approaches that of the open loop since $(1 + G) \to 1$ as $G \to 0$. Reducing the overall gain of the system reduces the height of the resonant peak and also the corresponding resonant frequency (ω_r).

The phase lag for the closed loop is much less than that of the open loop at all frequencies up to ω_c where the two curves cross, but approach each other again at higher frequencies. With increasing overall gain, the phase curves become steeper at the critical frequency.

Transient and Frequency Response

As noted above, the frequency response of an open-loop under-damped second-order system shows a resonant peak similar to that now seen in the closed-loop frequency response. In the case of the second-order system, the height of the peak and the corresponding frequency are determined by the damping constant ζ; in general the peak height and resonant frequency increase as the damping is decreased (pages 249–51).

Increasing the overall gain of a closed-loop system towards the maximum value reduces the damping of the transient response and, as seen in

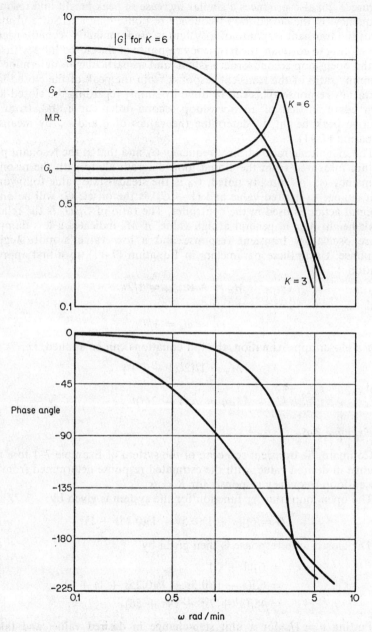

FIGURE 7-25. Bode closed-loop diagram for desired value changes (Example 7-5)

Figure 7–25, also produces a similar increase in peak height and resonant frequency on the closed-loop frequency response diagram. Since in many cases the transient response of a system is predominantly second-order, it is tempting to correlate the frequency response of the closed loop with that of the open-loop second-order system, and in particular to determine the damping ratio of the transient response from the peak of the closed-loop frequency response. This can be done by simply regarding the closed-loop gain curve as that of an open-loop second-order curve and, from the relative peak height, to determine the values of ζ and ω_n by means of Equation (7–11).

The magnitude ratio at zero frequency G_0 and that at the resonant peak G_p are measured from the closed loop diagram along with the resonant frequency ω_r. As already noted, G_0 is the steady-state value following a unit change in desired value and $(1 - G_0)$ is the offset; G_0 will be one if integral action is used in the controller. The ratio of G_p/G_0 is the relative peak height M_p; in general a high value of M_p indicates a less damped, more oscillatory transient response and a low value a more sluggish response. Using these parameters in Equation (7–11), to a first approximation

$$M_p \approx \text{A.R.}_{\text{max}} = 1/2\zeta$$

and

$$\omega_r \approx \omega_n = 1/T$$

For a closer approximation, the full equations can be applied, i.e.

$$M_p = 1/[2\zeta(1 - \zeta^2)^{\frac{1}{2}}]$$

and

$$\omega_r = \omega_n(1 - 2\zeta^2)^{\frac{1}{2}}$$

Example 7–6

Compare the transient response of the system of Example 7–1 to a unit change in desired value, with the estimated response determined from the closed-loop frequency response, for $K = 6$.

The open-loop transfer function for the system is given by

$$G = 6/[(s + 1)(0.5s + 1)(0.25s + 1)]$$

The closed-loop response is then given by

$$c/v = G/(1 + G)$$
$$= 6/[(s + 1)(0.5s + 1)(0.25s + 1) + 6]$$
$$= 48/(s^3 + 7s^2 + 14s + 56)$$

Putting $v = 1/s$ for a unit step-change in desired value, and taking factors of the denominator,

$$c(s) = 48/[s(s + 6.20)(s + 0.40 + 2.98\text{j})(s + 0.40 - 2.98\text{j})]$$

Expanding into partial fractions and inverting the terms in s gives, for the transient response,

$$c(t) = 0.86 + 0.18\,e^{-6.2t} + 0.88\,e^{-0.4t}\sin(171°t - 55°)$$

The term including the $\exp(-6.2t)$ will decay very quickly so that the response (Figure 7–26) is effectively second-order.

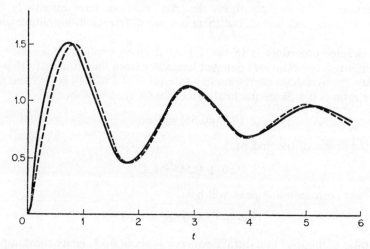

FIGURE 7–26. Transient response from closed-loop frequency response, comparison of theoretical and estimated responses

The response estimated from the closed-loop frequency response is obtained by the method outline above. From Figure 7–25,

$$G_0 = 0.86$$

$$G_p = 3.02$$

$$\omega_r = 2.88 \text{ rad/min}$$

Hence

$$M_p = G_p/G_0 = 3.52$$

For a first approximation,

$$\zeta = 1/2M_p = 0.142$$

(Alternatively from Figure 7–12, ζ may be estimated as about 0.15).
From Equation (7–11),

$$\zeta = 0.144$$

and

$$\omega_n = \omega_r/(1 - 2\zeta^2)^{\frac{1}{2}}$$
$$= 2.94 \text{ rad/min}$$

The transient response is obtained by substitution of the appropriate values in Equation (2–28) with a final value of 0.86, i.e.

$$c(t) = 0.86[1 + 1.01\ e^{-0.42t} \sin (2.88t - 35°)]$$

$$= 0.86 + 0.88\ e^{-0.42t} \sin (165°t - 35°)$$

The two results are compared in Figure 7–26, although it is evident from inspection of the two equations that the responses have essentially the same damping and period, but there is some difference in amplitude and phase.

A simpler procedure is to use Figure 2–26 to predict the amount of overshoot, decay ratio and damped frequency from the estimated value of ζ. Thus the overshoot corresponding to a value of ζ of 0.14 is 0.65 and the decay ratio is 0.4. Since the final value is 0.86, the overshoot is

$$(0.86)(0.65) = 0.56$$

and the height of the first peak is

$$0.86 + 0.56 = 1.42$$

The next corresponding peak will be

$$0.86 + (0.4)(0.56) = 1.08$$

The ratio of damped to natural frequency is about 0.98; hence the damped frequency is $(0.98)(2.94)$ rad/min, whence the period is 2.18 min. From this information an adequate sketch of the response curve can be drawn.

The conclusion to be reached is that the approximation is a useful method of predicting the general character of the transient response but it should not be used to estimate actual values, as the possibility of error is too great. The system used in this example was almost second order, which is the explanation of the quite reasonable agreement. For a system which is not close to second order, the agreement will be by no means as good and may even lead to erroneous conclusions in a particular case. The second-order approximation is a simple generalization which often can give reasonable results, but it should be used only as a guide to system design.

Effect of Measuring Lag

The system considered in Example 7–5 had unity feedback and is thus not typical of many process systems where a measuring feedback H is present. In such cases the closed-loop transfer function between the controlled variable and the desired value is

$$c/v = G/(1 + GH)$$

The closed-loop parameters, M and α, are derived from the closed-loop function $G/(1 + G)$, but the function with a measurement feedback can be put into this form by simple manipulation:

$$\frac{c}{v} = \frac{G}{1 + GH} = \frac{GH}{1 + GH}\left[\frac{1}{H}\right]$$

The values of the magnitude and phase of $GH/(1 + GH)$ can be derived from the open-loop characteristics of GH by the relationships already derived (Equations (7–15) and (7–16)) or from a Nichols chart, and the results divided by the feedback function H. If the latter is a measurement gain, this presents no problem but simply shifts the magnitude curve by a constant factor; if there is a measurement lag, e.g. if

$$H = K_m/(T_m s + 1)$$

the magnitude and phase of the measurement lag must be included into the closed-loop magnitude and phase. Since H is a divisor, a linear lag in measurement becomes a linear lead which must be combined with M and α. Examples of this are shown in the next section.

Load Changes

A similar procedure to the above must be followed for load changes since there is no general solution for values of M and α which are comparable to the Nichols chart, and it is necessary to resolve the equation for the response of the controlled variable to a load change to include the response to a change in desired value. The response to a load change, however, depends on the location of the disturbance, and each possibility gives a different solution. Considering a three-stage system, the load change may be introduced before each time-constant stage. For a load change before the first time constant,

$$c/u_1 = N_1/(1 + G)$$

where $N_1 = G_1 G_2 G_3$ and $G = G_c G_1 G_2 G_3$. N_1 thus differs from G only by the controller function G_c; hence

$$\frac{c}{u_1} = \frac{G}{1 + G}\frac{1}{G_c} = \frac{c}{v} \cdot \frac{1}{G_c}$$

In the simplest case of proportional control alone, $G_c = K_c$. The term $G/(1 + G)$ is the response to changes in desired value, and the appropriate values of M and α for a given frequency can be determined from the open-loop characteristics by Equations (7–15) and (7–16) or from a Nichols chart. It is only necessary then to divide the magnitudes by K_c to obtain the closed-loop magnitudes for the load change, the phase curve being unaltered.

If the controller should contain additional actions the gain and phase of the controller will be brought in to the equation, e.g. if PD control is used, $G_c = K_c(1 + T_d s)$, and a linear lag is introduced, i.e.

$$\frac{c}{u_1} = \frac{c}{v} \cdot \frac{1}{K_c(1 + T_d s)}$$

and the values of M and α for $G/(1 + G)$ must be divided by K_c and combined with the gain and phase of the $(1 + T_d s)$ term. This is now a linear lag since it appears in the denominator, but the term will already have been plotted as a linear lead in finding the open-loop frequency response curve. The effect on the phase angle is usually of little significance and the phase-angle curve is often not determined.

For load changes between the first and second time constants, the closed-loop transfer function is given by

$$c/u_2 = N_2/(1 + G)$$

where $N_2 = G_2 G_3 = G/(G_c G_1)$. Hence

$$\frac{c}{u_2} = \frac{G}{1 + G} \cdot \frac{1}{G_c G_1} = \frac{c}{v} \cdot \frac{1}{G_c G_1}$$

G_1 is the transfer function of the first time constant stage, i.e. $K_{p1}/(T_1 s + 1)$, and this introduces a linear lead term, e.g. for proportional control,

$$\frac{c}{u_2} = \frac{c}{v} \cdot \frac{(T_1 s + 1)}{K_c K_{p1}}$$

The gain and phase, M and α, for $G/(1 + G)$ are thus divided by $K_c K_{p1}$ and combined with the gain and phase of a linear lead which is the reciprocal of the linear lag of the first time-constant stage.

Similarly, a load change before the third time constant introduces a further linear lead from the second time constant;

$$N_3 = G_3 = G/(G_c G_1 G_2)$$

and hence

$$\frac{c}{u_3} = \frac{G}{1 + G} \cdot \frac{1}{G_c G_1 G_2}$$

$$= \frac{c}{v} \cdot \frac{(T_1 s + 1)(T_2 s + 1)}{K_c K_{p1} K_{p2}}$$

Example 7-7

Determine the closed-loop gains for load changes in the proportional control of a three-stage system with time constants of 0.25, 0.5, and 1.0 min and with process stage gains of 2, 0.8, and 2.5 when $K_c = 1.5$.

The system has the same three time constants as that of Example 7-1 and the open-loop Bode diagram (for $K = 1$) will be that of Figure 7-10.

The overall gain of the system is given by

$$K = K_c K_{p1} K_{p2} K_{p3} = (1.5)(2)(0.8)(2.5) = 6$$

which is the value of K used to plot the closed-loop response to desired value changes for the system with the same time constants (Figure 7–25).

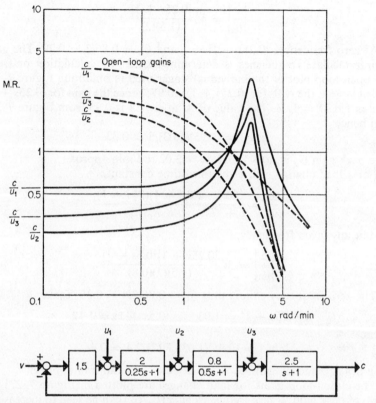

FIGURE 7–27. Bode closed-loop diagram for load changes and block diagram of system of Example 7–7

For load changes before the first time constant,

$$c/u_1 = [G/(1 + G)](1/K_c)$$

and at any given frequency,

$$\left| \frac{c}{u_1} \right| = \frac{M}{1.5}$$

Thus, the zero frequency gain $G_0 = 0.86/1.5 = 0.57$, and the resonant peak gain, $G_p = 3.02/1.5 = 2.01$. The gain curve for this case is that of Figure 7–25 displaced by the factor $(1/1.5)$ and is shown in Figure 7–27, which also shows the open-loop gain for (c/u_1).

For load changes before the second time constant,

$$\frac{c}{u_2} = \frac{G}{1 + G} \cdot \frac{(T_1 s + 1)}{K_c K_{p1}}$$

and at any given frequency,

$$\left| \frac{c}{u_2} \right| = M \frac{|0.25s + 1|}{(1.5)(2)}$$

At zero frequency, $|0.25s + 1| = 1$ and $G_0 = 0.86/3 = 0.29$. The gain at intermediate frequencies is determined either by calculation or from the open-loop plot of the individual stages used in obtaining Figure 7–10, e.g. at $\omega = 1$, the A.R. for $(0.25s + 1)$ is 0.97; hence the gain for $(0.25s + 1)$ lead is $1/0.97 = 1.03$. The value of M at $\omega = 1$ is 0.95 from Figure 7–25, and hence

$$|c/u_2| = (0.95)(1.03)/3 = 0.33$$

The peak gain G_p is now 1.6 at $\omega_r = 3.02$ rad/min approx.

For a load change before the third time constant,

$$\frac{c}{u_3} = \frac{G}{1 + G} \cdot \frac{(T_1 s + 1)(T_2 s + 1)}{K_c K_{p1} K_{p2}}$$

and at any given frequency,

$$\left| \frac{c}{u_3} \right| = M \frac{|0.25s + 1||0.5s + 1|}{(1.5)(2)(0.8)}$$

The zero frequency gain is thus $G_0 = 0.86/2.4 = 0.36$; for $\omega = 1$,

$$|0.25s + 1| = 1.03, \qquad |0.5s + 1| = 1.12$$

Hence

$$|c/u_3| = (0.95)(1.03)(1.12)/2.4 = 0.46$$

The peak frequency gain $G_p = 3.8$ at $\omega_r = 2.95$ rad/min approx.

The closed-loop gains for load changes are plotted in Figure 7–27; the ratio of peak gain to zero frequency gain (G_p/G_0) can be seen to increase as the load change is moved towards the end of the process, the values of M_p being 3.5 for load change u_1, 5.5 for u_2 and 10.6 for u_3. This is to be expected since there are fewer time constants after the load change to provide attenuation and damping of the disturbance before it reaches the controller.

At low frequencies, the load gain for the closed loop is $K_L/(1 + K)$ or the steady state value of the transient response; the numerical value will depend on the type and location of the load change. In the present case, the values of the load gain K_L for the three locations are: for u_1,

$$(2)(0.8)(2.5) = 4.0$$

for u_2,

$$(0.8)(2.5) = 2.0$$

and for u_3, 2.5. The values thus do not fall in numerical sequence. It is possible for c/u_3 to exceed unity at zero frequency if K_{L3} is sufficiently large; this would not imply poorer control since the open-loop gain would also be larger (Figure 7–27). The open-loop curves approach the value of K_L at low frequencies, and the control reduces the low frequency error by the factor $(1 + K)$.

For frequencies near to the resonant frequency ω_r, the closed-loop gain exceeds the open-loop gain, which implies that the error is greater than if the controller were not used to close the loop. This magnification of the disturbance by the controller is not necessarily a cause for concern since the major load changes in process control are not usually periodic but are random fluctuations at low frequency. The main function of a process system controller is to compensate for non-cyclical or at the least low frequency disturbances. The curves of Figure 7–27 are, in fact, typical of many process systems at the recommended controller settings, except that in most cases the zero frequency gain would be unity since integral action would be used in the controller.

It will be noticed that the arrangement of the time constants now assumes significance. In previous discussions of stability analysis, etc., the order of arrangement of the time constants had no effect on the transfer functions of the loop; but for consideration of load changes after the first time constant, both the order and the magnitude of the time constants are significant. It can be shown, by repeating the above example with the time constants arranged in a different order, that the results depend on the arrangement of the time constants. The difference between the response curves becomes greater if the smallest time constants are at the end of the system, thus making the peak height greater for a load change towards the end of the system. Fortunately the major time constant is usually the last in most processes (apart from the measuring lag which is often not significant) and this reduces the magnification of the load disturbance.

Use of the Closed-loop Response in Cascade Control

Cascade control is a complex multi-loop control in which the major control loop contains a subsidiary inner loop with a second controller. The desired value of this second controller is set by the output of the primary controller in the main outer loop, and the variable controlled by the inner loop controller is effectively the manipulated variable of the outer loop. The subject of cascade control will be discussed in more detail in a later chapter (see pages 351–5), but is introduced at this stage since, in order to determine the frequency response characteristics of the outer loop, it is necessary to include as one element the closed-loop characteristics of the inner loop. As will be seen in the later discussion, cascade control is most effective when the main disturbances affecting the system enter in the inner loop and if the latter is faster in response than the outer loop.

The use of an inner loop always tends to reduce the total lag of the system and so increase the critical frequency, as will be demonstrated in the next example.

Predicting the optimum controller settings for the two controllers of a cascade system is not as difficult as might at first appear. The inner loop is a conventional single-loop system, and the usual rules can be applied for the determination of recommended or optimum controller settings. Some authors recommend only proportional action for the inner loop [12, 29] on the grounds that any offset due to the omission of integral action will be small and eliminated anyway by the controller in the outer loop. Integral action is, however, worthwhile if the inner loop has a low gain and so will prevent large offsets. Derivative action is also useful if the reaction is slow due to a large number of time constants, although the improvement will be small compared to that of changing from a single controller to the cascade system.

The settings for the outer loop controller and the critical frequency of the whole system can be found from an open-loop Bode plot for the outer loop, the response of the elements of the outer loop being combined into the gain and phase $G/(1 + G)$ for the closed inner loop. As noted on page 278, the phase lag of a closed loop is always less than that of the open loop up to the critical frequency; the closed inner loop thus contributes less phase lag to the outer loop than if the inner loop controller were omitted. Since the phase lag is reduced, a higher frequency is required to attain the 180° phase lag corresponding to the critical frequency.

Example 7–8

Determine the critical frequency and maximum gain for a single loop controller and for the primary (outer) controller of a cascade system formed from the three-stage process of Example 7–7 with the addition of two further time constant stages with time constants of 2 min and unit gains, using proportional control in both cases.

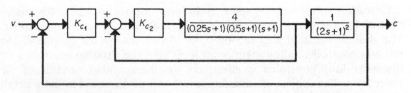

FIGURE 7–28. Block diagram of cascade control system of Example 7–8

The cascade system is shown in Figure 7–28. As a single loop system the second (inner loop) controller is omitted and the open loop function for the system is then

$$G = 4K_{c1}/[(0.25s + 1)(0.5s + 1)(s + 1)(2s + 1)^2]$$

This is plotted in Figure 7–29 as curve A and is the open-loop frequency response curve of Example 7–1 (Figure 7–10) with the addition of the two further time-constant stages. The critical frequency at 180° phase lag is 0.74 rad/min, the corresponding A.R. is 0.233, hence $K_{max} = 4.303$.

For the cascade system, curve B is the closed-loop frequency response of the inner loop with $K = 6$ (as determined in Example 7–5, Figure 7–25). This must now be combined with the open-loop response of the two additional time constants forming the outer loop (plotted as curve C) to give the open loop response of the outer loop as curve D. The critical frequency of the cascaded system is 1.82 rad/min, with A.R. $= 0.067$ and $K_{max} = 14.86$.

Comparing the single loop with the cascade system, the use of the inner loop controller more than doubles the critical frequency and trebles the maximum gain of the controller. Both factors contribute to increased controllability, and the error integral following a change in load in the outer loop (e.g. introduced after the controlled variable of the inner loop) would be reduced by 8–9 fold.

The increase in critical frequency is due to the phase curve of the closed inner loop being relatively flat until the critical frequency of this loop is approached. The curve becomes steeper if the gain of the inner loop controller is reduced and the critical frequency of the outer loop would be reduced; hence the general recommendation of a high gain in the inner loop. When there is only one time constant in the outer loop, the critical frequency will be much closer to that of the inner loop and may coincide with the resonant frequency of the inner loop; the A.R. may then be much larger and the gain of the outer loop correspondingly reduced.

The Nyquist Diagram

An example was given earlier of the Nyquist diagram method of plotting frequency response characteristics of the open loop (Figure 7–3). The Nyquist diagram is a plot of the magnitude and phase of the open-loop function G (or GH if a measurement feedback is included), on polar co-ordinates, i.e. a plot of $|G|$ *versus* $\angle G$, with the phase angle measured clockwise from the zero axis. Frequency is thus a parameter, specific values of which may be marked on the resulting curve, and the diagram is effectively a plot of $G(s)$ for values of $s = j\omega$ from $\omega = 0$ to ∞. It is thus the locus of the open-loop transfer function or simply the 'transfer locus' of the system. Since $G(s)$ is in general a complex number, $x + jy$, for values of s from j0 to j∞, the locus may also be regarded as a rectangular co-ordinate plot on a complex plane, the $G(s)$-plane, with real (x) abscissae and imaginary (y) ordinates.

Whilst the Nyquist diagram is basically the transfer locus of $G(s)$ for a range of values of s, which may be calculated from the open-loop transfer function, the two characteristics to be plotted are also the open-loop frequency response characteristics, and are usually more simply obtained

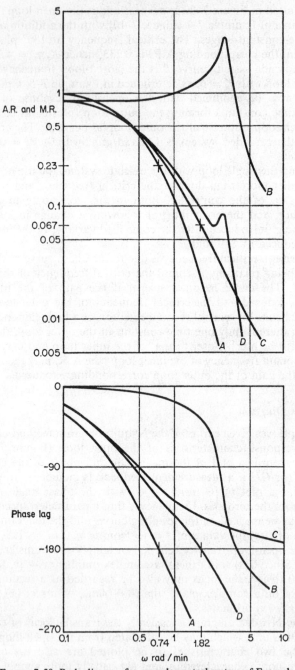

FIGURE 7–29. Bode diagram for cascade control system of Example 7–8

by combination of the characteristics of the individual elements by use of the Bode diagram. The curve is found to have a characteristic shape for individual elements or combinations of elements (Figure 7–30). Some general rules useful in sketching the curves are as follows:

(1) The locus approaches the origin as $s \rightarrow j\infty$ (i.e. high frequency) for all real systems with a finite source of power. If the locus does not approach zero at high frequencies then some terms in the transfer function which are important at high frequencies have not been included. It is usually permissible to merely extend the curve to end at the origin when the general direction of approach has been determined.

(2) The curve is tangent to one of the axes at the origin if no time delay is present; according to the difference $(n - m)$ between the number of poles and zeros in the transfer function, the tangent approaches the origin at an angle given by $-(n - m)\pi/2$.

(3) If the system contains a time delay the locus will spiral round the origin as $s \rightarrow j\infty$.

(4) The locus intersects and is perpendicular to the real axis at a magnitude which is equal to the steady-state gain of the function as $s \rightarrow j0$ (i.e. zero frequency), unless integrating elements are present, in which case the locus ends at infinity parallel to one of the axes.

(5) If the system is under-damped the curve shows a typical 'resonant bulge'.

The Nyquist diagram may be plotted from values which are determined analytically or obtained by experimental analysis. The power of the subsequent method of stability analysis based on the diagram lies in the fact that the mathematical form of the transfer function need not be known and it is not necessary to take factors of high order polynomials.

Sketching the Nyquist diagram is particularly useful when the frequency response characteristics depart from the usual pattern of a progressive increase in phase lag and decrease in amplitude with increasing frequency. Such irregular systems can often be confusing. As an example, there is the question of what angle to take in an experimental determination of frequency response, since a 90° phase advance is identical in appearance with a 270° phase lag, and so on. Similarly the calculation of $\tan^{-1}(y/x)$ can be confusing since $\tan^{-1} + 1$ can be $+45°$ or $-135°$, $\tan^{-1} - 1$ can be $-45°$ or $-225°$. In such cases it might be necessary to plot the appropriate vectors to determine the quadrant in which the resultant vector will lie.

In applying the Nyquist stability criterion (see below) it is necessary to follow the transfer locus of $G(s)$ from $s = +j\infty$ to 0 and from 0 to $-j\infty$. The negative transfer locus, i.e. $G(s)$ for negative frequencies (which have no real meaning), is always a mirror image of the curve for positive frequencies, and the two thus form a closed curve. In cases where the

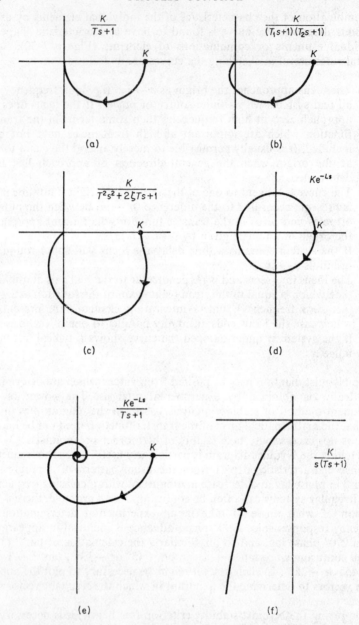

FIGURE 7–30. Typical Nyquist diagrams: (a) first-order lag, (b) second-order element (over-damped), (c) second-order element (underdamped), (d) pure time delay, (e) first-order lag and time delay of same order of magnitude, (f) first-order lag and integration

magnitude ratio is infinite at zero frequency, which only occurs when integration elements are present, the positive and negative curves apparently end at $\pm\infty$. Integrating elements are seen as an s^n term in the factors of the denominator of the transfer function, and in such cases the transfer locus is made into a closed curve by connecting the positive frequency end of the locus to the negative frequency end by $\frac{1}{2}n$ anti-clockwise rotations at infinite radius.

Stability and the Nyquist Criterion

The open-loop transfer locus or the Nyquist plot of the open-loop frequency response characteristics can be used to predict the stability or otherwise of the closed loop. As was seen previously, the limit of stability of the closed loop is given by solution of the characteristic equation, $1 + G(s) = 0$. The closed loop will thus be verging on instability if the transfer locus defining the value of $G(s)$ passes through the point at which $G(s) = -1$, which is the point $(-1, j0)$ on the $G(s)$-plane at which $|G| = 1$ and $\angle G = -180°$. Thus for simple systems, stable closed-loop operation will result if the open-loop transfer locus passes between the $(-1, j0)$ point and the origin in a clockwise direction with increasing frequency; unstable operation will occur if the locus passes outside or encloses the $(-1, j0)$ point. This statement will be recognized as equivalent to the Bode criterion that the magnitude ratio of the open-loop function should be less than 1 at 180° phase lag for stable response.

Changing the overall gain of the open loop does not change the shape of the transfer locus but only makes it expand or contract uniformly. It is not then necessary to replot the data for different gains, but it is more convenient to plot a single curve, usually for a normalized (unit) gain, and to imagine the scale of the axes to be changed. On this basis, from the curve for a unit gain, the value of the limiting gain is the reciprocal of the value of the locus at the intercept with the negative real axis, since multiplying the magnitude by this value would make the curve pass through the $(-1, j0)$ point.

The above statements are so far only an alternative statement of the Bode criterion. The Nyquist diagram is of greater value for loops containing unstable elements or integrations where the Bode criterion may not be applicable or ambiguous. In such cases the shape of the transfer locus is such that it is not usually possible to test for stability by the simple criterion outlined above, i.e. by simple enclosure of the $(-1, j0)$ point. The more general Nyquist stability criterion, of which the above, and also the Bode criterion, is an essential simplification, must then be employed. The derivation of the full Nyquist criterion is based on complex variable theory and is given in most texts on servo-mechanisms. The results are presented here without formal proof.

The basis of the method is to use the transfer locus to check whether the characteristic equation, $1 + G = 0$, has roots with positive real parts which will, of course, make the system unstable. Unlike the Routh test, it is not necessary to know the mathematical form of the transfer function, but only the transfer locus which may be plotted from experimental data. The Nyquist method also indicates how near the system is to stability or instability.

The roots of the characteristic equation are specific values of s which may be located on the complex s-plane as in Figure 7–31(a), where two real roots are shown as examples, i.e. r_1 (negative) and r_2 (positive). Suppose now that a vector is drawn from these roots to a point on the imaginary axis and that the tip of this vector follows a path from $+j\infty$ through zero to $-j\infty$ and then round a semi-circle of infinite radius from $-j\infty$ back to $+j\infty$. The value of s along this path thus changes from $+j\infty$ to 0 to $-j\infty$ and back again and makes a complete encirclement of the right (positive) real half-plane. For a root with a positive real part such as r_2, the vector $(s - r_2)$ makes one anti-clockwise rotation in following the path outlined above; for a root with negative real parts such as r_1, the vector $(s - r_1)$ makes no net revolution. It may be concluded that a root can be located in either the positive or negative half-plane by the number of revolutions of the test vector for $s = j\omega$ along the path which encircles the positive right half-plane. If, as may happen, a root lies on the imaginary axis or at the origin, the same procedure can be followed by altering the locus of s to include a semi-circle of infinitesimally small radius into the right half-plane around such a point.

For a particular function $G(s)$, for which $s = j\omega$, the problem can be transferred from the s-plane to the $G(s)$-plane by conformal mapping, since for any value of s there is a corresponding value of $G(s)$. Thus the transfer locus of Figure 7–31(b) represents $G(s)$ for all possible values of $s = j\omega$. A line from the origin to the point P represents a vector G; a line from the origin to the point Q at $(-1, j0)$ is the vector of -1 and by vectorial subtraction the line QP is the vector of $(1 + G)$. Thus the vector from Q, the point $(-1, j0)$ represents the denominator of the closed-loop transfer function. Taking the tip of this vector around the transfer locus from $\omega = +\infty$ to 0 to $-\infty$ is effectively the same as rotating a vector from a root of $(1 + G)$ around the s locus from $s = +j\infty \rightarrow 0 \rightarrow -j\infty$.

The Nyquist criterion follows directly from a consideration of the poles and zeros of the closed loop function $(1 + G)$. The theory shows that the number of values of s (i.e. the roots of $(1 + G)$) with positive real parts is the number of rotations of the vector $(1 + G)$ as the tip of the vector traces out the entire transfer locus, the closed curve for positive and negative frequencies. The trace is usually started from the origin along the positive locus (from $\omega = +\infty \rightarrow 0$) and back along the negative locus (from $\omega = 0 \rightarrow -\infty$) to the origin. The vector must make an integral (positive or negative) number of revolutions, clockwise rotation being counted as positive. The net number of revolutions is equal to the difference between the numbers of 'positive' poles and zeros of the closed-loop

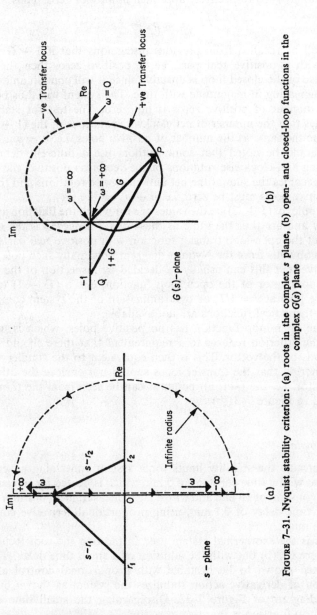

FIGURE 7–31. Nyquist stability criterion: (a) roots in the complex *s* plane, (b) open- and closed-loop functions in the complex *G(s)* plane

function, i.e. the values of s with positive real parts which make the function $(1 + G)$ respectively approach infinity or zero. Thus

$$n = p_+ - z_+$$

It will be recalled from previous discussions that if $1 + G = 0$ has a root with a positive real part, i.e. a positive zero, then the transient response of the closed loop is unstable since it will contain an exponential term increasing in magnitude with time. The system will thus be unstable if the number of positive zeros is not zero. The test for stability then becomes that the number of net clockwise rotations of the $(1 + G)$ vector must be the same as the number of positive poles, i.e. $n = p_+$.

It should be noted that some authors use a different definition; by counting anti-clockwise rotations of the vector as positive, the last statement becomes the sum of the net anti-clockwise rotations and the number of positive poles must be zero, i.e. $n + p_+ = 0$.

The poles of $(1 + G)$ are the values of s for which the function approaches infinity and are thus the same as those of G; hence p_+ is the number of poles of the open-loop transfer function with positive real parts. It is not always obvious from the Nyquist diagram how many such poles a system will have, but this can usually be decided by inspection of the factors of the denominator of the open loop function, e.g. a $(Ts - 1)$ term has a positive pole at $s = 1/T$, or by examination of the system components if the mathematical functions are not available.

If the open-loop function has no positive poles, which is usually the case, the criterion reduces to a requirement that there should be no net rotations of the vector. This is then equivalent to the simpler version of the criterion that the transfer locus should not enclose the other end of the vector, i.e. the $(-1, j0)$ point, as can be seen from the transfer locus plotted in Figure 7–31(b).

Example 7–9

Determine the stability limits for a system containing an exothermal reaction with a time constant of 5 min which is heated by an element with a time constant of 10 min and with a lag in measurement of 1 min combined with a time delay of 0.1 min, using proportional-derivative control with $T_d = 3$ min.

This is the exothermal system used in plotting the root locus diagram of Figure 6–8(b) but with the addition of a small time delay. The system was then shown to be unstable with proportional control alone. The addition of derivative action stabilizes the system as shown by the root locus diagram of Figure 7–32. Disregarding the small time delay, the open-loop function is

$$G(s) = K(1 + 3s)/[(5s - 1)(10s + 1)(s + 1)]$$

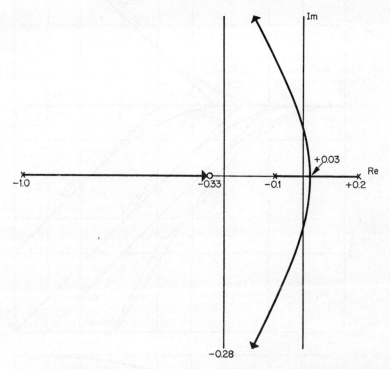

FIGURE 7–32. Root locus diagram for system of Example 7–9, omitting time delay

The poles are thus at -1, -0.1, and $+0.2$ min^{-1} and there is a zero at -0.33 min^{-1}. Two loci are asymptotic to $s = -0.28$ and leave the real axis at $+0.03$, these loci must then cross the imaginary axis and the corresponding complex roots will take on negative real parts. Hence a stable response can be obtained. The effect of the small time delay will be to bend the two loci back to the right after the first approach to the asymptotes, and so make the system response unstable at higher frequencies as the loci re-cross the imaginary axis and the roots take on positive real parts. The root locus plot in this instance is not well adapted to finding the second critical gain.

The transfer locus or Nyquist diagram is developed from the Bode plot of the open-loop responses of the separate elements. These are combined to give the Bode open-loop diagram for the system and the magnitude and phase transferred to a Nyquist plot on polar co-ordinates. As usual, the Bode plots are made for unit gain ($K = 1$), i.e. A.R.s are plotted and not $|G|$.

One advantage of using the Bode or Nyquist diagrams for stability analysis as compared to the root locus method is that empirical frequency-response data can be included for which the exact transfer function cannot

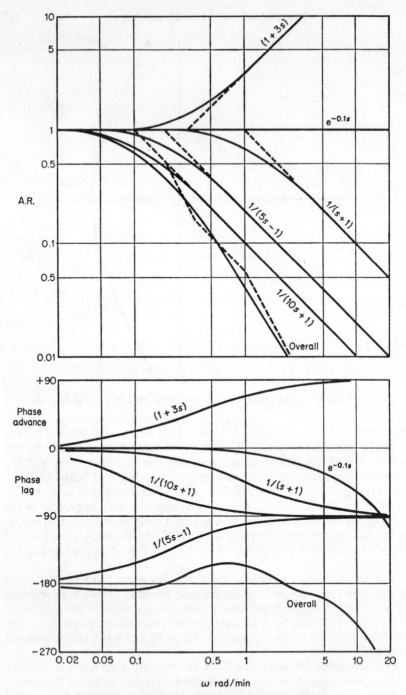

FIGURE 7–33. Bode diagram for system of Example 7–9

be defined. This example is made somewhat more realistic and in line with a practical system by including the small time delay to represent the small lags and delays which are always present in any real system. These small delays will have little effect at low frequencies, having a 'flat' frequency response ($|G| = 1$ and $\angle G = 0$) well beyond the corner frequencies of the major lags in the system. At higher frequencies these small lags will show the usual A.R. and phase behaviour as their corner frequencies are approached and passed. The effect on the A.R. curve is not likely to be great but the additional phase lag might be critical. The effect is therefore simulated by the small time delay, which has no effect on the A.R. curve but produces an increasing phase lag as the frequency increases.

The Bode diagram for the individual elements are shown in Figure 7–33 and are combined into overall curves for the open-loop response of the system in the usual way. The overall A.R. curve is fairly normal, falling away steeply as the frequency is increased. The phase angle curve is most unusual, being $-180°$ at zero frequency (due to the unstable exothermal element) and falling below 180° as the frequency is increased (due to the two other lags), but then crossing the $-180°$ value (due to the phase advance of the derivative action), and finally falling back below $-180°$ as the time delay takes effect. The stability of the system is thus difficult to analyse by the Bode criterion. The unusual features can be explained by the Nyquist diagram which is plotted from the values of A.R. and phase angle from the Bode diagram overall curves; plotting the A.R. as magnitude against the corresponding phase angle for a given frequency produces Figure 7–34(a). If the factors of the open-loop function are known, these values can be computed as the product of the individual A.R.s and the sum of the phase angles.

Since the Bode diagram is plotted for unit gain, the curve of the Nyquist diagram will meet the real axis at the $(-1, j0)$ point for zero frequency due to the exothermal lag of 180° at zero frequency. This is of little consequence since it is not necessary to replot the diagram for different gains; it is only necessary to effectively change the scale of the diagram by moving the point about which the $(1 + G)$ vector will be rotated. Moving this point towards the origin is equivalent to plotting the diagram on a larger scale and thus increasing the overall gain, as shown in Figures 7–34(b) and (c).

As the transfer locus is shown in Figure 7–34(a), the vector from $(-1, j0)$ makes one anti-clockwise rotation as the tip traces the curve from $\omega = +\infty$ to $-\infty$. The number of positive poles in the system (p_+) is known to be one because of the pole at $+0.2$ min^{-1} in the denominator of $G(s)$. Hence the number of positive zeros, z_+, given by $(p_+ - n)$ is $(1 - -1)$ or 2. The conclusion is that $G(s)$ has two positive zeros, and the characteristic equation, $1 + G = 0$, will then have two roots with positive real parts. The closed loop is therefore unstable for the particular value of the gain for which the diagram was plotted, i.e. $K = 1$.

The centre of rotation is then moved along the real axis in the direction of the origin, and the number of rotations of the vector is counted at a

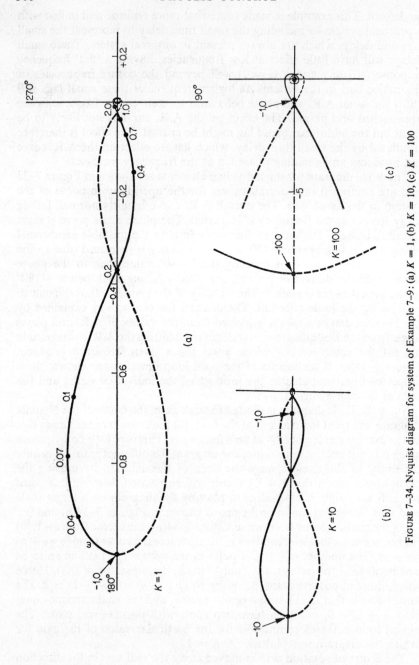

FIGURE 7-34. Nyquist diagram for system of Example 7-9: (a) $K = 1$, (b) $K = 10$, (c) $K = 100$

number of trial points. It will be found that when the point is taken to the right of -0.36, which is the point at which the positive and negative transfer loci cross the real axis, the number of rotations of the vector becomes once clockwise, i.e. $n = +1$. The number n is thus equal to p_+ and $z_+ = 0$; the closed-loop function thus has no zeros with positive real parts and the system is stable. To make the centre of rotation the $(-1, j0)$ point on the diagram, the scale of the real axis must be increased by a factor of $1/0.36 = 2.78$, and this is thus the minimum value of the overall gain K at which the system is stable. In Figure 7–34(b) the transfer locus is replotted for a gain greater than the minimum value ($K = 10$) and a trial vector rotation about the $(-1, j0)$ point shows that the system is stable. These findings are confirmed by the root locus diagram of Figure 7–32; the two dominant roots have positive real parts and the system is unstable at low gains, but there is a critical minimum value of the gain at which the loci cross the imaginary axis, and the roots take on negative real parts to stabilize the system. The transfer locus shows that this occurs at a minimum gain of 2.78.

If the gain is further increased, the transfer locus will again pass through the centre of rotation of the vector since there are two frequencies at which the phase lag of the system is $-180°$. This second intersection occurs at the point -0.011 on the real axis. At this point the rotation of the vector becomes once anti-clockwise, i.e. $n = -1$, and the system is again verging on instability. This then represents a maximum gain at which the system becomes unstable, the value of K being $1/0.011$ or 90.9. The small time delay has now caused the dominant roots to take on positive parts. Figure 7–34(c) shows the relevant parts of the diagram replotted for a gain greater than the maximum value ($K = 100$).

The range of open-loop gains for stable response is now clearly shown on the open-loop Bode diagram once the Nyquist diagram has been examined. The A.R. curve of the Bode diagram must be multiplied by a factor between 2.78 and 90.9 for a stable response. This will make the magnitude ratio greater than one at frequencies below about 0.20 rad/min (for which the phase angle curve is rising to 180° lag) and less than one for frequencies between about 0.20 and 2.25 rad/min (for which the phase lag is less than 180°). The minimum phase lag shown by the Bode diagram is about 151° at a frequency of 0.70 rad/min, giving a maximum phase margin of about 30°. This would require a gain of $1/0.074$ or 13.51.

The transfer locus is a concise graphical representation of the dynamic characteristics of a control system. For a control loop which is effectively linear and made up of many or complex elements, the Nyquist stability criterion is a graphic and unambiguous method of interpreting the stability characteristics and predicting the range of values of loop gain to give a stable response. In practice the Bode diagram is prepared first, because of the graphical facility of combination of elements to give the overall relationship, and for the inclusion of experimental data if required. The Nyquist diagram and criterion are normally only used if the system is one which produces an unusual Bode diagram for which the simple Bode criterion is ambiguous or not applicable.

Gain and Phase Margin

The Nyquist diagram has the additional advantage of showing how close a system is to stability for a particular value of the gain by the proximity of the intersections of the $G(s)$ locus with the real axis to the centre of rotation of the $(1 + G)$ vector, i.e. the $(-1, j0)$ point. Gain and phase margins were introduced previously as measures of the safety factors

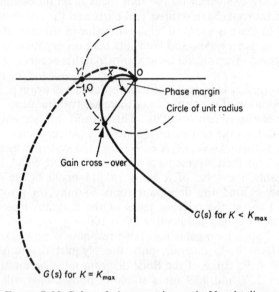

FIGURE 7–35. Gain and phase margins on the Nyquist diagram

of a stable system from the condition of instability, and these may also be interpreted on the Nyquist diagram. Considering Figure 7–35, the transfer locus of a simple system passes between the $(-1, j0)$ point and the origin in the clockwise direction of increasing frequency, thus showing that the system is stable. Changing the gain of the system does not change the shape of the curve but simply makes it expand or contract, and it will be apparent that if K is increased, the value of the magnitude $|G|$ will be greater at each phase angle including $-180°$. The intercept with the negative real axis will then be moved towards the $(-1, j0)$ point. At the limiting value of K_{max}, the intercept with the real axis X will coincide with $Y(-1, j0)$ and the system will be on the verge of instability. Conversely, when X is to the right of Y, the system will be stable; the further X is from Y, the greater is the degree of stability. This is then a measure of the gain margin, usually defined as OY/OX or $1/OX$; since $OX = K(A.R.)\omega_c$ and

$$OY = K_{max}(A.R.)\omega_c = 1$$

$$1/OX = K_{max}/K$$

The phase margin can also be defined on the Nyquist diagram by considering the previous definition, i.e. the difference between 180° and the phase lag corresponding to unit loop gain, the latter now becoming $|G| = 1$. The unit value of the magnitude is defined by a circle of unit radius centred on the origin which will pass through the $(-1, j0)$ point Y and intersect the $G(s)$ curve at Z, the latter being the gain cross-over point. The angle made by OZ with the negative real axis is then the difference between 180° and the phase lag at unit loop gain, i.e. the phase margin.

Closed Loop Analysis

The complex plane plot of the open-loop transfer function locus can also be used to investigate other characteristics of the closed loop response in addition to the stability. For example, the closed-loop magnitude ratio is defined by the closed-loop transfer function as

$$M = \left|\frac{G}{1 + G}\right| = \frac{|G|}{|1 + G|}$$

Since G is, in general, a complex number such as $(x + jy)$, M is the magnitude of a further complex number:

$$M = \frac{|x + jy|}{|1 + x + jy|} = \frac{(x^2 + y^2)^{\frac{1}{2}}}{[(1 + x)^2 + y^2]^{\frac{1}{2}}}$$

which, by squaring and re-arrangement, gives

$$[x + M^2/(M^2 - 1)]^2 + y^2 = M^2/(M^2 - 1)^2 \qquad (7\text{--}17)$$

This equation is of the form $(x - a)^2 + (y - b)^2 = r^2$, which is the equation of a circle on rectangular co-ordinates (x, y) with the centre at the point (a, b) and with a radius of r. Equation (7–17) thus defines a circle in the complex plane with the centre at the point given by

$$a = -M^2/(M^2 - 1)$$

on the real axis since $b = 0$, and with a radius of $M/(M^2 - 1)$, for a specific value of M. There is thus a family of circles each representing a constant value of M (Figure 7–36); for $M = 1$ the circle is the limiting case of infinite radius and is a straight line passing through the point $(-0.5, j0)$ parallel to the imaginary axis. As the centres of the circles approach the $(-1, j0)$ point, the radii approach zero and the value of M increases. Thus the stability of the system decreases as the value of M increases.

The intersection of the open-loop transfer locus $G(s)$ with the M-circles identifies the corresponding closed-loop magnitude ratios at specific frequencies according to the point of intersection on the $G(s)$ curve. It will be seen that there is a maximum value of $M(M_p)$ when the transfer locus is tangential to the particular M-circle; the frequency is then the resonant frequency ω_r, corresponding to the peak of the closed-loop frequency response curve. It is possible then to determine, from the open-loop transfer locus, the maximum amplitude of the steady-state sinusoidal response and the frequency of oscillation for the closed loop.

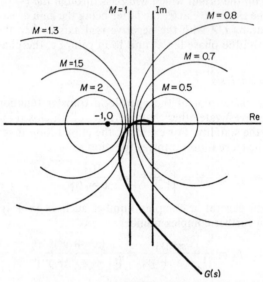

FIGURE 7–36. Constant M-circle on the Nyquist diagram

Certain empirical rules can be used to relate the maximum value of M to the transient response; a value between 1.3 and 2.0 results in a transient response which is generally acceptable for most control systems, the lower value of 1.3 being preferred for servo-systems and the higher for process controls [23]. A value of 2.0 for the maximum closed loop magnitude ratio corresponds to a value of ζ for the dominant second-order response of about 0.25.

The finding of the value of controller gain which causes the transfer locus to be tangential to a given M-circle in order to obtain a given maximum magnitude ratio is obviously a trial and error procedure. The following graphical procedure requires rather less effort than that of plotting the locus with different gains until it becomes tangent to the specified circle. Suppose the circle with centre at A on the real axis is an M-circle (Figure 7–37). The distance of A from the origin (OA) is then $M^2/(M^2 - 1)$ and the radius (AB) is $M/(M^2 - 1)$. A line from the origin tangential to the circle at B makes an angle with the negative real axis, $\angle AOB$, whose sine is AB/OA or $1/M$. A perpendicular from the point of tangency B to the real axis will intersect the latter at point C and

$$\angle ABC = \angle AOB = \sin^{-1}(1/M)$$

But $\sin \angle ABC = AC/AB$; hence

$$AC = AB \sin \angle ABC = 1/(M^2 - 1)$$

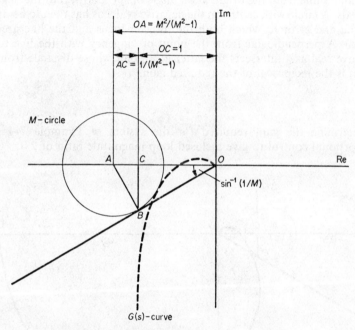

FIGURE 7-37. Determination of gain for specified $M_{\max}$

The distance of point C from the origin is

$$OC = OA - AC$$
$$= M^2/(M^2 - 1) - 1/(M^2 - 1) = 1$$

Thus the point C is the $(-1, j0)$ point and the ratio of OC/OA is

$$1/[M^2/(M^2 - 1)]$$

This relationship will apply irrespective of the scale to which the diagram is drawn.

If now a transfer locus with a gain of K is tangent to an M-circle (the value of M must then be M_p), reducing the gain by the factor K, i.e. to unity, reduces the scale of the diagram by the factor K. If a circle is now

drawn with centre on the negative real axis and tangential to both the
unit gain curve and to a line OB at an angle to the negative real axis of
$\sin^{-1}(1/M_p)$, this circle is the M_p-circle also reduced by the factor K. The
perpendicular from the point of tangency (B') with the line OB will intersect the real axis at a point C' where OC' is OC reduced by the factor K,
i.e. $OC' = OC/K = 1/K$; hence $K = 1/OC'$.

Thus to find the gain required to produce a specified maximum closed-
loop magnitude ratio M_p, the transfer locus is drawn for the usual unit
gain and a line from the origin at an angle of $\sin^{-1}(1/M_p)$ to the negative
real axis. A circle with centre on the negative real axis has then to be drawn
by trial and error to which both the transfer locus and the line are tangential. A perpendicular from the point of tangency with the line to the
negative real axis intersects the latter at a point whose distance from the
origin is the reciprocal of the required gain.

Example 7–10

Determine the gain required for the system of Example 7–1 with
proportional control to give a closed-loop magnitude ratio of 2.0.

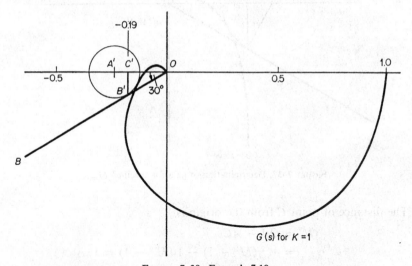

FIGURE 7–38. Example 7.10.

The Nyquist diagram for the three stage process is plotted with unit
gain ($K = 1$) in Figure 7–38, from the Bode diagram of Figure 7–10. For
a value of $M_p = 2.0$, $\angle BAC = \sin^{-1} 0.5$, or 30°. The line OB is then drawn
at an angle of 30° to the negative real axis. By trial and error, a circle is
drawn with centre on the real axis and tangential to both the curve and the
line OB. A perpendicular from the point of tangency (B') intersects the

real axis (C') at a value of -0.19; hence the required gain to achieve a maximum closed-loop magnitude ratio of 2.0 is $1/0.19 = 5.26$.

In a similar way to the development of the constant M-circles, the loci of constant closed-loop phase angles can also be identified as a family of circles on the $G(s)$-plane. These pass through the points 0 and -1 on the real axis with centres lying on a line parallel to the imaginary axis and passing through the point $(-0.5, j0)$.

PROBLEMS

7-1 Sketch the Bode open-loop frequency response diagrams for the following transfer functions and calculate the gain and phase angle for $\omega = 10$ rads/min in each case (s in min^{-1}).

(a) $50/[(10s + 1)(2s + 1)]$
(b) $10s/[(s + 1)(0.2s + 1)^2]$
(c) $(s + 1)/[(10s + 1)(3s + 1)]$
(d) $(s - 1)/[(10s + 1)(2s + 1)]$
(e) $(1 + s)^2$
(f) $(1 - 0.5s)/(1 + 0.5s)$

7-2 Sketch the Bode diagram for the system shown in Figure P7–1 (s in min^{-1}) and determine the maximum value of K_c for proportional control and that required to give 30° phase margin.

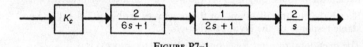

FIGURE P7–1

7-3 Sketch the Bode diagram for a process and measuring element with an overall transfer function of $e^{-0.5s}/(2s + 1)^2$ and determine the maximum value of K_c for proportional control (s in min^{-1}).

7-4 Plot the open-loop Bode diagram for the control loop of Figure P7–2 (s in min^{-1}) and determine the gain and phase margins.

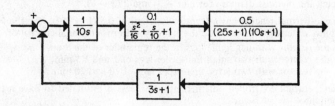

FIGURE P7–2

7-5 A control loop contains first-order lags with time constants of 30 and 10 s and a time delay of 3 s. Plot the open-loop Bode diagram and determine the value of K_c to give 30° phase margin. What is the gain margin?

7-6 The system of Problem 2–4 is to be controlled with a proportional controller. Use a Bode open-loop diagram and the Ziegler-Nichols rules to determine the value of K_c.

7-7 The system of Problem 5–11 is to be controlled with a proportional-integral controller. Determine the controller adjustments using an open-loop Bode diagram and the Ziegler-Nichols rules.

7-8 Sketch the open-loop Bode diagram for the system of Problem 6–10 and compare the Ziegler-Nichols recommended settings with those from the root locus plot.

7-9 Plot the open-loop Bode diagram for the outer loop of the system of Figure P7–3. The gain is increased until the system oscillates continuously at a frequency of 3 rads/min. Determine the magnitude of the time delay L.

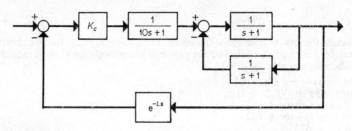

FIGURE P7-3

7-10 Plot the closed-loop Bode diagrams for the systems of Problems 7–6 and 7–7 and use the results to characterize the loops as second-order systems.

7-11 Sketch the Nyquist diagrams for the transfer functions of Problem 7–1.

7-12 Sketch the Nyquist diagrams for the systems of Problems 7–4 and 7–5 and test for stable response by the Nyquist criterion.

7-13 The open-loop transfer function of a three-stage process is

$$G(s) = 2/[(10s + 1)(5s + 1)(2s + 1)]$$

the time constants being in minutes. Determine by use of the Nyquist diagram if the system is stable

(a) with a proportional controller with a gain of 3.0,
(b) with a proportional-integral controller with a proportional gain of 3.0 and an integral time of 5 min.

7-14 Sketch the Nyquist diagram for $2/(s − 3)$ and $2/(3 − s)$.

7-15 An exothermal reactor has a transfer function of $0.5/(1 − 15s)$. Use the Nyquist stability criterion to investigate the stability with proportional control of

(a) the reactor with negligible lags in the remainder of the loop,
(b) the reactor with two small first-order lags of 1 and 1.5 min,
(c) the reactor with two large first-order lags of 10 and 20 min.

7-16 A temperature controlled polymerization process is estimated to have a transfer function of

$$G_p = e^{-5s}/[(s − 40)(s + 80)(s + 100)]$$

the time constants being in seconds. Plot open-loop Bode and Nyquist diagrams and determine the limiting gain for proportional control.

7-17 A process plant is characterized by an 8 s time delay and a second-order lag with time constants of 10 and 20 s. The plant is to be controlled by a proportional-integral controller set to give a gain margin of 2.5, but the integral action is to be used only if the steady-state offset with proportional control exceeds 5 per cent of the load change. Determine the required controller adjustments.

7–18 Two possible control schemes for a spray drier are shown in Figure P7–4; the control of the outlet temperature may be effected by regulation of the steam supply or the liquid feed. The air heater consists of two non-interacting first-order lags of

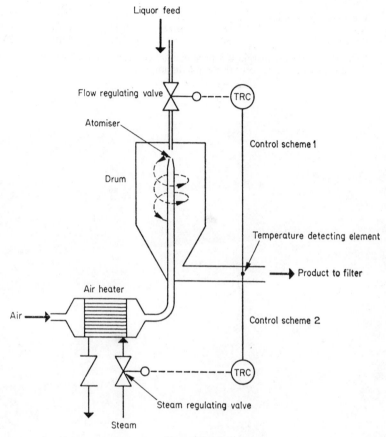

FIGURE P7–4

100 s each. The drier behaves as three first-order lags of 10 s each with a distance-velocity delay of 2 s. There is a further distance-velocity delay of 3 s between the air heater and the atomizer and the measuring element has a lag of 5 s. The control valves have no dynamic lag.

Draw block diagrams for the alternative schemes with appropriate values of the parameters and showing the points of entry of disturbances in the feed and steam supply. Select the control scheme which will give the better results by consideration of the steady-state offsets and determine the value of K_c for proportional control with adequate gain and phase margins.

7–19 A cascade control system has an inner loop comprising a proportional controller with a gain of 5.0 and a process stage with a single time-constant lag of 5 min and unity measurement feedback. The outer-loop process consists of two non-interacting first-order lags of 2 min each and also with unity feedback. Compare the proportional gains of the outer-loop controller for the cascade and as a single-loop system.

7-20 Plot the Bode diagram for an experimental test on a process which gives the following results:

Frequency (rads/min)	2.5	4.0	6.3	10.0	15.8
A.R.	0.87	0.72	0.57	0.35	0.14
Phase lag (°)	30	60	100	150	200

Determine the phase margin if a proportional controller is used with the proportional gain set at 60 per cent of the maximum value.

Chapter 8: Controller Mechanisms

The particular manner by which a controller produces its output signal has little bearing on the actual design of a control system. It is useful, however, to consider the basic operation of the controller, particularly when some departure from development of the ideal transfer function occurs.

Historically, the predominant type of control instrument used in the process industries has been the pneumatic instrument, and it is only within quite recent years that the electronic instrument has found increasing application. The predominance of the pneumatic controller has been principally due to the rugged construction, low maintenance costs and greater safety in the potentially imflammable and explosive atmospheres encountered in the chemical and process industries. The use of the pneumatic diaphragm valve as the final control element is also a considerable convenience. In recent years most of the advantages of the pneumatic instrument have been overtaken by developments in electronic instruments and most manufacturers now offer both types as alternatives; it is still necessary, however, to employ the pneumatic control valve as the final element with an electronic controller, except in the relatively few cases in which an electrically operated final control element can be used.

Most of the commercially available control instruments and the other associated hardware such as transmitters, valves, etc. are fully described in the literature [9, 12, 17, 29]. All that will be included here is a brief discussion of the operation of a typical pneumatic controller whose transfer functions will be derived for comparison with the ideal functions used in the previous chapters.

Pneumatic Control

The motion-balance, as distinct from the force-balance, type of pneumatic instrument is selected as being probably the simpler mechanism to explain. The basic element of all pneumatic instruments is the baffle-nozzle system, which is essentially a controlled leakage of compressed air from an otherwise closed system containing the diaphragm valve and supplied from a constant pressure source. The baffle is linked directly to the error or deviation signal, usually generated by a differential lever system connected to the reference input (desired value setting) and the measurement feedback element; the movement of the baffle is thus a direct measure of the error or deviation in the control system. The action of the baffle-nozzle arrangement is simply that the baffle impedes the flow of air through the nozzle and the pressure behind the latter is very nearly proportional to the nozzle-baffle separation over a very limited range of movement (Figure 8–1). As the error signal changes, so also does the nozzle back-pressure, and this provides the controller output signal through a suitable relay valve. The restriction behind the nozzle is an essential part of the arrangement to prevent the system filling to the pressure of the supply; obviously the restriction (usually a capillary tube) must present a greater restriction to the flow than the nozzle so that the pressure behind the latter can fall to a minimum value when the baffle presents the minimum impedance to the flow.

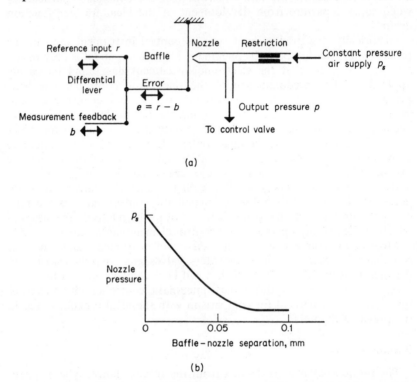

(a)

(b)

FIGURE 8–1. Pneumatic controller: (a) baffle-nozzle system, (b) nozzle pressure characteristics

The baffle-nozzle system is extremely sensitive, a change in the separation of less than 0.05 mm (0.002 in) is sufficient to cause a full change in the output pressure signal delivered by the instrument. Even with amplifying lever linkages in the measurement feedback, the bandwidth as a proportional controller would be extremely small ($\ll 2$ per cent) and the instrument would function effectively with an on-off action. It is necessary, therefore, to reduce the sensitivity of the baffle-nozzle system if the instrument is to function as a wide-band proportional controller.

Proportional Controller

The sensitivity of the baffle-nozzle system is reduced by introducing a pneumatic feedback bellows to oppose the movement of the baffle provided by the deviation linkage. The bellows is connected to the nozzle back pressure (usually via a relay valve which is not shown in Figure 8–2) so that the bellows expands as the baffle approaches the nozzle; owing to this opposing movement, the actual movement of the baffle with respect to the nozzle is very much less than if the end of the baffle were fixed, and a simple analysis shows that proportionality can be obtained with a much reduced sensitivity.

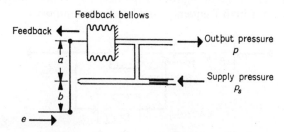

FIGURE 8–2. Schematic diagram of pneumatic proportional controller

Assuming that the system is initially in equilibrium with a constant output pressure at zero error and that an error signal e is introduced, the change in the baffle-nozzle separation x is given by

$$x = [a/(a + b)]e - [b/(a + b)]gp \qquad (8-1)$$

where a/b is the lever ratio of the baffle, g is the spring constant (extension per unit change of pressure) of the bellows, and p is the change in output pressure resulting from the error signal e. The pressure change is, however, proportional to the change in the baffle-nozzle separation, i.e.

$$p = cx \qquad (8-2)$$

Combining these two equations and re-arranging:

$$p[(a + b)/(bcg) + 1] = (a/bg)e \qquad (8-3)$$

The gain of the baffle-nozzle system ($p/x = c$) is very large since a baffle movement of less than 0.05 mm (0.002 in) with respect to the nozzle produces the full change of output pressure from the instrument. In British-American units the latter is from 3 to 15 lbf/in^2, in SI units the output pressure change will be about 80 kN/m^2 (probably from 20 to 100 kN/m^2). The gain c is thus of the order of 1600 kN/m^2/mm (6000 lbf/in^2/in); hence, in Equation (8–3),

$$bcg \gg (a + b)$$

If the gain is assumed to be infinite then

$$(a + b)/(bcg) \to 0$$

and Equation (8–3) reduces to

$$p = (a/bg)e \qquad\qquad (8\text{–}4)$$

This is effectively the transfer function of a proportional action generator, i.e. $p = K_c e$, where the proportional sensitivity is $K_c = a/(bg)$, and the latter parameter can be adjusted by changing the lever ratio a/b.

The above analysis assumes that the pressure in the bellows is equal to that at the nozzle at any instance; this is a valid assumption if sufficient time is available for the pressures to equalize. This is normally so, but may not be the case if high frequency oscillations are imposed.

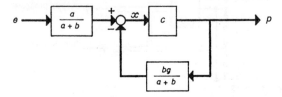

FIGURE 8–3. Block diagram of proportional controller of Figure 8–2

It is instructive to represent the action of the mechanism by means of a conventional block diagram; using Equations (8–1) and (8–2), the block diagram of Figure 8–3 results, which shows that the bellows introduces a negative feedback into the mechanism. The baffle-nozzle system thus acts as a high-gain amplifier in a negative feedback circuit, and any non-linearity in the baffle-nozzle relationship will have little effect on the linearity of the overall feedback system, i.e. the characteristics of the proportional action generator are essentially independent of the gain c so long as this is sufficiently large.

Proportional-derivative Controller

To add a derivative component to the proportional action, it is only necessary to delay the negative feedback by introducing a restriction in the connection of the output pressure to the feedback bellows. The reasoning here is that if a linear change of error is applied, the required response is a sudden rise in pressure (the derivative component) followed by a linear increase in pressure (the proportional component). The latter is provided by the proportional action generator already discussed, the effect of a restriction such as R_d in Figure 8–4 (usually a needle valve so that the resistance can be varied) is to hold back the operation of the feedback bellows by creating a pressure difference over the restriction between the output pressure p and that in the feedback bellows p_1. The amount of negative feedback applied is thus less than if the restriction were absent and a larger change in output pressure results. From this rather crude argument it can be seen that the response resembles the required PD action, and a simple analysis shows that the change in output pressure is given by

$$p = [a/(bg)](1 + T_d s)e \qquad (8\text{--}5)$$

if it is assumed that the nozzle gain is infinite. A separate treatment is required if the gain is regarded as finite (see pages 319–20).

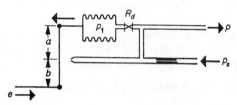

FIGURE 8–4. Schematic diagram of pneumatic proportional-derivative controller

Proportional-integral Controller

For a proportional-integral controller a further addition to the instrument is necessary, a positive feedback bellows connected via a restriction R_i to the nozzle pressure or the negative feedback bellows (Figure 8–5).

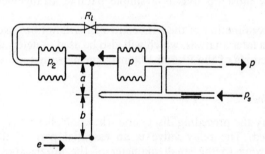

FIGURE 8–5. Schematic diagram of pneumatic proportional-integral controller

The need for the positive feedback can be seen by considering the output now required from a step change in error signal. The proportional generator develops the required proportional step-change in the output pressure; the additional integral action now calls for a linear increase in the output pressure whilst ever the error input remains constant. This can only be accomplished by bringing the baffle closer to the nozzle to increase the pressure; one end of the baffle, however, is virtually fixed in position by the constant error displacement, and the movement can only be accomplished by a movement opposing the negative feedback at the other end of the baffle. The positive feedback must be delayed by the restriction, otherwise the pressures in the two bellows would be equal and the negative feedback thereby completely impeded; the initial increase in pressure due to the proportional action creates a pressure difference over the restriction between the output pressure p and that in the positive feedback bellows p_2. The slow leakage of air through the restriction due to the pressure difference causes the bellows to expand and so move the baffle towards the nozzle. This increases the output pressure and so maintains the pressure difference over the restriction; the action thus continues and the output pressure increases until the error signal is removed or the limiting pressure is reached.

The above argument shows that the response is qualitatively that required from a PI controller and, with the assumption of an infinite nozzle gain, a simple analysis gives a response equation of

$$p = [a/(bg)] \, [1 + 1/(T_i s)]e \tag{8-6}$$

Proportional-integral-derivative Controller

As may be anticipated, the mechanism of the PID controller is a simple combination of the PI and PD generators. In most instruments of this type the positive feedback bellows and the integral restrictor are added to the basic PD generator of Figure 8–4, with the two restrictors in a series arrangement as shown in Figure 8–7. A number of variations are possible, of which the most obvious is a simple parallel arrangement of the restrictors.

The transfer function of the PID controller will not be developed at this stage but in a later analysis, which will also be used to determine the actual transfer functions of the PI and PD instruments with a finite nozzle gain.

Relay Valve

To simplify the preceding discussion, the controller relay valve has not been included. The relay valve is an essential part of the pneumatic controller owing to the small diameter of the nozzle and the capillary restriction in the air supply, both of which seriously limit the rate at which the nozzle pressure can change, particularly since the latter is effectively connected directly to the large capacitance offered by a transmission line and the diaphragm of the control valve, as implied in the preceding argument. Invariably the nozzle pressure is applied to a bellows or diaphragm capsule of very small capacity situated in close proximity to the baffle-nozzle system, so that a fast reaction to changes in the nozzle pressure are obtained. This relay unit operates a three-port valve, with a vent to atmosphere, which regulates the flow of air to and from the final control valve and also to the negative and positive feedback bellows. The rate of flow of air through the relay valve is very much greater than that through the nozzle and capillary since the diameters are much larger, and the supply is taken directly from the constant pressure supply to the instrument. Hence the output pressure applied to the diaphragm control valve responds relatively quickly to changes in the nozzle pressure caused by a change in the baffle-nozzle separation.

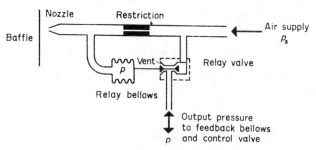

FIGURE 8–6. Pneumatic controller relay valve

The relay valve shown in Figure 8–6 is 'direct-acting', i.e. the output pressure from the valve increases with an increase in the input pressure from the nozzle; 'reverse-acting' relay valves, in which the output pressure decreases with an increase in the input pressure, are also used. The latter only requires that the direction of operation of the negative and positive feedback bellows be reversed in order to provide the same direction of movement of the baffle, but otherwise the unit design is basically the same. It was pointed out in an earlier discussion (pages 115–6) that the action of the controller itself must be reversible, i.e. it must be capable of direct or reverse action, as required to suit the direction of operation of the control valve. The latter must be arranged to 'fail safe' in the event of failure of the operating medium, and the controller action must be adjusted accordingly. This is accomplished by providing a reversing linkage in the mechanism between the differential lever developing the error signal and the linkage therefrom to the baffle, so that the direction of the baffle movement is effectively reversible with respect to the error.

If the relay valve has a unit gain, the operation of the controller mechanism is essentially that of the proportional controller diagram of Figure 8–3 with the addition of a further internal feedback loop due to the presence of the relay valve. The latter is often used also as an additional amplifier with a gain of up to about 5 : 1; this allows the nozzle pressure to vary over a rather smaller range with a consequent improvement in the linearity of the baffle-nozzle pressure system.

Controller Transfer Functions with Finite Gain

Consider the three-term controller (PID) of Figure 8–7 to be at a steady-state with zero error input. The output pressure and that in each of the feedback bellows will then be equal at some reference pressure (p_0) corresponding to the zero error. Assume that an error signal of e is now introduced through the differential linkage to the baffle; this will cause a change in the output pressure of p and a change in the bellows pressures of p_1 (negative feedback) and p_2 (positive feedback). The equation previously developed for the baffle movement of the proportional controller (Equation 8–1) will apply but must be modified for the different pressure changes and the presence of the positive feedback, i.e.

$$x = [a/(a + b)]e - [b/(a + b)]g_1p_1 + [b/(a + b)]g_2p_2 \qquad (8\text{–}7)$$

It will be assumed, for simplicity, that the two bellows units in the instrument are identical so that the spring constants are equal (i.e. $g_1 = g_2 = g$)

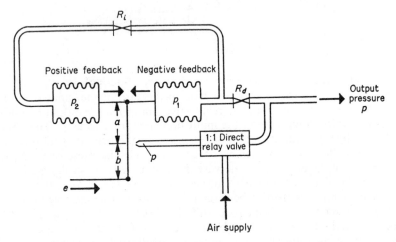

FIGURE 8–7. Schematic diagram of pneumatic proportional-integral-derivative controller

and the capacitances (volumes) are also equal. In practice one bellows unit is often used differentially to develop both negative and positive feedback by applying the negative feedback pressure inside the bellows and the positive feedback outside, or vice versa. The spring constant is then obviously the same on each side; the capacitances may differ but this would only introduce a constant numerical factor (the ratio of the volumes) into the subsequent analysis. The motion of the bellows is in general so small that any change in volume due to expansion or contraction is entirely negligible. Equation (8–7) can thus be simplified to

$$x = [a/(a + b)]e - [b/(a + b)]g(p_1 - p_2) \qquad (8\text{–}8)$$

The pressure changes in the two bellows may be related by taking a mass or volume balance over each; the bellows together form an interacting dead-end system of two first-order resistance-capacitance stages.

Thus for the positive feedback bellows with a pressure drop of $(p_1 - p_2)$ over the resistance R_i,

$$C_2(dp_2/dt) = (p_1 - p_2)/R_i$$

where C_2 is the capacitance of the bellows and the system has a time constant of $T_2 = R_iC_2$. This equation can be transformed to

$$p_2 = p_1/(1 + T_2s) \qquad (8\text{--}9)$$

The negative feedback bellows has an inflow due to the pressure drop $(p - p_1)$ over the resistance R_d and an outflow due to the pressure drop $(p_1 - p_2)$ over the resistance R_i. Hence

$$C_1(dp_1/dt) = (p - p_1)/R_d - (p_1 - p_2)/R_i$$

The time constants are now $T_1 = R_dC_1$ and T_{12} (an interdependence constant due to the interaction) $= R_iC_1$ but since it is assumed that the bellows capacitances are the same,

$$T_{12} = R_iC_2 = T_2$$

Transforming this equation yields

$$sp_1 = (p - p_1)/T_1 - (p_1 - p_2)/T_2 \qquad (8\text{--}10)$$

Combining Equations (8–9) and (8–10) with Equations (8–2) and (8–8) will yield the transfer function of the PID controller. Before developing this, the transfer functions of the PD and PI controllers will be derived.

Proportional-derivative Controller

For the PD controller, $R_i = \infty$ and $p_2 = 0$, since there is no positive feedback; Equation (8–9) does not therefore apply and Equation (8–10) reduces to

$$sp_1 = (p - p_1)/T_1$$

or

$$p_1 = p/(1 + T_1s)$$

Combining this with Equations (8–2) and (8–8),

$$p = c\{[a/(a + b)]e - [b/(a + b)]g[p/(1 + T_1s)]\}$$

which can be re-arranged to

$$p\left[\frac{(a + b)(1 + T_1s)}{bcg} + 1\right] = \left[\frac{a}{bg}\right](1 + T_1s)e \qquad (8\text{--}11)$$

If the nozzle gain c is now regarded as infinite (or very large),

$$bcg \gg (a + b)(1 + T_1s)$$

and Equation (8–11) reduces to

$$p/e = [a/(bg)](1 + T_1s)$$

which is equivalent to the ideal transfer function for the PD controller (Equation 8–5), i.e.

$$p/e = K_c(1 + T_d s)$$

and the time constant T_1 of the negative feedback system is thus identified as the derivative time T_d.

Since c is, however, finite in a practical instrument, a more realistic response equation is obtained by retaining the term containing c and re-arranging Equation (8–11) to

$$p/e = [a/(bg)](1 + T_d s)\left[\frac{1}{1 + (a + b)(1 + T_d s)/(bcg)}\right]$$

This may be simplified by writing $bcg/(a + b) = A$, whence

$$p/e = [a/(bg)](1 + T_d s)\left[\frac{1}{1 + (1 + T_d s)/A}\right]$$

$$= [a/(bg)](1 + T_d s)\left[\frac{A}{A + 1 + T_d s}\right]$$

$$= [a/(bg)](1 + T_d s)\left(\frac{A}{A + 1}\right)\left[\frac{1}{1 + T_d s/(A + 1)}\right]$$

The term A contains the gain c in the numerator; c is finite but is usually sufficiently large that A is generally greater than 100, hence $A + 1 \approx A$. The last equation can then be written with reasonable accuracy as

$$p/e = K_c\left[\frac{1 + T_d s}{1 + (T_d/A)s}\right] \tag{8–12}$$

It will be seen that the linear lead term of the derivative action is now opposed by a linear lag term in the denominator, the time constant of the latter being the derivative time divided by the parameter A. Since A is large, the time constant of the denominator term is very much smaller than that in the numerator. At low frequencies, $1 + (T_d/A)s \rightarrow 1$ and the approximate transfer function is that of the ideal function (Equation 8–5).

At high frequencies the phase lead of the numerator is opposed and ultimately cancelled by the phase lag of the denominator; consequently the phase lead of the derivative action does not reach the maximum of 90° at high frequencies as predicted by the ideal transfer function but passes through a maximum value of 40–60° (depending on the values of T_d and A) and then returns to zero. Similarly the gain is not now infinite at high frequencies but is limited to a value of $K_c A$, or an amplitude ratio of A. A typical Bode diagram for a value of $A = 100$ is shown in Figure 8–8, which should be compared with Figure 7–11(b) for the ideal PD controller. At very high frequencies the inertia of the moving parts becomes important and the gain and phase curves fall away as shown in Figure 8–8.

It will be noted that A is not strictly constant but varies with the gain

K_c since a change in the latter requires a movement of the pivot point of the baffle mechanism to alter the lever ratio a/b.

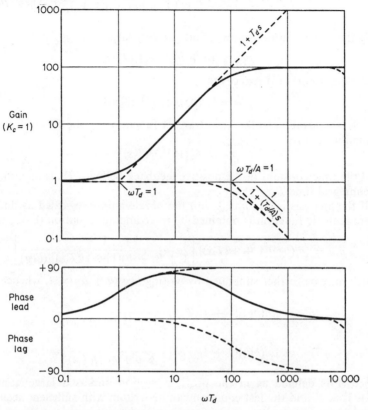

FIGURE 8-8. Bode diagram for proportional-derivative controller with finite nozzle gain

Proportional-integral Controller

For the PI controller, $R_d = 0$ and $p = p_1$ in Figure 8-7 and thus Equations (8-8) and (8-9) become

$$x = [a/(a + b)]e - [b/(a + b)]g(p - p_2)$$

and

$$p_2 = p/(1 + T_2 s)$$

Combining these equations with Equation (8-2),

$$p = c\{[a/(a + b)]e - [b/(a + b)]g[p - p/(1 + T_2 s)]\}$$

$$= c\left\{[a/(a + b)]e - [b/(a + b)]g\left[\frac{T_2 s}{1 + T_2 s}\right]p\right\}$$

which can be re-arranged to

$$p\left[\frac{(a+b)[1+1/(T_2s)]}{bcg}+1\right]=[a/(bg)][1+1/(T_2s)]e \qquad (8\text{--}13)$$

If the nozzle gain c is now infinite or very large,

$$bcg\gg(a+b)[1+1/(T_2s)]$$

and Equation (8–13) reduces to

$$p/e=[a/(bg)][1+1/(T_2s)]$$

which is equivalent to the ideal transfer function for the PI controller in Equation (8–6)

$$p/e=K_c[1+1/(T_is)]$$

and the time constant of the positive feedback bellows, T_2, is identified as the integral time T_i.

If the analysis is continued with the nozzle gain c regarded as finite, a more realistic function is obtained by re-arranging Equation (8–13) to

$$p/e=[a/(bg)][1+1/(T_is)]\left[\frac{1}{1+(a+b)[1+1/(T_is)]/(bcg)}\right]$$

which may be further simplified by writing $bcg/(a+b)=A$, whence

$$p/e=[a/(bg)][1+1/(T_is)]\left[\frac{1}{1+[1+1/(T_is)]/A}\right]$$

$$=[a/(bg)][1+1/(T_is)]\left[\frac{A}{A+1+1/(T_is)}\right]$$

Since A is defined as in the previous section and is a large number, $A+1\approx A$, and the last equation can be written with sufficient accuracy as

$$p/e=[a/(bg)][1+1/(T_is)]\left[\frac{A}{A+1/(T_is)}\right]$$

whence

$$p/e=K_c\left[\frac{1+1/(T_is)}{1+1/(AT_is)}\right] \qquad (8\text{--}14)$$

The ideal phase lag of the integral action is now opposed by a phase lead due to the time constant AT_i and the phase lag has a maximum value between the corner frequencies of $\omega=1/T_i$ and $1/(AT_i)$, falling to zero at low frequencies. Similarly, as shown in Figure 8–9, the gain is limited to K_cA at low frequencies instead of being infinite as with the ideal transfer function. At higher frequencies the effect of the additional time constant becomes negligible and the gain and phase curves are those of the ideal transfer function.

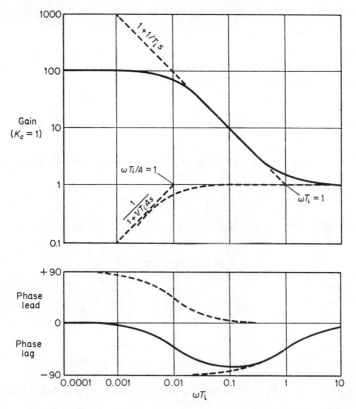

FIGURE 8–9. Bode diagram for proportional-integral controller with finite nozzle gain

Proportional-integral-derivative Controller

By analogy with the PI and PD controllers discussed above, it would be expected that the transfer function of the PID controller with a finite nozzle gain would be modified from the ideal function in a similar way to Equations (8–12) and (8–14), i.e. the ideal function would appear in the numerator of the actual function with a denominator of similar form but with action times modified by the parameter A, as follows:

$$p/e = K_c \left[\frac{1 + T_d s + 1/(T_i s)}{1 + (T_d/A)s + 1/(AT_i s)} \right] \qquad (8-15)$$

If A is sufficiently large (i.e. an infinite nozzle gain) Equation (8–15) reduces to the ideal transfer function for the PID controller; if $T_d = 0$ the equation reduces to that of the PI controller (Equation 8–14), and if $T_i = \infty$ to that of the PD controller (Equation 8–12). With a finite gain the PID function would show a limited gain at both high and low frequency

$(=K_cA)$ and a phase angle of zero at these frequencies, passing through a maximum lag at a low frequency and a maximum advance at a high frequency.

In practice the function does take this approximate form, but an additional complication now arises as a result of interaction between the integral and derivative generators.

Integral-derivative Interaction

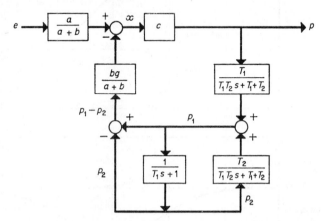

FIGURE 8–10. Block diagram for three-term pneumatic controller of Figure 8–7

If Equation (8–10) is re-written in the usual form of a transfer function, i.e.

$$p_1 = [T_2/(T_1T_2s + T_1 + T_2)]p$$
$$+ [T_1/(T_1T_2s + T_1 + T_2)]p_2 \qquad (8\text{–}10a)$$

it may be combined with Equations (8–2), (8–8), and (8–9) into a conventional block diagram (Figure 8–10). This representation of the controller operation is a fairly complex multiple loop system; the negative feedback loop contributing to the baffle displacement (x) can readily be identified, as also can the positive feedback of the pressure p_2 which makes a negative contribution to the negative feedback pressure p_1. The third loop in the diagram, however, is a positive contribution from p_2 to the value of p_1, and this may be identified as an interaction between the two resistance-capacitance stages of the negative and positive feedback bellows systems which are the derivative and integral function generators. As will be shown in the following analysis, effectively the derivative component of the response is integrated by the integral generator; since

$$\int(dx/dt)\,dt = x$$

this introduces an additional term into the proportional response.

The complete transfer function for the PID controller is obtained by combining Equations (8–2), (8–8), (8–9), and (8–10a).

Substituting the value of p_2 from Equation (8–9) in Equation (8–10a),

$$p_1/p = \frac{1 + T_2 s}{T_1 T_2 s^2 + (2T_1 + T_2)s + 1}$$

and

$$p_2/p = \frac{1}{T_1 T_2 s^2 + (2T_1 + T_2)s + 1}$$

whence

$$p_1 - p_2 = \left[\frac{T_2 s}{T_1 T_2 s^2 + (2T_1 + T_2)s + 1}\right] p$$

$$= \left[\frac{1}{T_1 s + (1 + 2T_1/T_2) + 1/(T_2 s)}\right] p$$

$$= \left[\frac{1}{T_1 s + I + 1/(T_2 s)}\right] p \qquad (8\text{–}16)$$

where $I = 1 + 2T_1/T_2$.

Substituting the value of $p_1 - p_2$ from Equation (8–16) into Equation (8–8), and combining with Equation (8–2):

$$p = c\left\{[a/(a + b)]e - [b/(a + b)]g\left[\frac{1}{T_1 s + I + 1/(T_2 s)}\right] p\right\}$$

whence

$$p\left[\frac{(a + b)[T_1 s + I + 1/(T_2 s)]}{bcg} + 1\right]$$

$$= [a/(bg)][T_1 s + I + 1/(T_2 s)]e \qquad (8\text{–}17)$$

If an infinite gain is assumed, this equation reduces to

$$p/e = [a/(bg)][T_1 s + I + 1/(T_2 s)]$$

and the response obviously contains the required proportional, integral and derivative terms, but the function is not directly comparable to the ideal transfer function of the PID controller without adjustment to

$$p/e = [a/(bg)]I[(T_1/I)s + 1 + 1/(T_2 I s)] \qquad (8\text{–}18)$$

Comparing Equation (8–18) with the ideal transfer function

$$p/e = K_c[T_d s + 1 + 1/(T_i s)]$$

it can be seen that the control parameters have been altered significantly from the values of the two previous cases of PD and PI control. The proportional gain K_c in those cases was $a/(bg)$ but is now $[a/(bg)]I$; the integral and derivative times T_i and T_d are not now identifiable with the time constants T_1 and T_2 but with modified values of T_1/I and $T_2 I$. The parameter I is the *interaction factor* which, for this particular design of mechanism, is $(1 + 2T_1/T_2)$ and must then be greater than unity if both integral and derivative actions are used in the controller. As a result, the proportional gain and the integral time are increased by a factor greater than one and the derivative time is decreased by the same factor. Note that $K_c = a/(bg)$, $T_d = T_1$ and $T_i = T_2$ when the integral and derivative components are used singly with proportional control. The adjustments for these three parameters are calibrated in terms of the gain and the two time constants (T_1 and T_2, or T_d and T_i) when the controller is operated in the proportional, proportional-derivative, or proportional-integral modes. When used in the three-term mode the calibrations are effectively changed by the interaction factor as noted above; i.e. the gain and action times actually developed by the three-term instrument are not the nominal values set by the adjustment of the control parameters.

The effect is most pronounced on the controller gain since the value of K_c developed for P, PD, or PI control is now multiplied by a factor (I) which is greater than one. An increase in the controller gain tends to make the control system less stable and it is, in fact, possible to make a system become unstable by a large increase in gain due to injudicious adjustment of the action times. The effect on the controller gain of the interaction is to produce a minimum value equal to the interaction factor, as seen in the Bode diagram of Figure 8–11; this must be borne in mind when setting the value of proportional gain with three-term control.

The important practical feature of interaction is that the factor I includes the ratio of the time constants, T_1 and T_2. A change in either of these parameters will thus change the value of the factor and so alter the values of all three of the control parameters, although, as already noted, the change in controller gain is likely to have the major effect. It is therefore advisable to maintain a constant ratio of the nominal integral and derivative action times when making any adjustment, in order to maintain a constant value of the interaction factor so that the gain is not affected. The value of the ratio chosen should be such as to permit the derivative component to develop the maximum phase advance. In the present case the ratio of the actual derivative to integral times developed is given by

$$T_d/T_i = (T_1/T_2)(1/I^2) = (T_1/T_2)(1 + 2T_1/T_2)^{-2}$$

and the ratio, $T_d:T_i$, has a maximum value of $1:8$ when $T_1:T_2 = 1:2$. This corresponds to a maximum phase advance of about $30°$ when the integral time is set equal to the period of oscillation as suggested by Aikman [1].

It should be noted at this point that the interaction factor of $(1 + 2T_1/T_2)$ applies only to pneumatic controllers of the basic design illustrated by the schematic diagram of Figure 8–7, and in which the bellows capacitances are equal. Any modifications to this basic design will develop a different interaction factor. The remarks above as to the general effects of inter-action will still apply, but the maximum ratio of $T_d:T_i$ and the maximum phase advance will almost certainly be different. The interaction factors for several different designs of instrument are determined by Aikman and Rutherford [2] and by Young [29] who also shows plots of the phase lead generated by several commercial designs of instrument for different ratios of the action times.

The limitation set by interaction in three-term pneumatic controllers is not in general a serious disadvantage in present practice, although there will be some applications in which a pneumatic controller might prove unsuitable on this account. The principal disadvantage has already been mentioned, that when a controller is adjusted on the plant a change in only one of the action times will change the ratio of T_1/T_2 and so alter the value of the interaction factor. This will lead to a change in the proportional sensitivity and in the overall controller gain; difficulties may then arise if due allowance for this has not been made, since an increase in the controller gain will always reduce the stability of the system.

The uncertainties in operating conditions and the changes therein often make estimation of controller adjustments unprofitable, particularly at the design stage, but this does not reduce the value of knowing the characteristics of the control system and the way that these are likely to change with a changed condition. Thus if the relationships between the action times and the gain of the three-term controller are understood, some consideration can be given to the likely effects of the interaction on the gain and to the consequent necessity of adjusting the proportional sensitivity before any change is made in the ratio of the action times.

Interaction in the three-term pneumatic instrument can be eliminated by generating the proportional, integral, and derivative signals in separate units (in a somewhat more complex instrument) and combining these signals in a summating relay. There appears, however, to have been little demand for an instrument of this type. If obtaining an ideal three-term response without interaction is critical, it can be provided by the electronic instruments now available.

Three-term Controller with Finite Gain

If the nozzle gain of the three-term pneumatic instrument is regarded as finite, the transfer function is obtained from Equation (8–17), after writing $A = bcg/(a + b)$, as

$$p/e = [a/(bg)][T_1 s + I + 1/(T_2 s)] \left[\frac{1}{1 + [T_1 s + I + 1/(T_2 s)]/A} \right]$$

$$= [a/(bg)][T_1 s + I + 1/(T_2 s)] \left[\frac{A}{T_1 s + (A + I) + 1/(T_2 s)} \right]$$

$$= [a/(bg)] \left[\frac{AI}{A + I} \right] \left[\frac{(T_1/I)s + 1 + 1/(T_2 I s)}{T_1/(A + I)s + 1 + 1/T_2(A + I)s} \right]$$

Since $A + I \approx A$, this last equation reduces to

$$p/e = [a/(bg)]I \left[\frac{(T_1/I)s + 1 + 1/(T_2 Is)}{(T_1/A)s + 1 + 1/(T_2 As)} \right]$$

The denominator terms become significant only at high and low frequencies as shown in the Bode diagram of Figure 8–11. The controller gain is limited to a maximum value of $K_c A$, but at intermediate frequencies the response is basically that of the ideal transfer function modified by the presence of the interaction factor which gives a minimum gain value of I. Both the phase lead and phase lag exhibit maximum values, being zero at very high and very low frequencies.

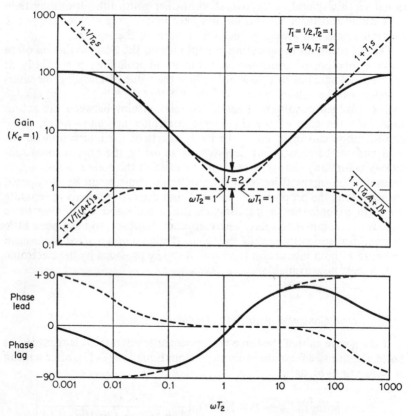

FIGURE 8–11. Bode diagram for three-term controller with interaction and finite nozzle gain

Non-linearities in Pneumatic Controllers

The response of pneumatic controllers with integral or derivative action tends to be slightly non-linear as the values of the resistances vary with the absolute pressure. If the air flow through the resistances is laminar (as tacitly assumed in the preceding sections) and the pressure differences are small, the mass flow rate is directly proportional to the pressure difference and the density, and the resistance (pressure difference per unit flow) is then inversely proportional to the absolute pressure. If the flow is turbulent and follows the usual orifice equation, the mass rate of flow varies with the square root of the pressure difference, and the density and the resistance would then vary with the square root of the absolute pressure.

Harriott [18] reports tests on a commercial pneumatic controller showing integral and derivative action times inversely proportional to approximately the 0.9th power of the absolute pressure, which indicates practically completely laminar flow. The change in time constants would not normally be of any importance, since fairly large changes in integral or derivative time must be made to have any significant effect on the system response.

If the error signal to a pneumatic controller is large or prolonged, the controller output pressure may reach a limiting value corresponding to the minimum or maximum of the fully open or fully closed control valve. In such conditions the controller is 'saturated' since an increase in the error can have no further effect on the controller output. If an integral component is present, the pressures may have time to equalize over the integral restrictor and an excessive positive feedback is applied. In such cases the baffle is moved completely out of the range of the movement within which the output pressure is determined by the baffle-nozzle separation, and the control action cannot change until the controlled variable regains the desired value and the error signal becomes of opposite sign. Saturation is a type of non-linearity which makes the normal definition of gain and phase lag invalid; the effect is essentially to decrease the gain as the input amplitude increases. The system therefore has excessive stability for large disturbances if the controller has been adjusted to deal with small disturbances. The major effect is that the peak and steady-state errors following large load changes are likely to be greater than those predicted from a linearized analysis. In most process operations saturation is most likely to occur during the start up period when large deviations may exist for a considerable time; it is often advisable, therefore, to carry out the starting up phase with manual control and to change over to automatic control as the desired values are approached.

Electronic Controllers

Since the early 1950s, several new types of electronic controllers have been developed which are fully competitive with pneumatic instruments for process applications. The major advantages of these instruments are negligible transmission lags, the absence of dead zones due to friction or hysteresis in moving parts, capability of operation at low temperatures, and compatability with other electrical devices such as electronic measurement transducers, data processing equipment and computers. The major disadvantage in process work is the need for intrinsically safe electrical equipment in hazardous areas, but some of the newer electronic instruments can meet this requirement. It is, however, usually necessary to convert the final electrical output of the instrument into an air pressure signal to operate a conventional pneumatic diaphragm valve in most applications since there is, as yet, no really satisfactory electrically actuated equivalent to the diaphragm valve.

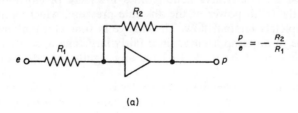

(a)

$$\frac{p}{e} = -\frac{R_2}{R_1}$$

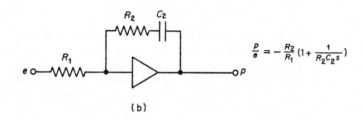

(b)

$$\frac{p}{e} = -\frac{R_2}{R_1}\left(1 + \frac{1}{R_2 C_2 s}\right)$$

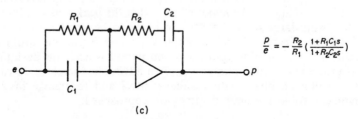

(c)

$$\frac{p}{e} = -\frac{R_2}{R_1}\left(\frac{1 + R_1 C_1 s}{1 + R_2 C_2 s}\right)$$

FIGURE 8–12. Basic electronic controller circuits

There is no basic design of electronic controller, although certain basic electronic circuits can be identified in commercial instruments. The input and output signals are usually direct currents of a few milli-amperes in magnitude although no standard range has yet been agreed. The principal component of these instruments is a high gain operational amplifier, with a gain usually exceeding 1000:1. The gain of the proportioning circuit does not depend on the gain of the amplifier but upon the ratio of the input and feedback resistors or capacitors, and can thus be very accurately calibrated. The error signal is generated by passing the input current from a suitable measuring instrument through a resistor of about 500 ohm, so generating an input voltage which is subtracted from a desired value voltage set by a voltage-dividing resistor. The integral and derivative components can be added in a number of ways by modifications or additions to the basic proportional circuit; a derivative component can be added by delaying the negative feedback over the amplifier or by advancing the input error signal; an integral component can be generated by adding a delayed positive feedback, by advancing the negative feedback, or by delaying the error input.

Some typical simplified circuits are illustrated in Figure 8–12, along with the appropriate transfer functions. The negative sign in the latter arises from the reversal of polarity by the amplifier. It will be noted that the PD function is not entirely ideal but can be made very nearly so by suitable choice of the circuit parameters. The integral and derivative terms can also be combined without interaction.

PROBLEMS

8–1 The three control functions of a pneumatic controller are generated in the separate elements shown in the block diagram of Figure P8–1, the integral and derivative

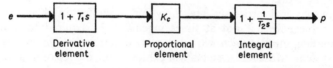

Derivative element   Proportional element   Integral element

FIGURE P8–1

generators having a fixed unit gain. Show by determination of the overall transfer function (p/e) that the controller exhibits integral-derivative interaction. Determine the interaction factor and hence the optimum ratio of the nominal action times for the maximum effective derivative action.

8–2 Plot on semi-logarithmic co-ordinates the gain and phase relationships for a three-term pneumatic controller with an interaction factor of $(1 + 2T_1/T_2)$ for different values of T_1 when the integral time (T_i) is set

(a) equal to the Ziegler-Nichols recommendation of $P_u/2$, and
(b) equal to the period of the oscillation.

8–3 Repeat Problem 8–2 for the controller of Problem 8–1.

8–4 The three-term controller discussed in the text has a series arrangement of integral and derivative resistances. Repeat the derivation of the transfer function, assuming infinite nozzle gain, for the parallel arrangement of integral and derivative resistances and hence determine the interaction factor.

8–5 The system of Problem 5–11 is to be controlled with a proportional-integral controller. Determine the controller adjustments using an open-loop Bode diagram and the Ziegler-Nichols rules assuming that a non-ideal controller is used in which the parameter A has a low value of 10 and a normal value of 100.

Chapter 9: Complex Control Systems

In the preceding chapters a number of relatively simple processes have been examined to illustrate the general application of process control theory. It has been seen that the quality of control available in such systems is determined by the characteristics of the process and by the nature, magnitude and point of entry into the control loop of disturbances in the input variables. If a simple system is defined as a single loop with linear parameters and with only one controlled output variable and one input manipulated variable, any system differing from this definition will be complex. The complexity may arise in the process due to non-linearities or multiple feedback loops, or in the control equipment requiring additional measuring, regulating or controlling elements. This latter situation arises when a simple control system cannot hold the deviation of the controlled variable within the specified limits; this is generally caused by large disturbances usually combined with large time lags in the process. Such complex control systems include cascade, feedforward, ratio, and various selective systems.

Many physical processes are, however, complex in themselves and considerable effort, if not at times ingenuity, is required to develop a mathematical model which will adequately describe the dynamic response of the system. Most real systems will exhibit some degree of non-linearity, and the linearization technique of Chapter 2 must be employed to obtain a suitable linear equation. In many cases interaction between parts of the system or the inter-dependence of two related variables may introduce internal feedback, both negative and positive, which makes the system effectively multi-loop. In many such cases the system can be reduced to a single equivalent loop by employment of block diagram algebra, and it will often be found that the system can be handled by a simple control.

Most process plants represent a combination of control loops; a number of output variables can be distinguished and measured separately, and there are also a number of different input variables which require manipulation to maintain optimum operating conditions. Such multi-variable systems are usually handled by constructing control loops between a particular pair of input and output variables so that the whole system becomes a combination of several single loops, some of which may interact with each other. In many applications of this type the complex control actions may be usefully employed.

The purpose of the present chapter is to discuss some representative complex arrangements in both the process and control elements and to consider, albeit briefly, the further problems of non-linearity and optimization. The systems to be described represent advances towards a fully integrated control system in which each condition or variable is measured and maintained at such a value relative to every other condition that the whole system continuously achieves its optimum performance. Such systems are the ultimate aim, particularly for processes requiring more consistent operating conditions than can be obtained with present techniques.

Theoretical Analysis of a Complex Process

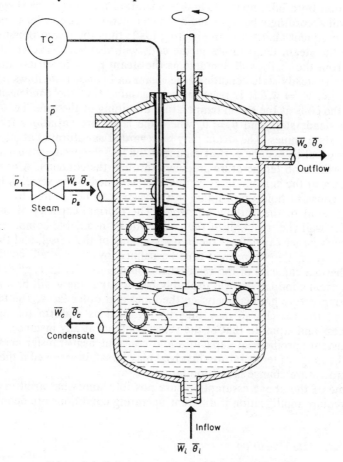

FIGURE 9–1. Temperature control of steam-heated vessel

The system shown in Figure 9–1 consists of a steam-heated agitated vessel through which flows a liquid at a variable rate of flow and with a variable inlet temperature. The liquid is heated by steam condensing in the coils, and the outlet temperature of the liquid would be controlled by measuring the temperature in the vessel and using the controller output signal to regulate the flow of steam to the coils. The major problem in the design of the control system is the dynamic response of the vessel temperature, since the system is actually non-linear. To obtain a linear model, a number of simplifying assumptions must be made.

The actual plant would most probably be designed on the basis of a steady-state condition with the required normal operating conditions of the input and output variables. The design or steady-state values of the variables can thus be defined and are indicated in Figure 9–1. During operation both inlet flow rate and temperature may vary; the steam flow rate will accordingly be regulated by the temperature controller to balance the consequent changes in the heating load; thus the steam pressure and hence the steam temperature in the coils will also vary. The condensate flow from the coils will fluctuate as the steam pressure varies, although under the steady-state conditions the steam and condensate flows must be equal, i.e. $\bar{w}_s = \bar{w}_c$. It is reasonable to assume that the condensate will leave the coils at the steam temperature obtaining at the time, i.e. $\bar{\theta}_s = \bar{\theta}_c$ at the steady-state and $\theta_s = \theta_c$ where the latter are variations from the steady-state values. The steam may be regarded as saturated at all times.

It is usual to assume perfect mixing if the vessel is well-agitated so that there will be no temperature gradients through the contents. A negligible heat loss to the surroundings can also be assumed if the vessel is lagged, and a constant hold-up of liquid in the vessel can be ensured by suitable arrangement of the outflow pipework. The thermal capacity of the heat transfer wall (i.e. the wall of the coils) will be negligible compared with the much larger capacity of the liquid contents of the vessel, and the heat transfer from the steam may thus be defined by an 'overall' coefficient. The thermal capacity of the vessel shell will not usually be negligible, but the thermal conductivity will probably be such that there will be a negligible temperature gradient between the inner and outer faces. The temperature of the vessel wall may thus be regarded as uniform through the thickness and equal to that of the vessel contents at any instance.

Constant specific-heat capacities and constant heat-transfer coefficient over the range of temperatures involved may be safely assumed if the range of temperature change is not too large.

Some of the above assumptions are possibly somewhat arbitrary and may require modification if different operating conditions are defined.

Steady-state Condition

The physical design of the apparatus, e.g. heat transfer area required, etc., will be based on the usual steady-state heat balances on the steam and liquid sides, i.e.

$$\bar{w}c_p(\bar{\theta}_o - \bar{\theta}_i) = UA(\bar{\theta}_s - \bar{\theta}_o) = \bar{w}_s(H_s - H_c) \qquad (9-1)$$

where $\bar{w}$ is the design inlet liquid flow rate, in kg/s,

$\bar{w}_s$ is the design steam flow rate, in kg/s,

$\bar{\theta}_i$ is the design inlet temperature of liquid, in °C,

$\bar{\theta}_o$ is the design outlet temperature of liquid, in °C,

$\bar{\theta}_s$ is the design steam temperature, in °C, at pressure $\bar{p}_s$ in kN/m²,

c_p is the specific heat capacity of liquid, in kJ/kg K,

U is the heat transfer coefficient, in kW/m² K,

A is the area of heat transfer surface, in m²,

H_s is the specific enthalpy of steam, in kJ/kg,

H_c is the specific enthalpy of condensate, in kJ/kg, both at temperature $\bar{\theta}_s$ and pressure $\bar{p}_s$.

The dynamic response of the system is based on the heat balances of the unsteady-state following changes in the input variables.

Unsteady-state Heat Balance, Liquid Side

The net heat gain of the contents and the vessel shell will be equal to the accumulation of heat in the vessel, thus:

Heat gain = Heat in − Heat out

= Sensible heat of inflow + Heat from steam

− Sensible heat of outflow

i.e.

$$(Mc_p + M_w c_{pw})(d\theta_o/dt) = wc_p\theta_i + UA(\theta_s - \theta_o) - wc_p\theta_o \quad (9\text{–}2)$$

where w, θ_i, θ_o, and θ_s are the deviations from the steady state values of the liquid flow rate and temperatures of the inflow, outflow, and steam, respectively,

M is the mass hold-up capacity of the vessel, in kg,

M_w is the mass of the vessel shell, in kg,

c_{pw} is the specific heat capacity of the vessel shell, in kJ/kg K.

The terms $w\theta_i$ and $w\theta_o$ are the products of the deviations in two variables and are thus non-linear; the other quantities in the heat balance are constant (or assumed to be so). To obtain a transfer function from Equation (9–2), the non-linear terms must first be linearized by the usual method:

$$d(w\theta_i) = [\partial(w\theta_i)/\partial w]_{\bar{\theta}_i} \, dw + [\partial(w\theta_i)/\partial\theta_i]_{\bar{w}} \, d\theta_i$$

$$= \bar{\theta}_i \, dw + \bar{w} \, d\theta_i$$

whence

$$w\theta_i = \bar{\theta}_i w + \bar{w}\theta_i$$

Similarly

$$w\theta_o = \bar{\theta}_o w + \bar{w}\theta_o$$

Substituting in Equation (9–2),

$$(Mc_p + M_w c_{pw})(d\theta_o/dt)$$

$$= (\bar{\theta}_i - \bar{\theta}_o)c_p w + \bar{w}c_p(\theta_i - \theta_o) + UA(\theta_s - \theta_o)$$

Transforming and solving for θ_o gives

$$[(Mc_p + M_wc_{pw})s + \bar{w}c_p + UA]\theta_o = wc_p\theta_i + (\bar{\theta}_i - \bar{\theta}_o)c_pw + UA\theta_s$$

whence

$$\theta_o(s) = G_1\theta_i - G_2w + G_3\theta_s \qquad (9\text{--}3)$$

where $G_1 = \dfrac{wc_p/(\bar{w}c_p + UA)}{T_1s + 1}$

$$G_2 = \dfrac{(\bar{\theta}_o - \bar{\theta}_i)/(\bar{w}c_p + UA)}{T_1s + 1}$$

$$G_3 = \dfrac{UA/(\bar{w}c_p + UA)}{T_1s + 1}$$

and $\quad T_1 = \dfrac{Mc_p + M_wc_{pw}}{\bar{w}c_p + UA}$

(N.B. $\bar{\theta}_o > \bar{\theta}_i$, and hence the gain of G_2 is positive.)

Unsteady-state Heat Balance, Steam Side

The heat balance on the steam side must allow for possible differences between the steam and condensate flow rates accompanying a change in the pressure p_s in the coils. A change in the steam pressure will cause a change in the internal energy content of the steam as well as in the steam temperature, and the unsteady-state heat balance is thus

Energy gain of steam = Heat in steam entering

$$\qquad\qquad\qquad - \text{ (Heat in condensate} + \text{Heat transferred)}$$

i.e.

$$d(M_s\varepsilon)/dt = w_sH_s - w_cH_c - UA(\theta_s - \theta_o) \qquad (9\text{--}4)$$

where w_s, w_c are the changes in steam and condensate flow rates from the steady-state values ($\bar{w}_s = \bar{w}_c$),

M_s is the mass of steam in the coils, in kg,
ε is the specific internal energy, in kJ/kg, of steam at temperature $\bar{\theta}_s + \theta_s$ and pressure $\bar{p}_s + p_s$.

The mass M_s of the steam is not constant owing to the pressure change and is thus not a convenient term to handle, but $M_s = V\rho_s$, where V is the volume of the steam space, in m^3, and ρ is the density of the steam, in kg/m^3, under the conditions obtaining. $d(M_s\varepsilon)$ may then be replaced by $V\,d(\varepsilon\rho_s)$, since V is constant, and Equation (9–4) becomes

$$V[d(\varepsilon\rho_s)/dt] = w_sH_s - w_cH_c - UA(\theta_s - \theta_o) \qquad (9\text{--}5)$$

The properties of the steam in the above equation, i.e. ε, ρ_s, H_s, and H_c, are all non-linear functions of the steam pressure, but the steam is assumed to be saturated at all times and they are thus equally non-linear functions of the steam temperature θ_s. Coughanowr and Koppel [10] show, in an extended analysis of a similar problem, that for small deviations in the steam temperature such as are likely to be encountered in practice (i.e. of the order of $\pm 5°C$), only changes in the density are of any significance, and negligible error is involved if the energy and enthalpy terms are regarded as constant.

The change in density with temperature of the steam must be linearized in the usual way about the design value $\bar{\theta}_s$, i.e.

$$d(\rho_s) = [\partial\rho_s/\partial\theta_s]_{\bar{\theta}_s} \, d\theta_s = \alpha \, d\theta_s$$

The constant α can be evaluated from the saturated steam tables as the net change in density per unit change in temperature $(\Delta\rho_s/\Delta\theta_s)$ for the specified change in temperature about the design value $\bar{\theta}_s$.

Equation (9–5) can now be written

$$V\alpha\varepsilon(d\theta_s/dt) = w_s H_s - w_c H_c - UA(\theta_s - \theta_o) \qquad (9\text{–}6)$$

To relate w_s and w_c, a mass balance is taken on the steam side, the rate of accumulation equals the difference in the flow rates, i.e.

$$d(M_s)/dt = w_s - w_c$$

But, as derived above,

$$d(M_s) = V \, d(\rho_s) = V\alpha \, d\theta_s$$

Hence

$$w_s - w_c = V\alpha(d\theta_s/dt) \qquad (9\text{–}7)$$

Substituting the value for w_c from Equation (9–7) in Equation (9–6):

$$V\alpha\varepsilon(d\theta_s/dt) = w_s H_s - H_c[w_s - V\alpha(d\theta_s/dt)] - UA(\theta_s - \theta_o)$$

Transforming and re-arranging,

$$[V\alpha(\varepsilon - H_c)s + UA]\theta_s = w_s(H_s - H_c) + UA\theta_o$$

whence

$$\theta_s(s) = G_4 w_s + G_5 \theta_o \qquad (9\text{–}8)$$

where $G_4 = \dfrac{(H_s - H_c)/UA}{T_2 s + 1}$

$$G_5 = \frac{1}{T_2 s + 1}$$

and $\quad T_2 = \dfrac{V\alpha(\varepsilon - H_c)}{UA}$

Control Valve

The flow of steam through the control valve is determined by the valve stem position and the pressure drop across the valve. The former may be taken as directly proportional to the controller output signal p, since the pneumatic diaphragm valve has an almost completely linear relationship between stem position and inlet pressure. Thus

$$w_s = f[p, (p_1 - p_s)]$$

where p_1 is a change in steam supply pressure from the design value of $\bar{p}_1$. The valve function will, in general, be non-linear but can be linearized by using an analytical expression such as the orifice flow equation. Alternatively a linearized form can be obtained by experimental testing of the valve; by plotting the steam flow rate against each of the three variables in turn whilst the other two are held constant at the design values, the slopes of the resulting curves can be measured at the design value of the steam flow rate $\bar{w}_s$. These slopes are the partial differential terms relating w_s to the particular variable in the linearized expansion

$$dw_s = [\partial w_s/\partial p]_{\bar{p}_1, \bar{p}_s}\, dp + [\partial w_s/\partial p_1]_{\bar{p}, \bar{p}_s}\, dp_1 + [\partial w_s/\partial p_s]_{\bar{p}, \bar{p}_1}\, dp_s$$

whence

$$w_s = K_v p + K_6 p_1 - K_7' p_s \qquad (9\text{-}9)$$

The negative sign is introduced since the plot of w_s *versus* p_s (the downstream pressure) will have a negative slope, in order to make the coefficient, K_7', positive.

The experimental approach to linearization of the flow characteristics of a control valve is always possible in principle, but it must be emphasized that the linear relationship only applies for small deviations from the normal operating load. If the latter is changed considerably, by a large load change on the system, the values of the coefficients in the above equation may be changed significantly.

It is a valid assumption in this instance to conclude that there will be no dynamic lag of any consequence between the controller output pressure and the valve stem position since the time constants associated with the vessel will be much larger than any possessed by the control valve.

To include Equation (9–9) in an overall transfer function with the previously derived equations, it is necessary to relate the steam pressure to the steam temperature. Since the steam is assumed to be saturated, there is a functional relationship between θ_s and p_s which can be derived from the steam tables. Thus

$$p_s = f(\theta_s)$$

which can be linearized about the design steam temperature:

$$p_s = [\partial p_s/\partial \theta_s]_{\bar{\theta}_s}s = k\theta_s$$

and Equation (9–9) becomes

$$w_s = K_v p + K_6 p_1 - K_7 \theta_s \qquad (9\text{-}10)$$

where $K_7 = kK_7'$.

Block Diagram and Overall Transfer Function

A block diagram for the complete system can now be drawn from the transfer function relationships of Equations (9–3), (9–8), and (9–10), along with the controller and measuring element equation

$$p = G_c(v - H\theta_o) \qquad (9\text{–}11)$$

It will be seen from the block diagram of Figure 9–2 that the apparently simple system is multi-loop and that the internal loops overlap. However,

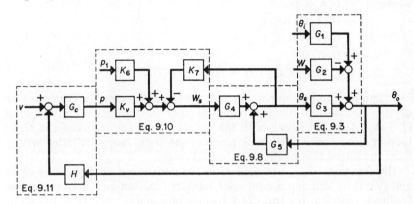

FIGURE 9–2. Block diagram of control system of Figure 9–1

so long as there is only one principal controlled variable and no other complications, any multi-loop system can be reduced to an equivalent single loop by applying the usual block diagram algebra. Thus an overall transfer function between the output θ_o and the input variables can be derived. The four individual equations are

$$\theta_o = G_1\theta_i - G_2w + G_3\theta_s \qquad (9\text{–}3)$$

$$\theta_s = G_4w_s + G_5\theta_o \qquad (9\text{–}8)$$

$$w_s = K_vp + K_6p_1 - K_7\theta_s \qquad (9\text{–}10)$$

$$p = G_c(v - H\theta_o) \qquad (9\text{–}11)$$

Combining Equations (9–10) and (9–11) gives

$$w_s = G_cK_v(v - H\theta_o) + K_6p_1 - K_7\theta_s$$

Substituting in Equation (9–8),

$$(1 + G_4K_7)\theta_s = G_cK_vG_4v - G_cK_vG_4H\theta_o + G_4K_6p_1 + G_5\theta_o$$

Substituting for θ_s in Equation (9–3),

$$(1 + G_4K_7)\theta_o = G_1(1 + G_4K_7)\theta_i - G_2(1 + G_4K_7)w$$
$$+ G_3(G_cK_vG_4v - G_cK_vG_4H\theta_o + G_4K_6p_1 + G_5\theta_o)$$

whence

$$(1 + G_cK_vG_3G_4H + G_4K_7 - G_3G_5)\theta_o$$
$$= G_1(1 + G_4K_7)\theta_i - G_2(1 + G_4K_7)w + G_cK_vG_3G_4v + G_3G_4K_6p_1$$

which can be re-written in the form of a closed-loop transfer function:

$$\theta_o(s) = \frac{N_1}{D}v + \frac{N_2}{D}\theta_i - \frac{N_3}{D}w + \frac{N_4}{D}p_1$$

where $N_1 = G_cK_vG_3G_4$,

$N_2 = G_1(1 + G_4K_7)$,

$N_3 = G_2(1 + G_4K_7)$,

$N_4 = G_3G_4K_6$,

$D = (1 + G_cK_vG_3G_4H + G_4K_7 - G_3G_5)$.

As can be seen by comparison with the block diagram, the denominator of the closed-loop function contains that of the main control loop $(1 + G_cK_vG_3G_4H)$, along with the negative feedback terms of the G_4K_7 loop and the positive feedback terms of the G_3G_5 loop, the latter arising from the interaction of the two thermal stages.

The functions G_{1-5} are defined in the derivations of Equations (9–3) and (9–8); by defining a controller transfer function and a measurement feedback function, the transfer function for a change in any of the input variables can be derived. Thus for proportional action, $G_c = K_c$, and with no measurement lag, $H = 1$, the transfer function for a change in desired value is given by

$$\frac{\theta_o}{v} = \frac{K_cK_vG_3G_4}{1 + K_cK_vG_3G_4 + G_4K_7 - G_3G_5}$$

$$= \frac{K_cK_vK_3K_4/[(T_1s + 1)(T_2s + 1)]}{1 + K_cK_vK_3K_4/[(T_1s + 1)(T_2s + 1)]}$$
$$+ K_4K_7/(T_2s + 1) - K_3K_5/[(T_1s + 1)(T_2s + 1)]$$

$$= \frac{K_cK_vK_3K_4}{(T_1s + 1)(T_2s + 1) + K_cK_vK_3K_4 + K_4K_7(T_1s + 1) - K_3K_5}$$

$$= \frac{K_cK_vK_3K_4}{T_1T_2s^2 + (T_1 + T_2 + K_4K_7T_1)s + 1 + K_cK_vK_3K_4 + K_4K_7 - K_3K_5}$$

$$= \frac{K}{T^2s^2 + 2\zeta Ts + 1} \tag{9–12}$$

where K_{1-5} are the steady-state gains of the transfer functions G_{1-5}, as defined previously.

The response of the system is thus second-order when proportional control is used, and neither the measuring element or control valve have any dynamic lag. The parameters of Equation (9–12) are seen to be

$$K = K_c K_v K_3 K_4 / D$$

$$T = (T_1 T_2 / D)^{\frac{1}{2}}$$

$$\zeta = \tfrac{1}{2}(T_1 + T_2 + K_4 K_7 T_1)/(T_1 T_2 D)^{\frac{1}{2}}$$

$$D = 1 + K_c K_v K_3 K_4 + K_4 K_7 - K_3 K_5$$

K, T, and ζ are all positive since the defining parameters are positive and $K_3 K_5 < 1$.

The present example is a thermal system in which the capacities on both sides are lumped. In the more usual tubular heat exchangers used in process heat transfer, a characteristic feature is that the capacities on one or both sides are distributed along the length of the exchanger. The theoretical analysis thus becomes even more complex and the reader is referred to the texts of Harriott [18], Buckley [3], and Campbell [5] for further discussion. These authors also deal with other complex examples in level, flow, pressure, etc. control systems.

Systems with Complex Control Elements

The performance of a simple single-loop feedback control system can often be improved by relatively minor changes in the process characteristics, such as the reduction of a time delay or a minor time constant, by use of a valve positioner to improve the response of the control valve, and so on. If the poor performance is, however, due to large uncontrolled load disturbances, it is often necessary to consider more complex schemes of control. These usually involve the use of two or more control instruments either interconnected, e.g. in feedforward, cascade and selective controls, or in separate loops, in which case the control becomes multivariable and may pose additional problems if the separate loops are not completely separate and have elements in common, leading to interaction between the loops.

Averaging Control

It is convenient to include this type of control at this stage, even though the control system is not necessarily complex since averaging control can be applied as a single-loop system. Averaging control functions as a regulator of the material balance across part of the process system, requiring inflows to be balanced against outflows at some point but with provision for smooth and gradual changes in rates of flow to avoid major disturbances entering the process.

It is very common in large process plants for the outflow or product stream of one processing stage to become the feed stream to a further stage in the overall process; any fluctuation in the outflow stream of the first unit then becomes a supply disturbance to the second unit. To prevent disturbances being passed on in this way from one process stage to another, it is usual to interpose between the two stages some type of storage vessel known variously as an accumulator, buffer or surge tank. The essential purpose of this additional storage capacity, which plays no part in the operation of either of the process stages, is to allow the fluctuating input to accumulate or to be drawn upon whilst the feed stream to the second process stage, i.e. the outflow from the storage capacity, is adjusted relatively slowly to prevent major disturbances entering the succeeding stage. For a liquid system the level in the accumulator, or the pressure in the case of a gaseous system, are direct indications of the state of balance between the inflow and outflow over the accumulator, and the particular variable is then controlled to regulate the outflow. However, if the control is adjusted in the normal way to give the usually required tight control of the level or pressure, the object of the accumulator is defeated since fluctuations of the inflow will be more or less immediately reflected in the outflow. To smooth out fluctuations in the inflow, the control must be made deliberately slower and more sluggish in response so that there is a relatively slow change in the outflow when the level or pressure changes following an inflow disturbance. The term 'averaging' implies that the outflow is averaged against the level or pressure and that the variable is controlled between upper and lower limits rather than at a fixed value.

The required response is thus critically or even over-damped, and can be obtained by use of proportional control with a much lower sensitivity than would normally be employed. Some authors also recommend the use of a very small integral component so that the level or pressure will be brought back to a definite value in due course of time. Buckley [3], however, points out that the addition of integral action in such cases of very low proportional sensitivity introduces a conditional stability and that a low frequency resonance can occur at the low values of gain required, with cycling at very long periods, up to twelve hours, having been reported. In addition, in response to a sustained change in inflow a controller with proportional action only will not permit a change in outflow greater than the change in inflow. The presence of an integral action component will always cause the change in outflow to exceed the change in inflow for some period of time in order to eliminate the offset and bring the level or pressure back to the desired value; this may then require the downstream plant to have some measure of additional load capacity (an over-design factor) to allow for the temporarily increased load.

Whilst essentially straightforward in concept, the design of an averaging control system has also an economic factor in the size of the accumulator. The latter must be sufficiently large to accomplish the required duty of smoothing out the flow fluctuations without risk of emptying or over-filling, yet it must not be too large that unused capacity is provided at additional capital cost. The problem is essentially one of determining the controller gain to give the required smoothness of outflow in terms of the dimensions of the accumulator.

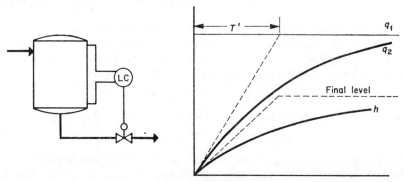

FIGURE 9–3. Averaging level control, response of outflow and level to step change in inflow

Consider the simple level control system with proportional control used as an averaging system, as shown in Figure 9–3(a). At a given time the inflow and outflow in volume/unit time will be q_1 and q_2, and the level, h, these values being measured from the normal steady-state operating conditions, i.e. inflow and outflow are equal with the level at the desired value h, and $h = 0$. A practical implication of the low sensitivity of an averaging control is that the dynamics of the valve and measuring element, and also compressibility effects in the case of a pressure control system, are negligible, and only the first-order lag of the vessel itself need be considered. As a first-order process, the closed-loop transfer function between the level and inflow is given (as seen in earlier sections on pages 26–7 and 140–1) by

$$h/q_1 = R/(Ts + 1 + K) \qquad (9–13)$$

where R is the flow resistance of valve and outflow piping,
    T is the time constant (RC) of the accumulator,
    C is the capacitance (cross-sectional area) of the accumulator,
    K is the overall gain ($K_c K_v R$).

The response of the level to a unit step change in inflow ($q_1 = 1$) is given by Equation (5–5):

$$h(t) = [R/(1 + K)][1 - \exp{(-t/T')}] \qquad (9–14)$$

where $T' = T/(1 + K)$.

The closed-loop transfer function between the outflow and inflow can be derived from the volume balance across the accumulator, i.e.

$$q_1 - q_2 = (Cs)h$$

and Equation (9–13), as

$$q_2/q_1 = (1 + K)/(Ts + 1 + K) = 1/(T's + 1)$$

From the last equation the response of the outflow to a unit change in inflow is given by

$$q_2(t) = 1 - \exp(-t/T') \qquad (9\text{--}15)$$

The responses of the level and outflow to the unit step change in inflow are shown in Figure 9–3(b) and are typical first-order responses. For an averaging control, the important variable after a change in inflow is the rate of change of the outflow (dq_2/dt), which from Equation (9–15) is

$$dq_2/dt = (1/T') \exp(-t/T')$$

which has a maximum value at $t = 0$ since the response is first-order, i.e.

$$(dq_2/dt)_{max} = 1/T' = (1 + K)/T \qquad (9\text{--}16)$$

To determine the size of vessel required, the overall gain must be found first; if the maximum and minimum flows are defined, a gain can be chosen such that the tank is almost full at maximum flow and almost empty at minimum flow. The time constant of the vessel, and hence the dimensions, can then be found from Equation (9–16) for a given maximum rate of change of the outflow.

Example 9–1

The normal liquid inflow to a process operating at 150 kN/m² is 20×10^{-3} m³/s, which may change suddenly by increments of up to 5×10^{-3} m³/s between minimum and maximum values of 10 and 30×10^{-3} m³/s. The liquid has a density of 10^3 kg/m³ and leaves the preceding stage of the process at a pressure of 300 kN/m². Determine the dimensions of an accumulator tank so that the inflow to the second process does not change at a rate greater than 5×10^{-6} m³/s².

Assume that the accumulator will be a vertical cylindrical tank, say 4 m in height. The normal operating level $(\bar{h})$ will then be 2 m. Determine the proportional gain so that the tank will be full at the maximum flow and empty at the minimum flow.

Outflow resistance,

$$R = (\partial h/\partial q) = 2(\bar{h} + \Delta p)/\bar{q}$$

$$\Delta p = 300 - 150 \text{ kN/m}^2$$

$$= \frac{150 \times 10^3}{10^3 \times 9.807} \frac{\text{N/m}^2}{(\text{kg/m}^3)(\text{m/s}^2)} = 15.3 \text{ m}$$

i.e. the pressure head over the liquid in the tank is equivalent to 15.3 m head of liquid.

$$R = 2(2 + 15.3)/(20 \times 10^{-3}) = 1.73 \text{ m}/(\text{m}^3/\text{s})$$

The steady-state change in level is found from Equation (9–14) at $t = \infty$, i.e.

$$h_\infty = R/(1 + K)$$

for unit step change in inflow

$$\text{Maximum change in inflow} = 30 - 20 \times 10^{-3}$$
$$= 10 \times 10^{-3} \, \text{m}^3/\text{s}$$

$$\text{Maximum change in level} = 4 - 2 = 2 \, \text{m}$$

Hence

$$1 + K = (R/h_\infty)\Delta q_{\text{max}}$$
$$= (1.73 \times 10^3/2)(10 \times 10^{-3})$$
$$= 8.65$$

Thus the required value of K for the tank to be full at the maximum rate of flow (and also empty at the minimum rate, since the normal condition is midway between the two) is 7.65. By defining a value of the valve sensitivity, K_v, the value of K_c can be found from $K = K_c K_v R$.

The required tank area is found from Equation (9–16); the largest sudden change in inflow is $\pm 5 \times 10^{-3}$ m³/s and the maximum permitted rate of change of outflow is 5×10^{-6} m³/s². Hence

$$T = [(1 + K)/(dq_2/dt)_{\text{max}}]\Delta q_1$$
$$= (8.65/5 \times 10^{-6})(5 \times 10^{-3})$$
$$= 8.65 \times 10^3 \, \text{s}$$

Since $T = RC$,

$$C = (8.65 \times 10^3)/(1.73 \times 10^3) = 5 \, \text{m}^2$$

and the tank diameter will thus be 2.52 m.

It will be noted that the volume occupied by the liquid under the normal operating conditions with the level at $\bar{h} = 2$ m will be $5 \times 2 = 10$ m³. With a normal flow of 20×10^{-3} m³/s, the residence or hold-up time is 0.5×10^3 s, whereas the time constant as found above is 8.65×10^3 s. This apparent discrepancy is due to the relatively high pressure over the liquid surface in the vessel, which is more than seven times the normal operating head. Consequently there is very little self-regulation in the process. If the pressure were smaller, the time constant would be nearer to the hold-up time; the size of the tank required would be the same for the same given conditions but a lower gain would be required.

If integral action is added to the controller in order to return the level to the normal operating value after any change in inflow, a somewhat similar analysis can be carried out, as discussed by Young [29]. The addition of the integral action makes the response second-order. It is then necessary to define a value of the damping ratio ζ (normally greater than one so that the response of the outflow change is over-damped) and the characteristic time, so that the peak disturbances in outflow do not exceed the specified maximum for the expected step changes in the inflow. Both the damping ratio and characteristic time can be identified in terms of the integral time, overall gain and vessel capacitance, and hence the latter can be determined. However, it is also necessary in this case to consider the steady-state following cyclical changes in the inflow and the particular conditions which might lead to resonance and consequent magnification of the inflow disturbances in the outflow.

There is a considerable similarity between pressure and level controls since both are defined by the same continuity equation. There is one practical distinction in that most pressure controls contain a significant amount of self-regulation, whereas in level controls there may be little or no self-regulation due to the effect of pressure heads over the liquid surface. Self-regulation opposes good averaging control, and in some cases the effects must be countered by the use of a cascade flow control in both level or pressure systems. A cascade system is also often necessary to eliminate the effects of downstream pressure disturbances which may affect the outflow from an accumulator.

Ratio Control

In process operations it is often necessary to maintain the values of two operating conditions in a constant proportion or ratio. The most common example is the ratio control of two flows to maintain a mixture of constant composition, e.g. to maintain a constant fuel/air ratio in combustion of a liquid or gaseous fuel, or the maintenance of a constant reflux ratio from the distillate of a fractionating column.

In these applications one of the two flow rates is independent of the ratio control system, being either 'free', i.e. effectively uncontrolled, such as the outflow from a preceding process, or alternatively regulated as the manipulated variable of a separate control loop, such as in the temperature control of a combustion process, as shown in Figure 9–4. The ratio control is applied by regulation of the second flow rate to maintain the required ratio of flows in spite of any changes which may occur in the first flow rate or in the second flow due to load disturbances influencing this flow. Control of the second flow is thus effectively a separate single-loop system and poses no special problems.

In practice it is necessary to measure both rates of flow, usually by means of flow transmitters, so that the values can be compared to determine the actual ratio which is the controlled variable of the ratio controller. It will be appreciated that it is not necessary to measure the flow rate in the other control loop since the flow is there the manipulated variable.

The adjustments of the ratio controller depend on the changes to be expected in the first flow rate. Essentially the controller must maintain the required precision at a constant desired value of the ratio when the 'free' flow is constant in spite of other disturbances which affect the second flow entering the loop. Additionally changes in the 'free' flow must not result in excessive overshoot or resonance in the corresponding changes produced in the second flow. The first requirement is satisfied by normal adjustment of the controller as a flow control loop, i.e. a low proportional sensitivity and integral time of the order of the operating period. To avoid excessive overshoot following a change in the first flow, the subsidence ratio of the response of the second flow must generally be somewhat larger than usual, e.g. of the order of 10:1 rather than the usual 4:1, and to avoid resonance in the case of cyclical disturbances in the first flow rate, the period of the closed loop system should differ by a factor of at least 3 from the period of any such cyclical disturbances.

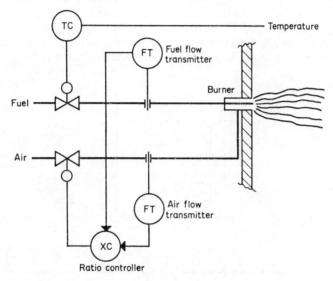

FIGURE 9–4. Fuel/air ratio control

Feedforward Control

In the early discussion of open-loop control, it was suggested that a method of overcoming the effect of input variable disturbances in such systems would be to measure the amount of disturbance in the variable and to adjust the input element of the open loop accordingly to limit the effect of the disturbance (see Figure 1–4). It is, however, difficult in practice to measure all possible input disturbances and to predict the quantitative effects on the process, and the more usual closed-loop 'feedback' type of system is invariably used. Nevertheless there are some process systems for which a 'feedforward' of certain signals may be a valuable addition to the normal feedback loop. If, as is often the case, the output of one process unit is the input stream to a succeeding unit, any disturbance in the output signals of the first unit will obviously affect the unit downstream, and it is thus advantageous to feed information of these disturbances forward to the downstream control loop and so 'anticipate' to some extent their arrival. Due to the fact referred to in the previous discussion that it is possible to apply open-loop load compensation to only a limited number of the input variables which affect a system, in practice rarely more than one or two such variables, the downstream unit must still be controlled by the usual feedback loop. Hence 'feedforward control' differs from open loop control with load disturbance compensation in that the feedforward system (the open loop) is invariably combined with a feedback loop.

One feature of the open loop which is of great advantage in this combination is that no matter how large the required correcting signal or however adverse are the transfer lags, the system cannot become unstable. Theoretically the open-loop correction can be made equal to the disturbance, and although temporary deviation may occur owing to the transfer lags, this should be followed by a non-oscillatory return to the desired value.

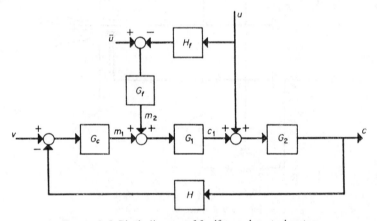

FIGURE 9–5. Block diagram of feedforward control system

A block diagram of a simple feedforward control system is shown in Figure 9–5, where the disturbed variable u might be a change in temperature, composition, or rate of flow of the feed stream to a reactor coming from a preliminary separation stage. The load variable is measured and fed forward, i.e. in a path parallel with actual passage of the disturbance, but the measurement is used to correct the output of the primary controller in the process loop and the signal therefore enters the loop earlier than the disturbance. It will be appreciated that the feedforward correcting element, with the transfer function G_f, must itself be a controller since it must produce a signal dependent on changes in the particular variable which is being measured from some normal operating value such as $\bar{u}$. The load compensator thus requires the elements of a typical controller, i.e. a desired value input element, an error discriminator and an output function generator. The outputs of the two controllers are then combined by a simple summation relay to operate the control valve.

The action of the system is thus that a change in the input variable u

produces a corrective signal in the control loop in advance of the entry of the disturbance into the loop. The primary closed-loop controller then adjusts the final position of the control valve to correct for any imperfections in the feedforward compensation as shown by the reactions of the final controller variable; in other words, the primary controller acts as a 'trimmer' to the feedforward compensator. Also, of course, it will deal with any other disturbances entering the closed loop which are not affected by the feedforward compensation.

The required transfer function of the feedforward controller may be readily identified from the transfer functions of the elements of the system; from Figure 9–5:

$$c_1 = G_1[G_f(\bar{u} - H_f u) + G_c(v - Hc)]$$

and $$c = G_2(c_1 + u)$$

For a constant desired value in the main loop ($v = 0$) and a constant normal value of the load variable ($\bar{u} = 0$), the overall transfer function between the final controlled variable and the disturbing variable is

$$\frac{c}{u} = \frac{G_2(1 - G_f G_1 H_f)}{1 + G_c G_1 G_2 H} \tag{9-17}$$

Comparing this with the closed-loop function with the feedforward compensator omitted, i.e.

$$\frac{c}{u} = \frac{G_2}{1 + G_c G_1 G_2 H}$$

it can be seen that the numerator of the transfer function is reduced by the presence of the additional terms due to the feedforward compensation; the deviation for a given step change in the load variable will consequently be less. In fact, according to Equation (9–17), perfect compensation would be achieved by making $G_f G_1 H_f = 1$, thereby making the closed-loop transfer function zero; there would then be no response from the controlled variable irrespective of any changes in the load variable.

If there is no dynamic lag in the measuring element, i.e. $H_f = 1$, the ideal compensator will have a transfer function of $G_f = 1/G_1$, i.e. the reciprocal of the process transfer functions between the control valve and the point of entry of the disturbance into the loop. For the simplest case of a single time-constant stage with a transfer function of $K_1/(T_1 s + 1)$, perfect compensation is obtained by use of a controller function of $(1/K_1)(T_1 s + 1)$, which is the transfer function of a proportional-derivative controller. A gradual increase in the load variable would lead to a gradual decrease in the intermediate variable c_1, so that the combined signal $(c_1 + u)$ would remain constant and there would be no change in the output variable c. Complete compensation is theoretically possible for all types of change in the load variable apart from a step change, which would call for an infinite output from a controller with a derivative component.

Unfortunately it is not possible in practice to achieve this ideal solution due to the uncertainties in measurement or estimation of time constants and transfer functions, and the non-linearities always present in items of plant, which may lead to changes in gain by factors of two or more over the usual range of inputs. In a practical system, compensation can only be imperfect but nevertheless may well reduce the net effect of a disturbance to some 20 per cent of the load change without compensation. A five-fold reduction in deviation is quite appreciable but cannot compare with the much larger reductions often possible with cascade control systems.

If there are two time constants between the load compensator and the point of entry of the disturbance into the loop, perfect compensation would then require a controller with first and second derivative actions. The difficulty in determining the system parameters and the additional problems of 'noise' with a second derivative (d^2e/dt^2) action have prevented the application of such devices. Using a conventional PD controller, Young [29] suggests setting the derivative time equal to the sum of the two time constants; the factor of improvement in the response then depends on the ratio of the lags in the final elements (G_2) to those in the earlier elements (G_1). In the example quoted by Young, G_1 is two time constants of 1 min each and G_2 is a single time constant of 2 min; proportional load compensation was found to have very little effect, but a PD compensation with a derivative time of 2 min produced a three-fold reduction in error.

When the last time constant in the system is much larger than the others, the dynamic effect of the lags in the compensating loop can be neglected and a proportional controller serves as well as any other.

The design of the feedback loop of a feedforward system is the same as that of a simple single loop system; the whole system can be reduced to a single loop if the load variable is introduced through an element with a transfer function of $(1 - G_f G_1 H_f)$, using the nomenclature of Figure 9–5. As already noted, the stability of the feedback loop is independent of the subsidiary compensation loop; the denominator of the closed loop function (Equation (9–17)) does not contain any terms apart from those of the closed loop.

The alternatives to feedforward control are the provision of storage capacity between the plant units to smooth out fluctuations in the load variable, using an averaging control where applicable, or the use of a cascade system. The provision of additional storage capacity is generally uneconomic in both capital and running costs if satisfactory control can be obtained by feedforward compensation. A cascade system will generally give a greater reduction in error than a feedforward system, but requires the point of entry of the disturbance to be included in the inner loop of the cascade. When the disturbance enters late in the loop, e.g. before the last element of the process, cascade control may be less efficient than feedforward, owing to the larger number of lags in the inner loop.

A similarity between feedforward control and selective control with two controllers operating one control valve is noted in the following section on pages 358–9.

Cascade Control

Reference has already been made in Chapter 7 to this type of complex control in which an inner and an outer control loop are formed, each with an individual controller and measuring feedback. The output of the outer-loop controller (the primary or master controller) sets the desired value of the inner-loop controller (the secondary or slave controller). The subject was considered on pages 287–9, since to determine the frequency response characteristics it is necessary to determine first the closed-loop frequency response of the inner loop as this forms an element of the outer loop.

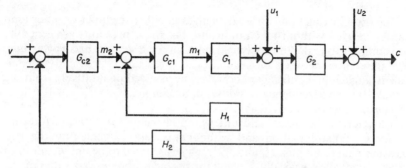

FIGURE 9–6. Block diagram of cascade control system

In the usual applications in process control the inner loop is almost invariably a flow control loop; the controlled variable of the outer loop is very often temperature but other variables, such as level and pressure in averaging control systems, are also encountered. Thus the block diagram of a typical cascade system shown in Figure 9–6 may represent temperature control in a fractionating column via the flow of steam to the reboiler. The steam flow is measured directly by the measuring element H_1, and controlled by the inner loop controller G_{c1}. The desired value of this controller is the manipulated variable of the outer loop temperature controller G_{c2}, which controls the temperature, measured by the element H_2 at some point in the column by varying the rate of flow of steam to the reboiler. If a conventional single-loop temperature control were used with the output of the temperature controller applied directly to the steam valve, any disturbance affecting the flow of steam (such as a change in supply pressure) would be subjected to the major lags in the column before reaching the temperature measuring element, and there would be a considerable lag before the change in temperature was detected and a correction applied; thus a large deviation would result. With the cascade system, using a separate flow control, any changes affecting the rate of flow of the steam are corrected very quickly by the fast reaction of the flow loop, and it is very unlikely that any effect of these disturbances would be detected at the temperature measuring element. The temperature control loop is, of course, still required to deal with any disturbance which arises outside the flow control loop and which would affect the temperature, e.g. changes in the feed composition or flow rate to the column which call for different rates of flow of steam to the reboiler. This is accomplished by the temperature controller re-setting the desired value of the flow controller.

Cascade control is especially effective in a system such as that described above where the main disturbances enter the inner loop and the reaction of the latter is much faster than that of the outer loop. Since the inner loop in process systems is usually a flow control loop, this last requirement is generally met.

The transfer functions for load changes in either the inner or outer loop can be derived without much difficulty, but the expressions are generally too complex to make the analytical solution for a step change worthwhile. The differences between the responses of a cascade system and an equivalent single loop control can, however, be assessed qualitatively from the steady-state and frequency responses quite simply.

Considering the block diagram of Figure 9–6, where the subscripts 1 and 2 refer to the inner- and outer-loop elements respectively, for convenience of manipulation and without any loss of generality the two measurement functions, H_1 and H_2, may be assumed to be unity, and since process control is basically a regulator problem, the desired value of the controlled variable may be taken as constant so that $v = 0$. The following equations can then be written by inspection from the block diagram:

$$m_2 = -G_{c2}c_2 \qquad m_1 = G_{c1}(m_2 - c_1)$$
$$c_1 = G_1 m_1 + u_1 \qquad c_2 = G_2 c_1 + u_2$$

Eliminating m_1, m_2, and c_1 from these equations yields

$$(1 + G_{c1}G_1 + G_{c1}G_{c2}G_1G_2)c = G_2 u_1 + (1 + G_{c1}G_1)u_2 \qquad (9\text{--}18)$$

For a load change such as u_1 within the inner loop, with $u_2 = 0$, the closed-loop transfer function from Equation (9–18) is

$$\frac{c_2}{u_1} = \frac{G_2}{1 + G_{c1}G_1 + G_{c1}G_{c2}G_1G_2} \qquad (9\text{--}19)$$

This function can be compared with that of a single-loop control where the inner-loop controller is omitted and $m_1 = m_2$, i.e.

$$\frac{c_2}{u_1} = \frac{G_2}{1 + G_{c2}G_1G_2} \qquad (9\text{--}20)$$

Comparing Equations (9–19) and (9–20), the effect of cascading the control is to introduce the inner-loop controller function into the denominator of the closed-loop transfer function at two points and so increase the magnitude of the denominator. The result is thus to reduce the magnitude of the change in the controlled variable c_2 for a given change in the load variable u_1. This can be demonstrated by assuming that the two controllers are both proportional, i.e. $G_{c1} = K_{c1}$ and $G_{c2} = K_{c2}$, and that the gains of the process elements G_1 and G_2 are unity. For cascade control (Equation (9–19)) the steady-state offset for a unit step change in u_1 is

$$c_2(\infty) = 1/(1 + K_{c1} + K_{c1}K_{c2}) \qquad (9\text{–}21)$$

If both controller gains are large, the offset will vary inversely with the product, $K_{c1}K_{c2}$.

For a single-loop control the steady-state offset for a unit step change in load is given by Equation (9–20) as

$$c_2(\infty) = 1/(1 + K_{c2}) \qquad (9\text{–}22)$$

and the offset varies inversely with the gain K_{c2}. However, as was demonstrated in a previous example, the value of K_{c2} for a single-loop control will be relatively much smaller than K_{c2} for a cascade control, and the offset for load changes in the inner loop will be very much smaller with the cascade control than with the equivalent single loop.

For disturbances such as u_2 occurring in the outer loop, the transfer functions corresponding to Equations (9–19) and (9–20) for inner-loop disturbances, and obtained from Equation (9–18) with $u_1 = 0$, are

Cascade
$$\frac{c_2}{u_2} = \frac{1 + G_{c1}G_1}{1 + G_{c1}G_1 + G_{c1}G_{c2}G_1G_2} \qquad (9\text{–}23)$$

Single loop
$$\frac{c_2}{u_2} = \frac{1}{1 + G_{c2}G_1G_2} \qquad (9\text{–}24)$$

The steady-state solution corresponding to Equation (9–21) for an outer-loop load change is thus

$$c_2(\infty) = (1 + K_{c1})/(1 + K_{c1} + K_{c1}K_{c2}) \qquad (9\text{–}25)$$

and the offset with cascade control will now vary inversely with K_{c2} when both gains are large. The steady-state solution for the single loop with load changes in u_2 is the same as that for changes in u_1 (Equation (9–22)). The gain of the inner-loop controller does not then help to reduce offset for load changes in the outer loop, and will have little effect on the peak deviation; the offset of the cascade system will, however, be less than that for the single loop due to the increase in K_{c2} permitted by the cascade. The advantage of cascade control for a primary (outer loop) load disturbance comes, in fact, from the higher frequency of oscillation of the system beyond that of the single loop system and thus the higher gain permitted at the critical frequency.

Example 9–2

Compare the offsets and error integrals for cascade control and the equivalent single-loop control of the system of Example 7–8.

For the cascade system the values of the proportional gains for the two controllers were $K_{c1} = 6$ and $K_{c2} = 14.9$, and for the single-loop system, $K_{c2} = 4.3$.

Substituting these values in Equations (9–21) and (9–22), the offsets following a unit step change in the inner loop are:

(a) Cascade system:

$$c_2 = 1/(1 + 6 + 89.4) = 0.0104$$

which is sufficiently small (1 per cent) to make integral action in the inner-loop controller unnecessary.

(b) Single-loop system:

$$c_2 = 1/(1 + 4.3) = 0.189$$

Thus, in this particular case, the cascade control reduces the offset for inner-loop load changes by a factor of nearly 20.

For a unit load change in the outer loop of the cascade, the offset is given by Equation (9–25):

$$c_2 = (1 + 6)/(1 + 6 + 89.4) = 0.073$$

The reduction in the offset is not as spectacular when the load change is in the outer loop but still shows an improvement by a factor of over 2.

The effect of cascading the control on the controllability of the system is shown by comparison of the error integrals for the two cases. The critical frequencies for the two systems were 0.74 rad/min for the single loop and 1.82 rad/min for the cascade. The error integral is inversely proportional to $\omega_c(1 + K)$; for the cascade system the denominator of the closed loop function corresponding to $(1 + K)$ for the single loop is $(1 + K_{c1} + K_{c1}K_{c2})$, assuming unit process stage gains. Thus

$$\text{Error integral ratio} = \frac{1.82(1 + 6 + 89.4)}{0.74(1 + 4.3)} \approx 45$$

Improvements in the error integral of up to a hundredfold are, in fact, quite common in cascade systems for disturbances within the inner loop compared to the equivalent single loop.

For outer-loop disturbances $(1 + K)$ in the error integral ratio must be replaced as a measure of offset by $(1 + K_{c1} + K_{c1}K_{c2})/(1 + K_{c1})$, giving

$$\text{Error integral ratio} = \frac{1.82(1 + 6 + 89.4)}{0.74(1 + 4.3)(1 + 6)} = 6.43$$

The effect of cascade control on the stability of the system can be assessed from the polynomial denominator of the closed-loop transfer function (which is the same for any type of disturbance) written as a characteristic equation, i.e.

$$1 + G_{c1}G_1 + G_{c1}G_{c2}G_1G_2 = 0$$

Dividing by $(1 + G_{c1})$ and re-arranging into the form used for definition of the limiting stability by the Bode criterion,

$$G_{c2}\left[\frac{G_{c1}G_1}{1 + G_{c1}G_1}\right]G_2 = -1 \qquad (9\text{--}26)$$

The limiting condition of stability thus occurs when the vector defined by the left-hand side of Equation (9–26) has a value of -1. The term in brackets will be recognized as the closed-loop transfer function for desired value changes of the inner loop of the cascade. If the latter is set to give an acceptable level of stability in itself, the maximum magnitude ratio will be of the order of 1.2 but the peak ratio will occur at a much higher frequency than the critical frequency of the outer loop so long as the response of the inner loop is faster than that of the outer. At the critical frequency of the outer loop, the magnitude ratio of the inner loop will be essentially unity and the angle contribution very small, as was demonstrated in Example 7–8.

This discussion implies that a cascade control system requires certain desirable features if the operation is to be completely successful. Initially the inner loop should include the major input disturbances and must be faster in reaction (i.e. a higher critical frequency) than the outer loop by a factor of at least 3, and preferably up to 5 or even 10. This condition is usually met in process control applications when the inner-loop controlled variable is a rate of flow. It is also advantageous to include within the inner loop as many disturbance inputs as can be accommodated without materially decreasing the response speed of the loop, since the correction for inner loop disturbances is so much better than for those in the outer loop.

Selective Control Systems

A selective control system also utilizes two controllers and has a basic similarity thereby to a cascade system, but the two controllers are now effectively in parallel rather than in series. The output of the primary controller is not now the desired value signal of the secondary controller, but instead the output of the latter has a 'selective' over-riding action on that of the main controller. The two major applications of this technique are to disturbances which occur at the extreme ends of the process system, i.e. close to the measuring element, either in a subsidiary variable or in the plant itself, or very close to or immediately after the control valve. In the former case both controllers use the same measuring element, with the secondary controller regulating a secondary supply into the process system in addition to that regulated by the main controller; in the latter case the controllers have separate measuring elements with the subsidiary controller virtually controlling an intermediate variable in the process loop (as with a cascade system), but both now operate the same supply valve through a summating relay.

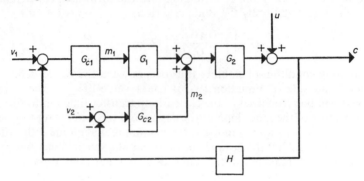

FIGURE 9–7. Selective control using single measuring element

As an example of the first type, Young [29] cites the case of an exothermic catalytic reactor which may develop a 'hot spot' due to erosion or channelling of the catalyst bed; the process may then become unstable and 'run away', and immediate corrective action is necessary to prevent a disastrous rise in temperature. In such a case the disturbance arises within the plant itself, due essentially to a mechanical failure; the control of such a reactor would almost invariably be based on the temperature of the catalyst bed, and the disturbance thus occurs in the very last stage of the process before the final measuring element. The temperature controller would manipulate the main supply condition, e.g. the feed stream to the reactor or possibly the steam supply to a feed preheater, and the main control loop would thus possess a number of transfer stages before the last one at which the temperature is measured and the disturbance occurs. In the circumstances as described, whilst the disturbance would be detected almost immediately by the measuring element, the correction of the controller would be delayed by all the other lags in the system before reaching the final stage, and the temperature could reach a dangerous level before the control becomes effective. A subsidiary controller would be used in such a case, with the same temperature measuring element as the main controller but operating a secondary supply valve such as an additional supply of feed or a diluent direct to the reactor, or a by-pass over the feed preheater, etc. The subsidiary loop would have fewer transfer stages than the main loop, as shown in Figure 9–7, so that the corrective action is developed much more quickly than that of the main controller; in this sense, the subsidiary controller 'over-rides' the action of the main controller in providing a much quicker corrective action if the temperature rises suddenly.

The function of the subsidiary controller in this instance is to correct quickly for disturbances which enter late in the control loop; the main controller takes essentially the same action by adjusting the main supply condition, but more slowly owing to the additional lags in the main loop. Once the correction of the main controller becomes effective and the controlled variable is driven back to the desired value, the subsidiary controller should return the secondary supply or by-pass to the normal operating level. This latter requirement implies that the secondary controller should not have an integrating action, since this would produce a permanent change in the output of this instrument after a load disturbance. The desired values of the two controllers need not necessarily be the same value of temperature; in fact there would be some advantage in offsetting that of the subsidiary controller so that the output signal is a relatively low value during normal operation. The proportional gain, and derivative time if such is necessary, of the secondary controller can be found by frequency response analysis of the secondary loop; the gain must not be too large to cause fluctuations in the secondary supply condition greater than can be tolerated for successful operation of the main loop. The controller in the latter loop may have any parameters required by the process, and these may be determined from the properties of the main loop in the usual way.

From the block diagram of Figure 9–7, the closed-loop transfer function for load changes at the end of the process (u) can be derived by the usual analysis; with constant desired values and negligible measurement lag:

$$\frac{c}{u} = \frac{1}{1 + G_{c2}G_2 + G_{c1}G_1G_2}$$

whereas for a single loop, omitting the secondary controller, G_{c2}, the function is

$$\frac{c}{u} = \frac{1}{1 + G_{c1}G_1G_2}$$

The introduction of the secondary controller thus introduces into the denominator of the closed loop function an additional term which is the transfer function of the subsidiary loop ($G_{c2}G_2$). This can only lead to a reduction in the response of the controlled variable to a given disturbance.

The effect on stability can be assessed from the characteristic equation of the system,

$$1 + G_{c2}G_2 + G_{c1}G_1G_2 = 0$$

which can be re-arranged to

$$\frac{G_{c1}}{G_{c2}} \left[\frac{G_{c2}G_2}{1 + G_{c2}G_2} \right] G_1 = -1$$

where the term in brackets is the closed-loop transfer function for desired value changes of the secondary loop. As the latter has fewer transfer stages, the response will be much faster than that of the main loop and the magnitude ratio will be essentially unity and the phase angle very small at the critical frequency of the main loop. The critical frequency of the double loop will thus be greater than that of the single loop since G_{c2} cannot have an integral action.

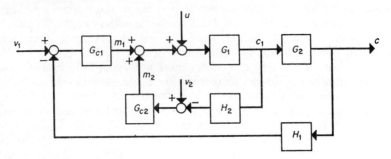

FIGURE 9–8. Selective control using single regulating element

The alternative form of selective control is for disturbances which enter very early in the control loop, usually just after the regulating unit, and the main process application is in the control of P.F. fired boilers. The principle is to measure the effect of the disturbance on an intermediate variable as close as possible to the entry of the disturbance, and to apply an immediate correction by means of the subsidiary controller without waiting for the disturbance to reach the final controlled variable. The main controller now acts as a 'trimmer' to adjust the final correction of the regulating unit on the basis of any residual deviations in the controlled variable. Thus two variables are measured, as in a conventional cascade system, but the output of the primary controller is combined with that of the secondary controller in a suitable summation relay to provide a single actuating signal to the control valve. A block diagram is shown in Figure 9–8 and a basic resemblance with feedforward compensation will be noticed if Figures 9–5 and 9–8 are compared. The essential difference between the two cases is that in feedforward compensation the subsidiary controller does not form part of an inner feedback loop but is part of an open loop outside the main feedback loop. In feedforward arrangements the load disturbance variable is measured outside the control loop; in the selective control, the effect of the disturbance on an intermediate variable within the loop is measured to generate the additional control action.

By making the usual assumptions that desired values are constant and measurement lag is negligible, the closed-loop transfer function can be determined from the block diagram as

$$\frac{c}{u} = \frac{G_1 G_2}{1 + G_{c1} G_1 G_2 + G_{c2} G_1}$$

whereas for the single loop, omitting the subsidiary controller G_{c2}:

$$\frac{c}{u} = \frac{G_1 G_2}{1 + G_{c1} G_1 G_2}$$

As in the previous case, the introduction of the second controller adds an additional term into the denominator of the closed-loop function and so produces a smaller deviation for a given change in the load variable. Re-arrangement of the characteristic equation of the system also produces virtually the same type of result, i.e.

$$\frac{G_{c1}}{G_{c2}} \left[\frac{G_{c2} G_1}{1 + G_{c2} G_1} \right] G_2 = -1$$

The control parameters for the two instruments may be found by the same techniques as used for a cascade system. The subsidiary controller (G_{c2}) is adjusted first using the frequency response characteristics of the inner loop. The critical frequency of this loop will usually be high, and a low proportional sensitivity will make the use of integral and/or derivative action unnecessary. The closed-loop response of the inner loop is then used to determine the response of the outer loop.

Sampled Data Control Systems

In the previous discussions of linear control systems it has been taken for granted that all signals are transmitted continuously from one element to the next, as is generally the case with the usual analogue type of measuring and controlling instruments where the output is a continuous function of time. However, in certain systems the signal at some point or points in the control loop is made available for transmission only at discrete separate intervals of time. This is usually due to the orderly sampling and measurement of a continuous signal at discrete intervals; hence the use of the term *sampled data*.

Two main types of sampled data systems can be identified; continuous control of a sampled data input signal to the controller, and sampled data control of either continuous or discrete inputs. In the latter case discrete impulses from the controller are used to position the control valve and the controller is basically digital rather than analogue in operation. A digital control computer can only accept signals at discrete intervals and a continuous input signal must be impulse-modulated or digitized. Although the discrete intervals are so short in this instance that for all practical purposes the measurement is continuous, it is more convenient to analyse the system by sampled data techniques and so to determine relatively more sophisticated algorithms for the digital control. This latter subject lies outside the scope of the present discussion.

The most usual application of sampled data control in process operations is that in which a particular variable, usually the output controlled variable, cannot be measured continuously but only intermittently. This occurs mainly with composition analysis measurements where a sample must be removed from a flow stream for analysis, either in the laboratory by manual or semi-automatic methods, or by automatic process analysers such as the gas chromatograph and similar instruments which are essentially intermittent in operation. In either case the final measurement of the composition analysis is available only at discrete intervals of time. When the interval is long, say more than 30 min, it is usual to apply the results to manual adjustment of the process, generally by manual adjustment of the desired values of the automatic control loops used on the process. The theory of sampled data control can, however, be used to evaluate a quantitative strategy for the manual operator as discussed by Buckley [3]. When the interval is relatively short, e.g. when using automatic process analysers, it is more usual to close the loop with a conventional continuous function controller so that a control signal is generated by sampling the process output at discrete intervals. With the continuous systems previously discussed, the input-output relations are described by differential equations; the discrete or sampled systems are now described by finite difference equations. Generalization is more difficult than with the continuous data systems, but the sampled data systems are usually sufficiently linear, or can be treated as such, to be studied and designed by an extension of the methods already discussed.

Principles of Sampled Data Systems

With a sampling of the controlled variable at discrete intervals, the measured value is now a series of pulses whose magnitude is proportional to the instantaneous value of the variable at the instant of sampling, and subject to any time delays which arise from the analytical or other measuring function carried out. These pulses are of sufficiently short duration compared with the sampling intervals to be regarded as 'impulses', i.e. of zero duration. A sequence of impulses is of no particular disadvantage when used as the input to a digital instrument, but is much too 'noisy' to accord with the continuous signal input required by the analogue type of controller. The basic principle in the use of intermittent signals as input to such instruments is to smooth the sequence of impulses into a series of steps by 'clamping' or 'holding' the output signal at the magnitude of the last impulse during the interval between it and the next impulse in the sequence. Thus if the continuous signal output of the process is $f(t)$ which is sampled at intervals of T, at a particular sampling instant of nT a new value of the measured variable $f(nT)$ is received by the sampler; the previous value, $f[(n-1)T]$ is discarded and the new value is held as a constant output over the interval T until the next sampling instant, $(n+1)T$. The sequence of impulses, each of magnitude $f(nT)$ where $n = 0, 1, 2, 3 \ldots$, is denoted by $f^*(t)$, where the asterisk denotes conventionally a sampled function. By means of the sample-and-hold device, the continuous function $f(t)$ is converted into the stepped or clamped signal, $f_c(t)$.

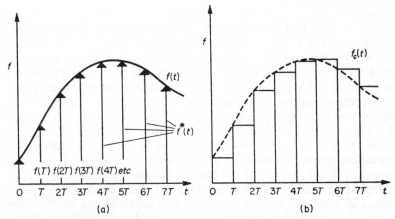

(a)   (b)

FIGURE 9–9. Sampled data: (a) impulse signals, (b) clamped signals

The impulse-modulation process is really a mathematical fiction as no physical device can produce a true sequence of impulses as an output signal. The important feature is that the input-output relationship of the sample-and-hold should be of the correct form so that the 'continuous' stepped signal leaving the 'hold' element can be compared continuously with a desired value signal to produce a continuous error signal to generate control action. The word continuous is used here not in the usual strict

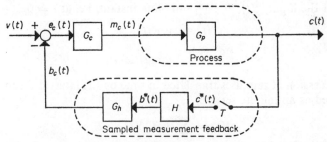

FIGURE 9–10. Block diagram of sampled data system with continuous control

mathematical sense, but rather to indicate that some value of an output signal is available at all times as opposed to $f^*(t)$ which only exists intermittently at the discrete sampling instants. As seen in Figure 9–10, the impulse-modulating (sampling) process is indicated by a switch with the sampling period T written near by. The sampled signal would usually then be measured (usual transfer element H) and the measured signal then passed to the holding element $G_h(s)$. The order of the last two elements may be reversed without affecting the discussion. It will be seen from Figure 9–10 that the error signal ($= v - b_c$) will be a stepped signal, as will also be the controller output.

It will also be seen from Figure 9–9 that a truly correct value of the input variable to the sampler is available only at the start of each sampling interval, and at other instants the value of the variable will not necessarily be correct if the variable is changing. Thus some information about the signal $f(t)$ is not contained in the output $f_c(t)$; there is an information loss due to the sampling process. The subjects of signal frequency and filtering are aspects of control theory not usually considered in process control applications since most process signals contain only very low frequencies. As is shown in the more specialized texts in this field [20, 21], the sampling frequency must be at least twice the greatest frequency for which reasonable signal recovery is required, i.e. if the sampling interval is T, a complete reconstruction of the input signal can be obtained from the sampler output for input frequencies up to $\omega = \frac{1}{2}(2\pi/T)$, or π/T rad/unit time. For process control systems, the highest frequency of interest would be the critical frequency of the control loop since the range of frequencies to be found in a disturbing signal would normally be well below the critical frequency. The choice of sampling interval may then be based on the critical frequency, T being chosen so that $\pi/T > \omega_c$.

Mathematical Analysis

Considering Figure 9–10, the sampler converts the continuous signal $c(t)$ into a train of uniformly-spaced impulses, $c^*(t)$. These output pulses may be regarded as the product of a unit impulse δ, and the instantaneous value of the input, $c(t)$, at each sampling instant, i.e. at $t = 0$,

$$c^*(t) = c(0)\delta(t)$$

at $t = T$,

$$= c(T)\delta(t - T)$$

The succession of impulses translated in time by the interval T can thus be expressed as an infinite series

$$c^*(t) = \sum_{n=0}^{n=\infty} c(nT)\delta(t - nT)$$

The Laplace transform of this series is also an infinite series

$$c^*(s) = \sum_{n=0}^{n=\infty} c(nT) \exp(-nTs) \qquad (9\text{–}27)$$

since $c(nT)$ is a constant in each term of the series; the transform of a unit impulse is itself unity (i.e. $F(\delta)(s) = 1$), and translation in time by nT gives a transform of $\exp(-nTs)$.

Assuming a measurement feedback of unity ($H = 1$), the signal $c^*(s)$

enters the holding element. This may take a number of possible forms but the simplest is the so-called *zero-order hold* where, as described above, the output variable is held constant over each sampling interval at the magnitude of the input pulse $c(nT)$ at the start of the interval. The output of the hold element thus consists of a positive step of magnitude $c(nT)$ followed by a negative step of the same magnitude at an interval of T later, i.e. at $(n + 1)T$. Thus

$$c_c(t) = c(nT)(t) - c(nT)(t - T)$$

and the output transform is

$$c_c(s) = c(nT)/s - [c(nT)/s] \exp(-Ts)$$

$$= c(nT) \left[\frac{1 - \exp(-Ts)}{s} \right]$$

However, $c(nT)$ is also the magnitude of the input pulse $c^*(t)$, i.e.

$$c^*(t) = c(nT)\delta(t)$$

whence

$$c^*(s) = c(nT)$$

The transfer function of the zero-order hold is thus given by

$$G_h(s) = \frac{c_c}{c^*}(s) = \frac{1 - \exp(-Ts)}{s} \tag{9-28}$$

The Laplace transform of the output signal of the hold element is then obtained from Equations (9–27) and (9–28) as

$$c_c(s) = G_h(s)c^*(s)$$

$$= \sum_{n=0}^{n=\infty} c(nT) \left[\frac{e^{-nTs} - e^{-(n+1)Ts}}{s} \right] \tag{9-29}$$

The z-transform

The calculation of system responses using Equation (9–29) requires a term-by-term inversion of the Laplace transform of each impulse and is thus a tedious procedure. The mathematical treatment is simplified by the introduction of a variable z, defined by

$$z = \exp(Ts)$$

Substituting z into Equation (9–27),

$$c^*(s) = \mathscr{L}[c^*(t)] = \sum_{n=0}^{n=\infty} c(nT)z^{-n} \tag{9-30}$$

The summation of Equation (9–30) is defined as the z-transform of the input signal $c(t)$, i.e.

$$c(z) = Z[c(t)]$$

$$= \sum_{n=0}^{n=\infty} c(nT)z^{-n}$$

$$= c^*(s)\big|_{z=e^{Ts}}$$

For most functions of interest the infinite series can be summed directly to give a simple expression in z, e.g. the Laplace transform of a unit step at successive intervals of T is

$$u^*(s) = 1 + e^{-Ts} + e^{-2Ts} + e^{-3Ts} + \ldots$$

which, in terms of z, becomes

$$u(z) = 1 + z^{-1} + z^{-2} + z^{-3} + \ldots$$

$$= \sum_{n=0}^{n=\infty} z^{-n}$$

$$= \frac{z}{z-1} \tag{9–31}$$

(for $z > 1$).

A more interesting example is given by the decaying exponential function, $f(t) = \exp(-at)$, which at sampling intervals of T has the values

$$f^*(t) = 1 + e^{-aT} + e^{-2aT} + \ldots$$

$$f^*(s) = 1 + e^{-aT} e^{-Ts} + e^{-2aT} e^{-2Ts} + \ldots$$

Substituting z for e^{Ts},

$$f(z) = 1 + (e^{aT}z)^{-1} + (e^{aT}z)^{-2} + \ldots$$

$$= \sum_{n=0}^{n=\infty} (e^{aT}z)^{-n}$$

$$= \frac{z}{z - e^{-aT}} \tag{9–32}$$

(for $z > 1$).

Since the z-transform is mathematically the Laplace transform with a change in variable, it may be concluded that there are no new restrictions imposed for the existence of a z-transform beyond these existing for the Laplace transform from which it is derived; z-transforms have been evaluated for most functions of interest and extensive tables with the corresponding Laplace transforms and time functions are available [3]. The use of these tables is basically similar to that of tables of Laplace transform pairs. The introduction of a z-transformed input signal into the z-transform of the transfer function of an element or series of elements (the so-called pulse transfer function) produces an output z-transform which may be inverted (by methods discussed below) to give the output signal as a function of time, either intermittent or continuous, but which in the latter case will, in general, have correct values only at the instants of sampling.

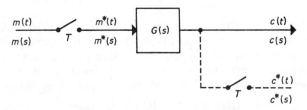

FIGURE 9–11. Pulse transfer function

The z-transform of a transfer function, either of a single element or a series, is known as a *pulse transfer function* and is defined as follows. If a continuous element with a transfer function of $G_1(s)$ is preceded by a sampler, the continuous input signal $m(t)$ becomes an impulse sequence $m^*(t)$. If it is desired to know the system output only at the instants of sampling then, in effect, the output signal $c(t)$ must also be sampled at the same instant as the input signal; this is often shown by a fictitious sampler perfectly synchronized with the input sampler, as shown in Figure 9–11. The output of this second sampler is, by definition, $c^*(t)$, and the determination of this signal is exactly the same as the determination of $c(t)$ at the sampling instants. By the usual transfer function relationship,

$$c(s) = G_1(s)m^*(s)$$

from which it follows that

$$c^*(s) = m^*(s) \sum_{n=0}^{n=\infty} \mathscr{L}^{-1}[G_1(s)](nT)\delta(t - nT)$$

where $\mathscr{L}^{-1}[G_1(s)]$ is the impulse response of the component $G_1(s)$. Hence

$$c^*(s) = m^*(s)G_1^*(s)$$

and substituting z,

$$c(z) = m(z)G_1(z)$$

where $G_1^*(s)$ or $G_1(z)$ is the pulse transfer function between the sampled input and output signals.

It should be emphasized that the z-transform is not directly comparable with the Laplace transform in all particulars, and that some of the manipulative conveniences of the latter are not applicable. The most important example occurs in the pulse transfer function of a series of elements. Unless the elements in the series are separated by synchronized samplers, i.e. the input and output signals to each element are impulse sequences, the

z-transform of the series is *not* given by the product of the z-transforms of the individual elements, as is the case for the Laplace transforms, i.e.

$$G(s) = G_1(s)G_2(s) \ldots$$

but

$$G(z) \neq G_1(z)G_2(z) \ldots$$

The reason that the z-transforms are not multiplicative when the signals between elements are not sampled is that the output depends not only on the inputs at the times of the sampling instants but also on the inputs between these instants. The z-transform of the series is given by the z-transform of the product of the individual Laplace transforms, i.e.

$$G(z) = Z\{\mathscr{L}^{-1}[G_1 G_2 \ldots (s)]\}$$

$$= \overline{G_1 G_2 \ldots (s)}(z)$$

$$= \overline{G(s)}(z)$$

A z-transform inversion into the time domain may be performed by a number of methods, of which the partial fraction expansion and the long division methods are most usually employed. The former method is analogous to Laplace transform inversion, and consists of simplifying the z-transform by a partial fraction expansion into terms which are sufficiently simple to be identified in the table of z-transforms and the corresponding time function found. As most z-transforms contain a z in the numerator, it is often more convenient to expand $c(z)/z$ by partial fractions and to multiply the expansion by z before inverting. As an example of this method, consider the transform

$$c(z) = z/[(z - a)(z - b)]$$

Expanding $c(z)/z$ by partial fractions gives

$$\frac{c(z)}{z} = \frac{1}{(z - a)(z - b)}$$

$$= \frac{1}{(a - b)}\left[\frac{1}{z - a} - \frac{1}{z - b}\right]$$

whence

$$c(z) = \frac{1}{(a - b)}\left[\frac{z}{z - a} - \frac{z}{z - b}\right]$$

From Equation (9–32), or a table of z-transforms, the inverse transform is recognized as

$$c^*(t) = \frac{1}{(a - b)}(a^n - b^n) \tag{9–33}$$

where $n = 0, 1, 2, 3. \ldots$

This method of inversion leads to an expression which is a continuous function of nT and which in general has correct values only at the instants of sampling, since no information is available between these instants. A modification of the z-transform is to introduce a fictitious delay element, with a delay time which is a fraction of the sampling period, between the system and the output sampler. As discussed by Koppel [20], this permits values of the output signal to be determined between sampling instants.

The alternative method of z-transform inversion is to expand $c(z)$ into a power series in z^{-n} by synthetic division of the numerator by the denominator. The coefficients of the z^{-n} terms are then the magnitudes of the impulse responses at the times, nT, i.e. if

$$c(z) = a_0 + a_1 z^{-1} + a_2 z^{-2} + \ldots + a_n z^{-n}$$

then

$$c^*(s) = a_0 + a_1 e^{-Ts} + a_2 e^{-2Ts} + \ldots + a_n e^{-nTs}$$

which, by inversion of the Laplace transform, yields

$$c^*(t) = a_0(0) + a_1(T) + a_2(2T) + \ldots + a_n(nT)$$

This method therefore leads to an intermittent function of time which gives the values of the variable at the sampling instants, i.e. $f^*(t)$ may be written as $f(nT)$. The division may be carried out to yield as many terms as desired, but the value found for any particular sampling instant depends on the values determined for all previous instants, and errors are therefore cumulative. The method does provide a reasonably quick procedure for finding the response at the first few sampling instants after a disturbance, which are often the most important, and the ease of inversion makes this a particularly useful technique.

Continuous Control with Sampled Data Signals

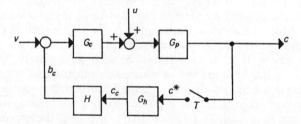

FIGURE 9–12. Block diagram of sampled data system with continuous control

Consider the block diagram of the control loop shown in Figure 9–12, where a continuous controller G_c is applied to a process with transfer function G_p, the output of the latter being sampled and measured at intervals of T, with a holding element G_h providing a clamped feedback signal to the controller. As in the study of continuous data systems, it is first necessary to determine the closed-loop transfer functions between the input and output variables. From the block diagram of Figure 9–12, it follows directly that

$$c(s) = G_c G_p(v - G_h H c^*) + G_p u$$

Solving this equation for $c(s)$ at the sampling instants when $c = c^*$, $v = v^*$ and $u = u^*$,

$$c^*(s) = G_c G_p v^* + G_p u^* - G_c G_p G_h H c^*$$

Introducing the z-variable, the closed loop relationship becomes

$$c(z) = \frac{\overline{G_c G_p v}^*}{1 + G}(z) + \frac{\overline{G_p u}^*}{1 + G}(z) \qquad (9\text{–}34)$$

where $G = \overline{G_c G_p G_h H}(z)$. [N.B. $G \neq G_c(z)G_p(z)G_h(z)H(z)$.]

Equation (9–34) is essentially identical in form with the corresponding result in the Laplace transform which would be obtained with a control loop with continuous signals, but there are some differences. It is not in this case possible to write expressions for the control ratios c/v or c/u, since the values of the input variables are now 'bound' with the process and controller transfer functions in the pulse transforms. The loop response cannot then be generalized but must be calculated for each particular input disturbance. The equation also defines the value of the output only at the sampling instants; if values between these instants are required, the modified z-transform must be used.

It will be seen that the denominators of the response terms of Equation (9–34) for both load and desired value changes are the same $(1 + G)$, as is the case for continuous data systems. By analogy with such systems, the stability of the response *at the sampling instants* will be determined by the roots of a characteristic equation formed by equating the denominator to zero, i.e. $1 + G(z) = 0$. Owing to the essential differences in the transforms different conditions for stability will now apply. As seen in the inversion of the transform in Equation (9–33), a factor such as $(z - a)$ of $G(z)$ will lead to a term in the output variable of a^n at the sampling instants, nT, where $n = 0, 1, 2, 3. \ldots$ It will be seen immediately that if a is negative, the response impulses are alternately positive and negative, and the response is oscillatory. If $a = -1$, the positive and negative impulses are of constant magnitude and the response is a continuous oscillation, i.e. the condition of limiting stability. If $1 > a > -1$, the impulses are of decreasing magnitude at successive sampling instants and the response is stable. The essential requirement for stability is then that the magnitude of all roots of $1 + G(z)$ should be less than $1(|z| < 1)$; for the system to be stable, all such roots must lie within the unit circle of $|z|$ in the z-plane.

The graphical methods of stability analysis developed for linear continuous data systems may be applied also to sampled data systems with only slight modifications. Thus a root locus diagram may be constructed for $G(z)$ using the usual rules for the poles and zeros of $G(z)$; the only difference in application is that limiting stability occurs when the root loci cross the unit circle in the z-plane rather than crossing the imaginary axis in the s-plane (for a continuous data system). The Nyquist method can also be applied, mapping a contour to encircle the unit z-plane circle rather than the right half of the s-plane. However, by changing the variable of $G(z)$ to w, defined by $z = (w + 1)/(w - 1)$, the unit circle of the z-plane is transferred to the left half of the w-plane, as can be demonstrated by conformal mapping. This change in variable allows the stability criterion of the root locus diagram as developed for continuous data systems to be applied directly on the w-plane for a sampled data system. When the root locus now crosses from the left half-plane into the right half-plane (equivalent to the loci crossing the imaginary axis in the s-plane), the roots in the z-plane lie outside the unit circle and the system is unstable. The Routh criterion can also be applied directly to the characteristic equation with w substituted for z, to determine whether or not the system is stable for given values of the parameters.

Proportional Control of a First-order Process

To illustrate the techniques discussed in the previous section, consider the loop of Figure 9–12 with a proportional controller, $G_c = K_c$, and a first-order process, $G_p = 1/(T_p s + 1)$. It may be assumed that the hold element is zero-order, i.e.

$$G_h = [1 - \exp(-Ts)]/s$$

and that there is negligible lag in the measurement function, i.e. $H = 1$.

The closed-loop function $G(s)$ is then given by

$$G(s) = \frac{K_c(1 - e^{-Ts})}{s(T_p s + 1)}$$

which may be written

$$G(s) = X(s) - X(s)\, e^{-Ts}$$

The Laplace transform inversion of the second term is the inversion of the first term but delayed by the time T; hence the z-transform of G will be that of the term X multiplied by $(1 - z^{-1})$. The z-transform may be found by inverting $G(s)$ and taking the z-transform of the resulting time function and summing the infinite series (as in the derivation of Equations (9–31) and (9–32)), or from a table of z-transforms by reference to the corresponding Laplace transform or its partial fraction expansion. Thus the z-transform corresponding to $f(s) = 1/s$ is $z/(z - 1)$, and that for $f(s) = 1/(s + a)$ is $z/(z - e^{-aT})$ (see Appendix). Expanding $X(s)$ by partial fractions:

$$X(s) = \frac{K_c}{s(T_p s + 1)} = K_c \left[\frac{1}{s} - \frac{1}{s + (1/T_p)} \right]$$

Hence

$$X(z) = K_c \left[\frac{z}{z - 1} - \frac{z}{z - \exp(-T/T_p)} \right]$$

Let $\exp(-T/T_p) = \alpha$; introducing the $(1 - z^{-1})$ term of the zero-order hold,

$$G(z) = K_c \left[\frac{z}{z - 1} - \frac{z}{z - \alpha} \right] (1 - z^{-1})$$

$$= K_c \frac{1 - \alpha}{z - \alpha} \qquad (9\text{-}35)$$

The characteristic equation thus takes the form

$$1 + K_c \frac{1 - \alpha}{z - \alpha} = 0 \qquad (9\text{-}36)$$

or $\qquad\qquad (z - \alpha) = -K_c(1 - \alpha)$

α is thus a real pole of $1 + G(z)$ and there are no zeros. Since Equation (9–35) is first-order, the root of $1 + G(z)$ can only be real and the locus will lie on the real axis; α is also real and hence

$$|z - \alpha| = |z| - |\alpha|$$
$$= K_c(1 - \alpha)$$

whence $\qquad\qquad |z| = K_c(1 - \alpha) + \alpha$

The stability constraint is that the roots of the characteristic equation $1 + G(z) = 0$ lie within the unit circle $|z| = 1$. Hence

$$K_c(1 - \alpha) + \alpha \leqslant 1$$

$$K_c \leqslant \frac{1 + \alpha}{1 - \alpha} \qquad (9\text{-}37)$$

Equation (9–35) can also be written

$$z - [\alpha - K_c(1 - \alpha)] = 0$$

which is of the form $(z - a) = 0$; the closed-loop response at the sampling instants will then contain a term of the form a^n. If a is negative the term will be oscillatory, and if K_c is set at the limiting value of $(1 + \alpha)/(1 - \alpha)$, then

$$a = \alpha - K_c(1 - \alpha) = -1$$

and the response is an oscillation of constant amplitude. It can readily be seen that if K_c is greater than the limiting value, the terms in a^n are alternately positive and negative and of increasing magnitude, i.e. the oscillation is unstable. It is interesting to note that an unstable response is possible from a simple first-order system with proportional control which is not, of course, the case with a continuous data loop.

The response of the sampled data loop to a load disturbance can be derived from Equation (9–34), thus:

$$c(z) = \frac{G_p u(z)}{1 + G(z)} \tag{9-34a}$$

For the first-order system with a unit step change in load,

$$G_p u(s) = 1/[s(T_p s + 1)]$$

whence

$$G_p u(z) = \frac{z(1 - \alpha)}{(z - 1)(z - \alpha)} \tag{9-38}$$

where $\alpha = \exp(-T/T_p)$, as before.

Combining Equations (9–34a) and (9–38),

$$
\begin{aligned}
c(z) &= \frac{z(1 - \alpha)/[(z - 1)(z - \alpha)]}{1 + K_c(1 - \alpha)/(z - \alpha)} \\[2mm]
&= \frac{z(1 - \alpha)}{(z - 1)[z - \alpha + K_c(1 - \alpha)]} \\[2mm]
&= \frac{1}{1 + K_c}\left[\frac{z}{z - 1} - \frac{z}{z - \alpha + K_c(1 - \alpha)}\right]
\end{aligned}
\tag{9-39}
$$

Using the z-transform inversion,

$$c(nT) = \frac{1}{1 + K_c}\{1 - [\alpha - K_c(1 - \alpha)]^n\} \tag{9-40}$$

Substituting the limiting value of K_c (i.e. $(1 + \alpha)/(1 - \alpha)$) in Equation (9–40) gives

$$c(nT) = \tfrac{1}{2}(1 - \alpha)[1 - (-1)^n]$$

and the values at the first sampling instants ($n = 0, 1, 2, 3, \ldots$) are

$$c(nT) = 0, 1 - \alpha, 0, 1 - \alpha, \ldots$$

In the following examples the effects of changing the value of K_c and the sampling period T on the responses are illustrated.

Example 9–3

Determine the system response at the first few sampling instants following a unit step change in load applied to a first-order sampled data loop (as Figure 9–12) with a proportional controller, using values of K_c equal to $\tfrac{1}{3}$, $\tfrac{1}{2}$, and $\tfrac{2}{3}$ of the limiting value, with a sampling period such that $\alpha = 0.8$.

The maximum value of K_c for stable response is given from Equation (9–37); for a value of $\alpha = 0.8$, $K_{c\,max} = 1.8/0.2 = 9$.

For $K_c = \tfrac{1}{3}K_{c\,max}$, $K_c = 3$.

The response equation is Equation (9–40), for $K_c = 3$ and $\alpha = 0.8$:

$$c(nT) = \tfrac{1}{4}\{1 - [0.8 - 3(1 - 0.8)]^n\}$$
$$= \tfrac{1}{4}[1 - (-0.2)^n]$$

Hence the values of $c(nT)$ for $n = 0, 1, 2, 3, 4$ are 0, 0.2, 0.24, 0.244, and 0.25; the final value ($nT = \infty$) is 0.25.

An alternative approach is to use the long division method of inversion from the value of the z-transform of the response given by Equation (9–39), i.e. for $K_c = 3$ and $\alpha = 0.8$:

$$c(z) = \frac{z(1 - 0.8)}{(z - 1)[z - 0.8 + 3(1 - 0.8)]}$$

$$= \frac{0.2z}{(z - 1)(z + 0.2)}$$

$$= \frac{0.2z}{z^2 - 0.8z - 0.2}$$

Dividing numerator by denominator, the following series in z^{-n} is obtained:

$$c(z) = 0.2z^{-1} + 0.24z^{-2} + 0.244z^{-3} + 0.25z^{-4} + \ldots$$

where the coefficients of z^{-n} are the magnitudes of the response variable at the appropriate value of nT.

Proceeding similarly for other values of K_c the following table of values of $c(nT)$ results:

TABLE 9–1

K_c	0	1	2	3	4	5	Final value
3	0	0.2	0.24	0.244	0.25	0.25	0.25
4.5	0	0.2	0.18	0.182	0.182	0.182	0.182
6	0	0.2	0.12	0.152	0.139	0.144	0.143
9	0	0.2	0	0.2	0	0.2	—
10	0	0.2	−0.04	0.248	−0.098	0.317	—

The response for K_c equal to and greater than the limiting value are included for comparison; Figure 9–13 is a sketch of $c(t)$ for values of K_c up to the limiting value.

Example 9–4

Compare the system response at the first few sampling instants following a unit step change in load for the system of the previous example for

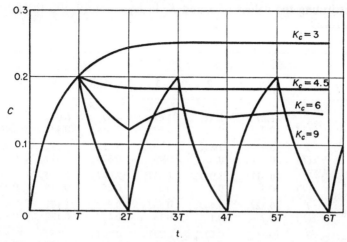

FIGURE 9–13. Response of sampled data proportional control for different values of K_c (Example 9–3)

different sampling periods when K_c is adjusted in each case to give a subsidence ratio of $4:1$ in the resulting oscillation.

Considering the response as given by Equation (9–40), if the term $[\alpha - K_c(1 - \alpha)]$ is made equal to $-\frac{1}{2}$, the successive peaks of the oscillatory response will be reduced by a factor of 4 every second sampling instant, thus giving the required subsidence ratio. The required value of K_c for a particular value of α is given by

$$K_c = (\alpha + 0.5)/(1 - \alpha)$$

The parameter α is determined by the ratio of the sampling period (T) to the time constant of the first-order process, T_p; hence different sampling periods for a given time constant yield different values of α and thus different values of $K_{c\,max}$ and K_c for the given subsidence ratio. Selecting a suitable range of values for α, the following table can be drawn up:

TABLE 9–2

$\alpha = \exp(-T/T_p)$	T/T_p (approx)	$K_{c\,max}$	K_c
0.3	1.20	1.857	1.143
0.4	0.92	2.333	1.5
0.6	0.51	4.0	2.75
0.8	0.22	9.0	6.5
0.9	0.11	19.0	14.0

Substituting the values of α and K_c in Equation (9–40), the following values result:

TABLE 9–3

α		1	2	n 3	4	5	Final value
0.3	T	$1.2T_p$	$2.4T_p$	$3.6T_p$	$4.8T_p$	$6.0T_p$	∞
	$c(T)$	0.7	0.35	0.525	0.438	0.481	0.467
0.4	T	$0.92T_p$	$1.84T_p$	$2.76T_p$	$3.68T_p$	$4.6T_p$	∞
	$c(T)$	0.6	0.3	0.45	0.375	0.413	0.400
0.6	T	$0.51T_p$	$1.02T_p$	$1.53T_p$	$2.04T_p$	$2.55T_p$	∞
	$c(T)$	0.4	0.2	0.3	0.25	0.275	0.267
0.8	T	$0.22T_p$	$0.44T_p$	$0.66T_p$	$0.88T_p$	$1.1T_p$	∞
	$c(T)$	0.2	0.1	0.15	0.125	0.138	0.133
0.9	T	$0.11T_p$	$0.22T_p$	$0.33T_p$	$0.44T_p$	$0.55T_p$	∞
	$c(T)$	0.1	0.05	0.075	0.063	0.069	0.067

These responses are sketched in Figure 9–14. As would be expected, it can be seen that increasing the frequency of sampling permits the use of a larger proportional gain, thus leading to smaller offset and initial peak deviation, with faster recovery.

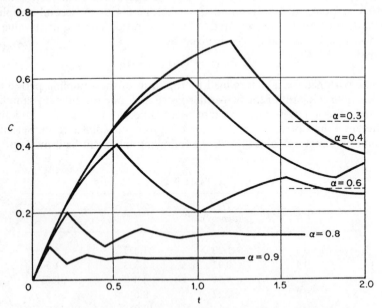

FIGURE 9–14. Response of sampled data control for different sampling periods
$(\alpha = \exp(-T/T_p))$

Although only a relatively simple example has been treated, it will be apparent that the technique can be extended to systems of higher order or to more complex controls, but with increased algebraic complexity. In general, the closed-loop transfer function of Equation (9–34) can be applied for determination of the controller function G_c, by testing the response at the first few sampling instants following a convenient forcing function, such as a step change in v or u, with likely values of the controller parameters, K_c, T_i, and T_d. The process is repeated until a satisfactory response is obtained.

Multi-variable Systems

The control systems which have been considered up to this point have been limited to systems in which there is effectively only one output variable of importance, and which is readily identified as the controlled variable. Although there may be a larger number of input variables, again one may be readily identified as having the most direct influence on the controlled variable and capable of being manipulated by the controller to regulate the controlled variable.

In general, however, process control systems are rarely so straightforward; usually there are a number of output variables associated with the process which must be held within certain limits if the process is to function efficiently and economically, and there will be a number of input variables which can be manipulated to regulate the output variables. A chemical reactor, for example, will respond to changes in the flow rate, composition and temperature of the inlet flow streams, and to the rate of heat addition or removal by heating or cooling streams. These changes will be shown by the dynamic response of the output variables which are principally the flow rates and compositions of the product stream or streams, and the pressure and temperature within the reactor. Usually in such systems, the transfer functions relating any particular pair of an input and an output variable will not be known with any accuracy. Present control theory is often inadequate to predict such functions accurately, and experimental determinations are usually costly and difficult to perform and interpret quantitatively. The practice of process control in such circumstances is largely based on past experience and simple tests of the process and, it must be admitted, to some extent on intuition. What is often done, in effect, is to appraise the system in terms of each possible pair of an input and an output, i.e. a manipulated and a controlled variable, and then to devise a system of control loops linking the most appropriate pairs of variables. Fortunately in many cases, only a relatively small number of pairs of variables are important, and these may often be directly identified by inspection of the process flow sheet.

The initial step is usually to regulate the principal input variables separately as far as is possible. Thus, in the reactor shown in Figure 9–15, the input flow rates are controlled directly, or possibly by ratio control, to maintain the required proportion. The feed temperature is regulated by individual temperature control of one or more feed preheaters. Control of input temperature and feed flow rates in this way basically maintains a constant material and energy input into the system, and with a reasonably tight control over these variables, a material and energy balance over the reactor could be maintained by regulating the temperature level by manipulation of the coolant flow, and by regulating the pressure by manipulation of the outflow. It is not usually practicable to maintain constant material and energy balances over the whole system by regulating both input and output flow rates, since slight errors in maintaining desired values would lead to either depletion or accumulation in the system.

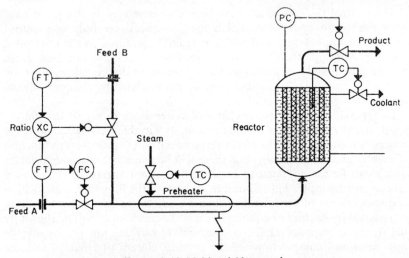

FIGURE 9–15. Multi-variable control

Maintaining a constant material and energy balance over a system would normally maintain a steady level of operation, but no provision has yet been made for changes in the input stream compositions or for changes in the process operation due to deterioration of catalyst, fouling of heat exchange surface, etc. These will generally require some change in the material and/or energy balance to suit the new operating conditions, which will be indicated by changes in the composition of the outflow stream. If the latter can be measured by a suitable analytical instrument, a further control impulse can be generated which can be used in a cascade system to adjust the material or energy balance via the feed flow and reactor temperature controllers.

The control scheme resulting from the above discussion may involve rather more instrumentation than is strictly necessary, but with adequate sizing of the reactor and cooling system, an effective overall control would be obtained.

In the above discussion a number of control loops have been specified for the multi-variable system. The fact that each individual loop may be designed and adjusted separately to operate satisfactorily *per se* is no guarantee that the overall operation of the whole system will be satisfactory; it is quite possible for each loop to be individually stable yet the whole system might be unstable or oscillatory in response. This is due to either a miscalculation of the degrees of freedom of the system or, more usually, to interaction between individual control loops.

Degrees of Freedom

Certain relationships between the variables in a system are specified by the usual physical laws governing the system, and setting the value of any particular one of these variables must necessarily place some limitations on the values of some of the others. This is equivalent to the mathematical principle that a system of n independent equations in n variables is required to define the value of each variable; defining an arbitrary value of any one variable means that one equation must be discarded to make the number of equations equal to the remaining number of variables. Specifying a controller for a particular process variable effectively amounts to setting that variable at an arbitrary value and thus reduces the degrees of freedom of the system. The state of a process is defined completely when each of its degrees of freedom is specified; the number of the latter is finite and is equal to the difference between the number of variables and the number of defining equations relating the variables.

The number of independently-acting controllers used in a multi-variable system may not exceed the number of degrees of freedom, since this will imply that two controllers are attempting to control one variable, usually by one controller acting indirectly through a second variable. In designing a complex system with multiple variables, it is necessary to observe that the number of degrees of freedom is not exceeded by the number of controllers. Such overinstrumentation is not uncommon; it is quite usual to use the temperature in the upper plates of a distillation column as an indirect indication of product composition to correct for changes in feed composition via the reflex ratio. At a constant column pressure this effectively fixes the temperature gradient in the column; but if now a temperature measurement in the lower half of the column is used to correct the boil-up rate following changes in feed composition, the two corrections are effectively made for the same disturbance, one can over-correct the other and lead to an oscillatory condition. Similarly the density and boiling point of an evaporator product should not be separately controlled, since both variables depend on concentration.

In assessing the number of degrees of freedom, a distinction must be made between system variables (flow rates, temperatures, etc.) and system parameters (hold-up volumes, etc. and properties such as specific heat which can be regarded as constant if disturbances are not excessively large).

Interaction Between Control Loops

When related variables are regulated by separate feedback loops, it is not unusual for interaction to occur between the loops, i.e. the action of one controller influences the feedback to the other controller(s), and the stability of all the loops may thus be affected. Fortunately, in many multi-variable processes there is considerable over-damping in this inter-action between loops, and the overall effects may be relatively small. The existence of interaction between loops must, however, be recognized, as it may well render a proposed scheme of control unworkable and it may be necessary to modify the design to minimize the effect. This particular field is open to numerous pitfalls which in general may be avoided only by past experience; when new and different applications arise, considerable difficulty may be experienced in finding a satisfactory solution. Here again there is some room for improvement in the present state of knowledge of process system design.

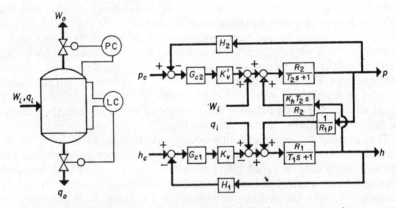

FIGURE 9–16. Interaction between control loops in level-pressure control system

To illustrate the principle of interaction between control loops, consider the relatively simple case of simultaneous control of pressure and liquid level in a vessel by individual control loops, where the liquid and gaseous outflows are the manipulated variables, as shown in Figure 9–16. An increase in the pressure over the liquid surface increases the rate of both gaseous and liquid outflows; the pressure change is thus a load variable on the liquid level control system. Similarly an increase in level produces an immediate increase in pressure, and the change in level is a load variable in the pressure control loop. The two individual control loops are thus interconnected, the controlled variable in one loop being a load variable in the other loop in each case. As a multi-loop system, it is always possible in principle to relate each controlled variable to the input variables by an overall transfer function; there will obviously be two such relationships since there are two controlled variables.

It is necessary to derive first the transfer equations for the individual loops; considering the level control, the continuity equation is given by the volume balance:

$$A(dh/dt) = q_i - q_o$$

where A is the sectional area and h, q_i, and q_o are changes in level and the volumetric inflow and outflow rates from the datum conditions of $\bar{q}_i = \bar{q}_o$ with the level at $\bar{h}$. Assuming a constant outflow back pressure, the outflow rate is determined by the liquid outflow valve position m_1, and the total pressure head in the vessel, i.e. $\bar{H} = \bar{p}/\rho g + \bar{h}$, at the datum condition, where ρ is the liquid density. For simultaneous changes in pressure and level, the change in pressure head is thus $H = p/\rho g + h$, and

$$q_o = f(H, m_1)$$

$$= \frac{\partial q_o}{\partial H}\bigg|_{\bar{m}_1} H + \frac{\partial q_o}{\partial m_1}\bigg|_{\bar{H}} m_1$$

$$= H/R_1 + K_v m_1$$

where R_1 is the flow resistance of the outflow valve $(= 2(\bar{p}/\rho g + \bar{h})/\bar{q}_o$ for turbulent flow) and K_v is the valve sensitivity in the operating range. Substituting for q_o in the continuity equation,

$$A(dh/dt) = q_i - (H/R_1 + K_v m_1)$$

$$= q_i - (p/\rho g + h)/R_1 - K_v m_1$$

Transforming and re-arranging,

$$h = (R_1 q_i - p/\rho g - R_1 K_v m_1)/(T_1 s + 1) \tag{9-41}$$

where T_1 is the time constant $(R_1 A)$ of the level control system.

An increase in level has a rather different effect on the pressure, since there is an immediate increase in pressure by compression of the gas space over the liquid. A change in pressure is related to the other variables by

$$p = [(w_i - w_o)/C]t + (\partial p/\partial h)h \tag{9-42}$$

where the first term is the accumulation in time t due to a difference in inflow and outflow, where w_i and w_o are changes in the mass flow rates from the datum values of $\bar{w}_i = \bar{w}_o$ at pressure $\bar{p}$, and C is the capacitance of the pressure vessel. The second term is the compression effect of the change in level, h; the partial differential may be evaluated by considering the compression under conditions of zero flow. An increase in level of ∂h from the datum $\bar{h}$ reduces the gas space by a volume of $A\partial h$ and increases the pressure from $\bar{p}$ to $\bar{p} + \partial p$. If the volume of gas over the liquid when the level is $\bar{h}$ is V, then

$$\bar{p}V = (\bar{p} + \partial p)(V - A\partial h)$$

whence, if the term $(\partial p \partial h)$ is taken to be negligible,

$$\frac{\partial p}{\partial h} = \frac{A}{V}\bar{p} = \frac{\bar{p}}{h_{max} - \bar{h}} = K_h$$

where h_{max} = total height of the vessel.

Substituting in Equation (9–42), and differentiating gives

$$C(dp/dt) = w_i - w_o + K_h C(dh/dt) \tag{9-43}$$

At a constant outflow back pressure, the outflow is a function of the valve position m_2 and the pressure within the vessel, i.e.

$$w_o = f(p, m_2)$$
$$= p/R_2 + K'_v m_2$$

where R_2 is the flow resistance $(= 2\bar{p}/\bar{w}_o$ for turbulent flow) and K'_v is the valve sensitivity in the operating range. Substituting in Equation (9–43),

$$C(dp/dt) = w_i - (p/R_2 + K'_v m_2) + K_h C(dh/dt)$$

Transforming and re-arranging,

$$p = (R_2 w_i - R_2 K'_v m_2 + K_h T_2 sh)/(T_2 s + 1) \tag{9-44}$$

where T_2 is the time constant $(R_2 C)$ of the pressure control system.

The block diagram of Figure 9–15 combines Equations (9–41) and (9–44) with control functions of G_{c1} and G_{c2} and measurement functions of H_1 and H_2, with no dynamic lag in the control valves. Assuming that proportional control is used in both loops and with unit measurement functions, the two closed-loop equations can be written by inspection from the diagram as

$$p = \frac{R_2}{T_2 s + 1}\left[\frac{K_h T_2 s}{R_2} h + w_i - K_{c2} K'_v (p_c - p)\right] \tag{9-45}$$

and

$$h = \frac{R_1}{T_1 s + 1}\left[\frac{1}{R_1 \rho} p + q_i - K_{c1} K_v (h_c - h)\right] \tag{9-46}$$

The overall equations for the double-loop system are obtained by eliminating p or h from Equations (9–45) and (9–46), which yields two equations of the form

$$p = \frac{N_1}{D} q_i - \frac{N_2}{D} h_c + \frac{N_3}{D} w_i - \frac{N_4}{D} p_c$$

and

$$h = \frac{N_5}{D} q_i - \frac{N_6}{D} h_c + \frac{N_7}{D} w_i - \frac{N_8}{D} p_c$$

The numerators of the transfer functions are collections of the various parameters; the denominator is the more interesting as determining the stability and is a quadratic term

$$D = T_1 T_2 s^2 + [T_1(1 - K_{c2}K_v'R_2) + T_2(1 - K_{c1}K_vR_1 - K_h/\rho g)]s$$
$$+ (1 - K_{c1}K_vR_1 - K_{c2}K_v'R_2 + K_{c1}K_{c2}K_vK_v'R_1R_2)$$

Thus, although each individual loop is a single time-constant system and will exhibit first-order behaviour if the controlled variable to the second loop is constant, the combination has the second-order response of a two-time-constants system. In the present case the overall system is more stable than a single loop system with the same two time constants, since the interactions contribute to self-regulation and are not delayed by additional lags. Nevertheless there are more terms to be considered in the combined equations than in the two individual loops. As a result the controller settings that are satisfactory for the individual loops may lead to excessive cycling when both loops are considered as a complete system. The effect is most pronounced when both loops have approximately the same critical frequency, and is much less marked if the critical frequencies are very different. The situation becomes much worse if there are additional lags before the interactions take effect.

Mathematical Analysis of Multi-variable Systems

Only brief reference can be made here to the mathematical treatment of multi-variable systems. The general analysis of such systems is basically an extension of the 'classical' control theory of single-variable systems involving the use of matrix algebra, the resulting stability equations yielding matrices corresponding to the sets of equations.

If a system has r inputs $m_1, m_2 \ldots, m_r$, and p outputs $c_1, c_2 \ldots, c_p$, column vectors of the Laplace transforms of the input and output signals can be defined by

$$\mathbf{M}(s) = \begin{pmatrix} m_1(s) \\ m_2(s) \\ \cdot \\ \cdot \\ \cdot \\ m_r(s) \end{pmatrix} \qquad \mathbf{C}(s) = \begin{pmatrix} c_1(s) \\ c_2(s) \\ \cdot \\ \cdot \\ \cdot \\ c_p(s) \end{pmatrix}$$

If now one particular input variable, say m_k, is varied whilst all others are held constant (i.e. at zero), a response will be induced in each of the p output signals; there will thus be a transfer function relating each output signal to each input signal. The usual transfer function relationship will apply between any pair of input and output signals, thus for input m_k and output c_j,

$$c_j(s) = G_{jk}(s)m_k(s)$$

where $G_{jk}(s)$ is the transfer function between m_k and c_j. If all the inputs are varied simultaneously, $c_j(t)$ must be determined by the principle of superposition, i.e. for a linear system, by summing the response of the output c_j to all the inputs $m_1, \ldots, m_r$, which yields

$$c_j(s) = \sum_{k=0}^{k=r} G_{jk}(s)m_k(s)$$

This equation will apply to each output signal such as c_j; to include the response of all output signals $c_1, \ldots, c_p$, the relationship is written in matrix notation as

$$\mathbf{C}(s) = \mathbf{G}(s)\mathbf{M}(s) \qquad\qquad (9\text{--}47)$$

where $\mathbf{G}(s)$ is a matrix of p rows and r columns in which $G_{jk}(s)$ is the element in the jth row and kth column. Equation (9–47) will possess an inversion in the time domain which will also appear in matrix notation. In principle these equations are identical in form to the single-variable counterparts, and the analysis of multi-variable systems offers no theoretical complexity over the single-variable system apart from the introduction of the matrix notation. Both analogue and digital computers can handle the matrix operations readily in principle.

Recent developments in multi-variable control theory are based on a concept of *state* and *state variables*, the system behaviour being studied through its state rather than through its output variables. In general terms, the state of a system may be defined as sufficient information about the system at some instant of time; this information, together with a specification of the input vector $\mathbf{m}(t)$ for subsequent time enables the output vector $\mathbf{c}(t)$ to be determined. The state of the system can be represented by a vector $\mathbf{x}(t)$; if the symbol $\mathbf{m}(t_o, t)$ is used as an abbreviation for all values of $\mathbf{m}(t)$ during the interval from t_o to t, the state concept can be expressed mathematically by

$$\mathbf{c}(t) = \mathbf{c}\{\mathbf{x}(t_o), \mathbf{m}(t_o, t)\} \qquad t > t_o$$

For further discussion of this subject reference should be made to the recent literature, e.g. Koppel [20].

Non-linear Systems

Systems considered previously in the text have been assumed to be linear or to operate sufficiently closely to a steady state to exhibit effectively linear behaviour, i.e. perturbations from the steady state are sufficiently small that a linear relationship between variables can be derived by using only the first-order terms from the Taylor expansion of the actual non-linear functions between the variables.

It is instructive to consider, albeit briefly, the position with systems which cannot be assumed to behave with a linear relationship. Initially two different types of non-linear behaviour can be distinguished; continuous and discontinuous. The latter occurs when the response of a variable is not a continuous function of the input, and in process systems arises mainly from mechanical malfunction or incorrect sizing of equipment. Such discontinuous non-linearities include saturation, due to under-sizing of control valves, etc. so that a limit to the magnitude of the response is imposed; backlash or dead zone, due to play in moving parts so that there is no response to small magnitude signals; and hysteresis, due to 'stiction' in valve glands, etc. These discontinuities may be treated mathematically but the best approach is by direct mechanical solution at the source. Saturation may be eliminated by correct sizing of equipment, backlash by replacement of worn parts, and hysteresis by use of valve positioners or by the application of a low amplitude oscillation to maintain continuous movement in parts subject to Coulomb friction.

With a continuous non-linearity, the response is a monotonic but non-linear function of the input variable. These may be handled readily enough by linearization techniques, but only so long as changes from the normal steady-state values are relatively small. When there are large changes in the process variables, such as occurs during plant start-up, linearizing procedures are not reliable and other methods must be used.

An obvious and direct approach is that of simulation by analogue or digital computers, the former being particularly suitable for rapid determinations of the effects of changes in system parameters. Systems of any complexity, however, require a large scale computer, and some experience is necessary to reduce the analysis to a tractable level.

Mathematical approaches to non-linear functions are presented by phase plane analysis and the describing function technique.

Phase Plane Analysis

Phase plane analysis of a non-linear dynamic system is based on a conceptual simplification of changing to a co-ordinate system (the phase plane) in which time no longer appears explicitly but is replaced by a differential function. When dx/dt is plotted against x, a curve is produced (known as a phase portrait) on which time is a parameter. In general, for a particular non-linear response function, there will be a number of different curves (or trajectories) specified by differing initial conditions, and the stability of the response for particular initial conditions may be shown by the direction and behaviour of the corresponding trajectory.

Several graphical techniques are available for sketching phase plane portraits [24], of which the method of isoclines, i.e. the loci of points at which different trajectories have the same slope, is usually the most suitable. Since the slope of a trajectory at any point is unique, trajectories can only intersect at critical points where the slope is indeterminate; such critical points are the steady states for the non-linear system. A major difference between linear and non-linear systems is that the latter may exhibit more than one steady state.

In general, if the trajectories for the different initial conditions converge *into* a critical point the system will be stable, since the phase portrait indicates that, irrespective of the initial condition (i.e. any applied disturbance), the system returns to a steady state as shown by the critical point. Conversely, if the trajectories *diverge* from a critical point, the system must be unstable. Certain trajectories, however, may form a closed figure, and this corresponds to a condition of limiting stability; the system will develop a sustained oscillation whose amplitude and frequency are determined only by the properties of the system and not by the initial conditions (or disturbance). Such oscillations are known as limit cycles, and may be stable when the closed figure on the phase plane is approached asymptotically by nearby trajectories, or unstable when the nearby trajectories diverge. A stable limit cycle is set up, for example, by the oscillation of two-position control which is a non-linear system since the relationship between correction and deviation is discontinuous. The amplitude and frequency of the oscillation of the variable in a two-position control system is dependent entirely on the properties of the system and independent of the initial value of the variable. Similarly, the continuous oscillation set up in a continuous control system when the controller gain is increased to a maximum critical value is actually a stable limit cycle. The controller gain in a real system cannot be set with the absolute accuracy required to the exact value to just sustain a continuous oscillation; in practice it will be slightly less or slightly greater than the critical value and, according to theory, the oscillation will be very slightly damped or just unstable. In practice a continuous oscillation would be set up under these conditions due to the small non-linearities present in any real system.

An interesting application of phase plane analysis is to the linearized equations for a non-linear system. It can be shown that in the vicinity of a critical point (the steady state), the linearized solutions give accurate information about the behaviour of the non-linear system. This essentially follows from the fundamental stability theorem of Liapunov; in the vicinity of the steady state of a non-linear system, the stability characteristics are the same as those of the linearized system. The proof of this theorem does not require the boundaries of the region of stability to be defined numerically, and it is relatively simple to establish the local stability (or instability) of the steady state for any non-linear system by systematic linearization. This is often adequate for qualitative phase-plane analysis since in many cases large areas of the phase plane have no physical significance (e.g. negative values of the system variables).

Linear control theory is successful in system design essentially because of Liapunov's theorem. However, a disadvantage of linear theory is that the stability of the linearized system refers to the stability of the actually non-linear system only in some region in the vicinity of the steady state. No information about the size of this region or of the system behaviour outside this region is given by linear theory. If the region is very small then almost any plant disturbance may result in control system failure if the controllers have been designed to operate only in the linear region.

Present research into non-linear control theory is mainly an attempt to determine the extent of the region in which the linearized equations will apply. There are virtually no useful generalizations for non-linear systems at present, although considerable effort is being expended on such systems.

Phase plane analysis is a powerful and convenient technique which can be used to obtain extensive information about the behaviour of a non-linear system, but the usefulness of the method is limited to systems which can be described graphically in two dimensional space, i.e. to systems which are effectively second order.

The Describing Function

The alternative mathematical approach to non-linear systems analysis is that of the describing function, which is an extension of frequency response analysis to systems containing a non-linear element followed by linear elements in series. The response of a non-linear element to an input sine wave is a distorted sinusoidal output with a fundamental frequency equal to that of the input, but with additional higher frequency harmonic components. By Fourier analysis of the response wave the fundamental component can be identified by a *describing function* which defines the magnitude and phase angle of the fundamental component with respect to the input wave. These quantities differ from the frequency response characteristics of a linear element by being functions not only of the frequency but also of the amplitude of the input sinusoid, thus an appropriate amplitude or range of amplitudes of the input signal must be chosen. Linear elements following in series after the non-linear element exercise a filtering action on the higher frequency harmonics due to the greater attenuation of the higher frequencies and in general the harmonic components are eliminated from the system response.

If the describing function for the non-linear element, ignoring harmonics, is N, the necessary condition for a stable limit cycle (i.e. a continuous oscillation of the response), is

$$|N||GH| = 1$$

and
$$\angle N + \angle GH = -180°$$

where G and H are the overall transfer functions of the forward and feedback linear elements in the loop. The magnitude and angle of N will in general depend on the amplitude A and frequency ω of the input; the magnitude and angle of GH depends only on ω. The above equations can then be rewritten

$$|GH(\omega)| = 1/|N(A, \omega)|$$

and
$$\angle GH(\omega) = -180° - \angle N(A, \omega)$$

As shown by Coughanowr and Koppel [10], for a simple case such as two position control where N is independent of frequency, these equations can be solved graphically by a gain-phase plot, i.e. a plot of log AR *versus* phase angle. The linear elements are plotted as $|GH|$ *versus* $\angle GH$ with ω as a parameter on the curve; the non-linear element is plotted as $1/|N|$ *versus* $-180° - \angle N$, with the amplitude A as a parameter. According to the above equations a limit cycle occurs at the intersection of the two curves, and the corresponding amplitude and frequency can be read directly from the parametric values on the curves.

The describing function technique is less laborious than phase plane analysis, particularly if a solution can be obtained from a gain-phase plot, and it is not restricted to second-order systems; in fact the higher the order of GH, the more accurate is the result due to the better elimination of the harmonics. The method, however, gives no information about the transient response, but the limit cycle amplitude and frequency are very often the quantities of primary interest.

Computer Control

The conventional pneumatic or electronic process controller is essentially an analogue computer in principle, containing summation, multiplication, integration and/or differentiation elements which produce an analogue form of correcting signal as a compressed air pressure or electrical signal output. There is, therefore, no sharp boundary between process control with conventional instruments and control using a general purpose analogue computer, although it is sometimes convenient to refer to the latter as 'computer control' since the purpose of the computer is to generate control action based on a more realistic appraisal of the system performance and objective than is possible with the measurement and control of a single variable as with conventional instruments.

Some simpler types of analogue computer control include the measurement of two or more variables which are combined by analogue computation into a property which is a more direct or realistic indication of the system performance than any of the single measured variables and on which a conventional control loop is operated, the control function generator normally being included in the computer. Thus the stream flow rates and temperatures over a heat exchanger may be measured and combined into a heat balance by the analogue simulation of the mathematical relationship between the variables. Such devices are in general special-purpose; they are designed for this specific duty and are equivalent in scope to a standard programme controller with the added facility of limited calculation.

The capability and flexibility of the digital computer in handling large volumes of information extremely quickly, combined with the additional facilities of rapid sequential operation and time sharing, have directed the development of computer control firmly into the digital field. The initial application of digital computers to process control recognized the limitations then existing in computer reliability, and the computer was used essentially to supervize the operation of standard control instruments. The data input to the computer was applied to a steady-state mathematical model of the process, and the necessary adjustments to the desired values of the process controllers were calculated to maintain the required level of operation of the whole system. The main advantage of this interim application was that the direct operation of the process was carried out by conventional instruments, and the process could thus continue to operate in spite of computer failure, although there would be some loss in efficiency. Advantages could be taken of the computer facility for handling multi-variable calculations, but there are obvious limitations in channelling the output through control loops using standard control instruments. With the greater speed and, more important, reliability of the modern digital machine, this technique has now been superseded by on-line or direct digital control (d.d.c.).

The conventional approach to direct digital control is the translation of the continuous proportional-integral-derivative algorithm of the conventional analogue process controller into an approximately discrete form of digital impulses. The variable to be controlled is sampled (measured) as frequently as is necessary to obtain an essentially equivalent performance from the sampled data as from the continuous data controller. Since, as considered in the section on sampled data control (page 362), the sampling theorem states, in approximate terms, that the output signal must be sampled at twice the highest frequency for which information recovery is required, taking direct advantage of the digital characteristics can yield superior performance whilst requiring less information, and a large number of variables can be sampled and controlled at relatively low sampling frequencies.

The data processing role is fundamental to any on-line computer application and is effectively the function of process monitoring by which the computer first establishes the status of the process variables, both input and output, as may be necessary. This requires that the outputs of conventional process measuring instruments be converted from analogue to digital signals and be continuously scanned by the computer at the selected frequency. The frequency of scanning is completely flexible and may differ for different measurements and at different times; the computer may be programmed to perform various checks on the measuring instruments, to indicate malfunctions, to round off, average or otherwise combine various input signals and to check the validity of spurious signals. Logging of the data at fixed or variable intervals may be performed by the usual print-out devices; selected variables may be displayed intermittently or continuously or on demand, and the trend of a variable may be shown by a chart plotter or other display. Alarm signals may be given whenever a variable exceeds an established limit; this latter may be pre-determined or may be varied in reference to operating conditions.

The output signals from the computer to the final control elements are computed explicitly and applied directly to the valves, etc. either as digital impulses or via digital-to-analogue converters. In applying regulatory control in this way, advantage may be taken of all aspects of classical control theory; the method of control is basically feedback, but the regulation may be based on output variables which are not directly measured but are computed from several measured variables, both input and output. For illustration, a single-loop system is shown in Figure 9–17; as many of the input and output variables are measured as required to compute the particular controlled output and this, by comparison with the desired reference input, provides the feedback control impulse to the manipulated variable. Computer application to a single-loop control would be grossly uneconomic but, with the rapid sequencing facility available, a number of loops can be handled sequentially in this way. The application of cascade and feedforward control is relatively straightforward.

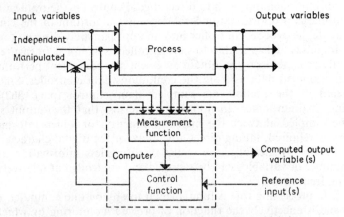

FIGURE 9–17. Single-loop computer control with multiple inputs

A further advanced level of control is achieved by the computer calculating the explicit control action on an integrated basis, i.e. by taking into account the multiplicity of input and output variables and the complexities of their inter-relationships, and applying this action as appropriate signals to a number of manipulated variables. Although the signals are derived and applied sequentially, the rapid sequencing produces a virtually continuous signal to each control valve. This is a true multivariable control with effectively simultaneous manipulation of as many input variables as are necessary to regulate the multiple output variables to attain the desired operating level of the process as a whole.

On-line computer application must be based on a dynamic mathematical model of the complete process; all the input data accumulated from the measuring instruments is applied, through any necessary sub-routines, to the equations which form the model and the computer then calculates the 'updated' values of its output signals to maintain the desired values of the output variables. The mathematical model is usually pre-determined by theoretical analysis of the process based on experimental data and testing, and is itself updated in the light of operating experience and statistical analysis of the operations. A further development which has been suggested is the use of the computer itself to update the model by analysis of the process operation; this would require the computer to apply some perturbations to the process in a pre-determined fashion and to use the resulting input data to compute and store the relationships between the input and output variables which constitute the dynamic model.

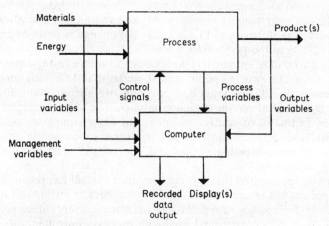

FIGURE 9–18. Multi-variable computer control

Probably the most powerful development of direct computer control is in *self-optimization* by which the overall implications (usually economic) are explicitly considered in the strategy of the control operation. Thus the computer is designed to control the ultimate objective of the process rather than controlling symptoms such as temperatures and pressures at particular points in the plant. The aim would then be to maximize the yield of the process, on a basis of both product quality and rate of production, whilst at the same time to minimize the production cost. The computer must now be provided with suitable criteria of optimality and, after taking account of all the significant process variables must compute the process conditions required to achieve such performance. In the simpler cases this may be done by the virtually trial and error method of 'hill climbing'; the computer calculates the economic advantage at the operating conditions at a particular time and directs the system to search for an improved performance. Each time a more nearly optimal condition is identified, this becomes the basis of a new optimizing search sequence. When more than two variables are involved, optimization rapidly becomes a complicated procedure requiring very powerful computing equipment. The basic economics of on-line optimization have been reviewed by Eliot and Longmire [14] and Williams [27] who have concluded that the best opportunities for its use lie in very large-scale production and in production-limited plant, i.e. a plant whose production capacity is the limiting factor on sales of the product.

A further development from optimization is *adaptive* control in which the system is capable of adapting its own parameters in response to changes in environment with the object of improving the system performance. The nature of the data processing used by the computer to generate the optimum response is now itself to be changed by variations in factors such as the process characteristics, uncontrolled input variables, the policy of operation or management, and so on, provided always that the observed performance indicates that the change is necessary to effect improvements in the performance.

There are no inherent barriers, apart from economic factors, in operating a fully automatic process plant, and some large installations, notably in the manufacture of ammonia and ethylene, are already being fully controlled by a central on-line digital computer. The same principles can be extended in theory to control of integrated plant complexes such as oil refineries and petrochemical plants, or even to corporate enterprises comprising manufacturing plants and the related operations of warehousing, marketing, etc.

It will be appreciated that on-line computer control has potentially very wide applications where the use can be economically justified. It does not necessarily follow, as interpretation of some current literature might suggest, that the conventional analogue process controllers are now to become obsolete. In the absence of exceptional factors, conventional process control instruments will continue to give acceptable and satisfactory results in very many applications. These instruments are designed for control of a single variable, and their particular weakness in multi-variable control has already been noted; in general detailed consideration cannot usually be given to the effects of interactions between the single-variable control loops in a multi-variable system, but this is due to the complexities of the process analysis. In a surprisingly large number of cases, interactions of this type are found to have a relatively minor effect on the system as a whole, and can often be virtually ignored at the design stage and later corrected by trial and error methods during commissioning of the operation. It is not, of course, possible with present practice to make conventional control systems self-optimizing although some techniques are being explored. Optimization is generally required only when major changes occur in conditions or requirements, and these tend to be relatively infrequent in many process operations.

A digital computer control installation will generally be somewhat more expensive in capital outlay than an equivalent multiple loop installation with conventional instruments, and it is essential that the increased capital cost is met by increased operating efficiency. Williams [26] has stated that a return on plant investment of some 0.5 per cent per annum is probably possible on any process control application using an on-line computer, arising solely from the better monitoring and long-term statistical optimizing ability of the computer. On a large-scale plant of high capital cost, a return of this magnitude might well justify the use of direct digital control.

PROBLEMS

9–1 A process system consists of three stages in series with time constants of 15, 5, and 60 s respectively, followed by a time delay of 10 s. Determine the critical frequency of the system using a single controller and a cascade system with the first two time-constant stages in the inner loop, assuming that each measuring element required has a time constant lag of 2 s.

9–2 The temperature (θ_o) of the oil leaving a gas-fired tube still is controlled by a proportional controller operating a diaphragm valve in the fuel supply to the furnace. The latter corresponds to a system with two time constants of 15 min and 25 s, with a dead time of 20 s. The steady state gain between valve pressure and outlet temperature per lbf/in² is 40°C. Determine the proportional gain (K_c) for 45° phase margin.

The major disturbance variable in this process is the oil inlet temperature (θ_i) and the open-loop relationship between θ_o and θ_i is represented by a time delay of 10 min and a time constant lag of 1 min with unit gain. Determine the perfect feedback function, assuming negligible lag in measurement of the inlet temperature and comment on the practical realization of this function.

9–3 Determine the transfer function relating x and x_c for the interacting system of Figure P9–1. Calculate the effective time constants and damping coefficient if

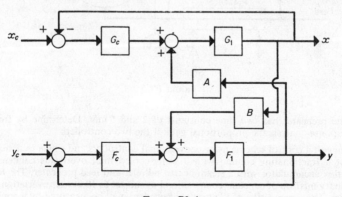

FIGURE P9–1

$G_1 = 5/(10s + 1)$, $F_1 = 5/(5s + 1)$, $A = 2$, and $B = 3$, with proportional control in both loops. What happens if A and/or B are negative?

9–4 A partial evaporator which separates a feed stream containing volatile components into vapour and liquid phases is shown in Figure P9–2. Three variables (pressure, level, and temperature) may be measured and three flow streams (vapour and liquid outflows and steam supply) can be manipulated, the feed stream being derived from a previous process and hence subject to uncontrollable variations in flow rate and

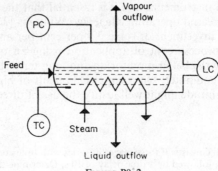

FIGURE P9-2

composition. Consider the possible controller arrangements and suggest the most efficient control scheme. Draw a block diagram showing all interactions.

9-5 A reactor with a selective control system is shown in Figure P9-3, the reactor having time constants of 1 and 4 min with a time delay in measurement of 5 s.

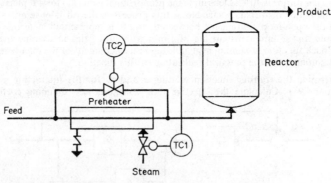

FIGURE P9-3

The preheater has two time constants of 2 and 5 min. Determine by frequency response analysis the proportional gain of the two controllers.

9-6 Suggest a control scheme using conventional controllers and single feedback loops for a fractionating column with a water-cooled total overheads condenser and reflux accumulator with a steam-heated reboiler and feed preheater. The feed is a binary mixture of constant composition but subject to 10 per cent variation in flow rate. The composition of the overhead product may be measured by a continuous analysing instrument.

9-7 Repeat Problem 9-6 using cascade and ratio controls to improve the design.

9-8 An error-sampled control system has the block diagram shown in Figure P9-4. The transfer function of the process is

$$G_p = K/s(s + 1)$$

and the sampling period is 1 s. Evaluate the open-loop z-transfer function and determine the marginal value of K for stability.

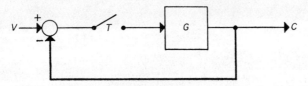

FIGURE P9-4

9-9 The open-loop z-transfer function for a sampled measurement control system is

$$GH(z) = \frac{0.792zK}{(z-1)(z-0.208)}$$

Determine the limiting value of K for stability and the response to a unit change in load when $K = 1.57$.

Appendix I

TABLE OF LAPLACE TRANSFORM INVERSIONS

Item	$f(s)$	$f(t)$
1	1	$\delta(t)$, i.e. unit impulse at $t = 0$
2	$\dfrac{1}{s}$	1, i.e. unit step from $t = 0$
3	$\dfrac{1}{s^n}$ $(n = 1, 2 \ldots)$	$\dfrac{t^{(n-1)}}{(n-1)!}$
4	$\dfrac{1}{s + a}$	e^{-at}
5	$\dfrac{1}{(s + a)^n}$ $(n = 1, 2 \ldots)$	$\dfrac{t^{(n-1)}\, e^{-at}}{(n-1)!}$
6	$\dfrac{s}{(s + a)^2}$	$e^{-at}(1 - at)$
7	$\dfrac{1}{s(s + a)}$	$\dfrac{1}{a}(1 - e^{-at})$
8	$\dfrac{1}{(s + a)(s + b)}$	$\dfrac{1}{b - a}(e^{-at} - e^{-bt})$
9	$\dfrac{s}{(s + a)(s + b)}$	$\dfrac{1}{a - b}(a\,e^{-at} - b\,e^{-bt})$
10	$\dfrac{1}{s(s + a)(s + b)}$	$\dfrac{1}{ab}\left[1 + \dfrac{1}{a - b}(b\,e^{-at} - a\,e^{-bt})\right]$
11	$\dfrac{(s + c)}{(s + a)(s + b)}$	$\dfrac{(c - a)\,e^{-at} - (c - b)\,e^{-bt}}{a - b}$
12	$\dfrac{1}{(s + a)(s + b)(s + c)}$	$\dfrac{e^{-at}}{(b - a)(c - a)} + \dfrac{e^{-bt}}{(a - b)(c - b)}$ $+ \dfrac{e^{-ct}}{(a - c)(b - c)}$

Item	$f(s)$	$f(t)$
13	$\dfrac{(s + d)}{(s + a)(s + b)(s + c)}$	$\dfrac{(d - a)\,\mathrm{e}^{-at}}{(b - a)(c - a)} + \dfrac{(d - b)\,\mathrm{e}^{-bt}}{(a - b)(c - b)}$ $+ \dfrac{(d - c)\,\mathrm{e}^{-at}}{(a - c)(b - c)}$
14	$\dfrac{1}{s^2 + a^2}$	$\dfrac{1}{a}\sin at$
15	$\dfrac{1}{s(s^2 + a^2)}$	$\dfrac{1}{a^2}(1 - \cos at)$
16	$\dfrac{s}{s^2 + a^2}$	$\cos at$
17	$\dfrac{1}{(s + a)^2 + b^2}$	$\dfrac{1}{b}\mathrm{e}^{-at}\sin bt$
18	$\dfrac{(s + a)}{(s + a)^2 + b^2}$	$\mathrm{e}^{-at}\cos bt$
19	e^{-as}	$\delta(t - a)$, i.e. unit impulse at $t = a$
20	$\dfrac{\mathrm{e}^{-as}}{s}$	$1(t - a)$, i.e. unit step at $t = a$

In the following inversions for damped oscillatory responses when $\zeta < 1$, $A = \sqrt{(1 - \zeta^2)}$; $B = \sqrt{(1 - \zeta^2)}/T$; $C = \zeta/T$.

21	$\dfrac{1}{T^2 s^2 + 2\zeta Ts + 1}$	$\dfrac{1}{AT}\mathrm{e}^{-Ct}\sin Bt$
22	$\dfrac{1}{s(T^2 s^2 + 2\zeta Ts + 1)}$	$1 + \dfrac{1}{A}\mathrm{e}^{-Ct}\sin(Bt - \phi)$ $\phi = \tan^{-1}(A/-\zeta)$
23	$\dfrac{(T_1 s + 1)}{T^2 s^2 + 2\zeta Ts + 1}$	$\dfrac{1}{AT}[1 - 2T_1 C + (T_1/T)^2]^{\frac{1}{2}}\,\mathrm{e}^{-Ct}\sin(Bt + \phi)$ $\phi = \tan^{-1}[T_1 B/(1 - T_1 C)]$
24	$\dfrac{(T_1 s + 1)}{s(T^2 s^2 + 2\zeta Ts + 1)}$	$1 + \dfrac{1}{A}[1 - 2T_1 C + (T_1/T)^2]^{\frac{1}{2}}\,\mathrm{e}^{-Ct}\sin(Bt + \phi)$ $\phi = \tan^{-1}[T_1 B/(1 - T_1 C)] - \tan^{-1}(A/-\zeta)$

Appendix II

SHORT TABLE OF Z-TRANSFORMS

Item	$f(s)$	$f(nT)$	$f(z)$
1	1	$\delta(t)$ (unit impulse at $t = 0$)	1
2	e^{-nTs}	$\delta(t - nT)$ (unit impulse at $t = nT$)	z^{-n}
3	$\dfrac{1}{s}$	1 (unit step)	$\dfrac{z}{z - 1}$
4	$\dfrac{1}{s^2}$	nT	$\dfrac{Tz}{(z - 1)^2}$
5	$\dfrac{1}{s + a}$	e^{-anT}	$\dfrac{z}{z - e^{-aT}}$
6	$\dfrac{1}{(s + a)^2}$	$nT e^{-anT}$	$\dfrac{Tz\, e^{-aT}}{(z - e^{-aT})^2}$
7	$\dfrac{a}{s(s + a)}$	$1 - e^{-anT}$	$\dfrac{(1 - e^{-aT})z}{(z - 1)(z - e^{-aT})}$
8	$\dfrac{b - a}{(s + a)(s + b)}$	$e^{-anT} - e^{-bnT}$	$\dfrac{(e^{-aT} - e^{-bT})z}{(z - e^{-aT})(z - e^{-bT})}$
9	$\dfrac{(b - a)s}{(s + a)(s + b)}$	$b\,e^{-bnT} - a\,e^{-anT}$	$\dfrac{(b - a)z^2 - (b\,e^{-aT} - a\,e^{-bT})z}{(z - e^{-aT})(z - e^{-bT})}$
10	$\dfrac{a}{s^2 + a^2}$	$\sin anT$	$\dfrac{z \sin aT}{z^2 - 2z \cos aT + 1}$
11	$\dfrac{s}{s^2 + a^2}$	$\cos anT$	$\dfrac{z(z - \cos aT)}{z^2 - 2z \cos aT + 1}$
12	$\dfrac{b}{(s + a)^2 + b^2}$	$e^{-anT} \sin bnT$	$\dfrac{z\,e^{-aT} \sin bT}{z^2 - 2z\,e^{-aT} \cos bT + 1}$
13	$\dfrac{s + a}{(s + a)^2 + b^2}$	$e^{-anT} \cos bnT$	$\dfrac{z(z - e^{-aT} \cos bT)}{z^2 - 2z\,e^{-aT} \cos bT + 1}$
14	$f(s + a)$	$e^{-anT}f(t)$	$f(e^{-aT}z)$

References

1. AIKMAN, A. R., *Trans. Soc. Inst. Tech.*, 1951, **3**, 2.
2. AIKMAN, A. R. and RUTHERFORD, C. I., in Tustin, A. (ed), *Automatic and Manual Control* (Butterworths, London, 1952).
3. BUCKLEY, P. S., *Techniques of Process Control* (Wiley, New York, 1964).
4. CALDWELL, W. I., COON, G. A., and ZOSS, L. M., *Frequency Response for Process Control* (McGraw-Hill, New York, 1959).
5. CAMPBELL, D. P., *Process Dynamics* (Wiley, New York, 1958).
6. CHU, Y., *Trans. A.I.E.E.*, 1952, **71**, 29.
7. COHEN, G. H. and COON, G. A., *Trans. A.S.M.E.*, 1953, **75**, 827.
8. COHEN, W. C. and JOHNSON, E. F., *Chem. Engrg. Prog. Symp. Ser. 36*, 1961, **57**, 86.
9. CONSIDINE, D. M., *Process Instruments and Controls Handbook* (McGraw-Hill, New York, 1957).
10. COUGHANOWR, D. R. and KOPPEL, L. B., *Process Systems Analysis and Control* (McGraw-Hill, New York, 1965).
11. DAY, R. L., in *Soc. Inst. Tech.* 'Plant and Process Dynamic Characteristics', (Butterworths, London, 1957).
12. ECKMAN, D. P., *Automatic Process Control* (Wiley, New York, 1958).
13. ECKMAN, D. P., *Mech. Engrg.*, 1953, **75**, 582.
14. ELIOT, T. Q. and LONGMIRE, D. R., *Chem. Engrg.*, 1962, **69**, 99.
15. EVANS, W. R., *Trans. A.I.E.E.*, 1948, **67**, 547.
16. FARRINGTON, G. H., *Fundamentals of Automatic Control* (Chapman and Hall, London, 1951).
17. GRABBE, E. N., RAMO, S., and WOOLDRIDGE, D. E. (eds), *Handbook of Automation, Computation and Control*, Vol. 3 (Wiley, New York, 1961).
18. HARRIOTT, P., *Process Control* (McGraw-Hill, New York, 1964).
19. JONSON, E. F., *Automatic Process Control* (McGraw-Hill, New York, 1967).
20. KOPPEL, L. B., *Introduction to Control Theory* (Prentice-Hall, Englewood Cliffs, N.J., 1968).
21. RAGGAZINI, J. R. and FRANKLIN, G. F., *Sampled Data Control Systems* (McGraw-Hill, New York, 1958).
22. ROUTH, J. F., *Dynamics of a System of Rigid Bodies* (Macmillan, London, 1877).
23. ST. CLAIRE, D. W., COOMBS, W. F., and OWENS, W. D., *Trans. A.S.M.E.*, 1950, **72**, 1135.
24. THALER, G. J. and PASTEL, N. P., *Analysis and Design of Non-linear Feedback Control Systems* (McGraw-Hill, New York, 1962).
25. TRUXALL, J. G., *Automatic Feedback Control Systems Synthesis* (McGraw-Hill, New York, 1955).
26. WILLIAMS, T. J., *Systems Engineering for the Process Industries* (McGraw-Hill, New York, 1961).
27. WILLIAMS, T. J., *I.S.A. Journal*, 1961, **8**, 50.
28. WILTS, C. H., *Principles of Feedback Control* (Addison-Wesley, Reading, Mass., 1960).
29. YOUNG, A. J., *An Introduction to Process Control System Design* (Longmans Green, London, 1955).
30. ZIEGLER, J. G. and NICHOLS, N. B., *Trans. A.S.M.E.*, 1942, **64**, 759.

Selected Bibliography

(1) *Mathematics*

CHURCHILL, R. V., *Complex Variables and Applications* (McGraw-Hill, New York 1960).

JENSEN, V. G. and JEFFERYS, G. V., *Mathematical Methods in Chemical Engineering* (Academic Press, London, 1963).

MICKLEY, H. S., SHERWOOD, T. E., and REED, G. E., *Applied Mathematics in Chemical Engineering* (McGraw-Hill, New York, 1957).

SAVANT, G. J., *Fundamentals of the Laplace Transformation* (McGraw-Hill, New York, 1957).

THOMSON, W. T., *Laplace Transformation* (Longmans Green, London, 1957).

(2) *General Control Theory*

BROWN, G. S. and CAMPBELL, D. P., *Principles of Servomechanisms* (Wiley, New York, 1948).

CHESTNUT, H. and MAYER, R. W., *Servomechanisms and Regulating System Design* (Wiley, New York, 1951–55).

EVANS, W. R., *Control System Dynamics* (McGraw-Hill, New York, 1954).

GIBSON, J. E., *Nonlinear Automatic Control* (McGraw-Hill, New York, 1963).

GRABBE, E. M., RAMO, S., and WOOLDRIDGE, D. E. (eds.), *Handbook of Automation, Computation and Control* (3 vols) (Wiley, New York, 1958–61).

GRAHAM, D. and McRUER, D., *Analysis of Nonlinear Control Systems* (Wiley, New York, 1961).

GUY, J. J., *Solutions of Problems in Automatic Control* (Pitman, London, 1966).

HOROWITZ, I. M., *Synthesis of Feedback Systems* (Academic Press, New York, 1963).

MERRIAM, C. W., *Optimisation Techniques and the Design of Feedback Control Systems* (McGraw-Hill, New York, 1964).

MESAROVIC, M., *Control of Multivariable Systems* (Wiley, New York, 1964).

MISHKIN, E. and BRAUN, L., *Adaptive Control Systems* (McGraw-Hill, New York, 1961).

OLDENBOURG, R. C. and SARTORIUS, H., *The Dynamics of Automatic Control* (A.S.M.A., New York, 1948).

OLDENBURGER, R., *Frequency Response* (Macmillan, London, 1956).

RAGGAZINI, J. R. and FRANKLIN, G. F., *Sampled Data Control Systems* (McGraw-Hill, New York, 1958).

SMITH, O. J. W., *Feedback Control Systems* (McGraw-Hill, New York, 1958).

TRUXALL, J. G., *Automatic Feedback Control Synthesis* (McGraw-Hill, New York, 1955).

TRUXALL, J. G. (ed.), *Control Engineers' Handbook* (McGraw-Hill, New York, 1958).

TUSTIN, A. (ed.), *Automatic and Manual Control* (Butterworths, London, 1952).

(3) *Process Dynamics and Control*

BUCKLEY, P. S., *Techniques of Process Control* (Wiley, New York, 1964).

CALDWELL, W. I., COON, G. A., and ZOSS, L. M., *Frequency Response for Process Control* (McGraw-Hill, New York, 1959).

CAMPBELL, D. P., *Process Dynamics* (Wiley, New York, 1958).

CEAGLSKE, N. H., *Automatic Process Control for Chemical Engineers* (Wiley, New York, 1956).

Very Good ✱ COUGHANOWR, D. R. and KOPPEL, L. B., *Process Systems Analysis and Control* (McGraw-Hill, New York, 1965).

ECKMAN, D. P., *Automatic Process Control* (Wiley, New York, 1958).

FARRINGTON, G. H., *Fundamentals of Automatic Control* (Chapman and Hall, London, 1951).

HADLEY, W. A. and LONGOBARDO, G., *Automatic Process Control* (Pitman, London, 1963).

HARRIOTT, P., *Process Control* (McGraw-Hill, New York, 1964).

HENGSTEBECK, R. J., *Distillation Principles and Design Procedure* (Reinhold, New York, 1961).

JOHNSON, E. F., *Automatic Process Control* (McGraw-Hill, New York, 1967).

McADAMS, W. H., *Heat Transmission* (McGraw-Hill, New York, 1961).

PERLMUTTER, D. D., *Introduction to Chemical Process Control* (Wiley, New York, 1965).

SAVAS, E. S. (ed), *Computer Control of Industrial Processes* (McGraw-Hill, New York, 1965).

SHILLING, C. D., *Process Dynamics and Control* (Holt, Rinehart, and Winston, New York, 1963).

SOCIETY OF INSTRUMENT TECHNOLOGY, *Plant and Process Dynamic Characteristics* (Butterworths, London, 1957).

STAINTHORP, F. P., 'Process Control', in Cremer, H. W. and Watkins, S. B. (eds.), *Chemical Engineering Practice*, Vol. 9 (Butterworths, London, 1965).

TRIMMER, J. D., *Response of Physical Systems* (Wiley, New York, 1950).

TUCKER, G. K. and WILLS, D. M., *A Simplified Technique of Control System Engineering* (Minneapolis-Honeywell, Philadelphia, 1958).

WILLIAMS, T. J., *Systems Engineering for the Process Industries* (McGraw-Hill, New York, 1961).

YOUNG, A. J., *An Introduction to Process Control System Design* (Longmans Green, London, 1955).

(4) *Measuring and Control Instruments, etc.*

BEARD, C. S., *Control Valves* (Instruments Publishing Co., Pittsburg, 1957).

CARROLL, G. S., *Industrial Process Measuring Instruments* (McGraw-Hill, New York, 1962).

CONSIDINE, D. M., *Process Instruments and Controls Handbook* (McGraw-Hill, New York, 1959).

ECKMAN, D. P., *Industrial Instrumentation* (Wiley, New York, 1951).

HOLZBOCK, W. G., *Instruments for Measurement and Control* (Reinhold, New York, 1962).

JONES, E. B., *Instrument Technology*, 3 vols. (Butterworths, London, 1953).

POLLARD, A. and CARRUTHERS, T. G., 'Measurement of Process Variables', in Cremer, H. W. and Davies, T., eds. *Chemical Engineering Practice*, Vol. 4 (Butterworths, London, 1957).

Index